Mastering Python for Bioinformatics

How to Write Flexible, Documented, Tested
Python Code for Research Computing

Ken Youens-Clark

Beijing · Boston · Farnham · Sebastopol · Tokyo

Mastering Python for Bioinformatics

by Ken Youens-Clark

Published by O'Reilly Media, Inc., 1005 Gravenstein Highway North, Sebastopol, CA 95472.

O'Reilly books may be purchased for educational, business, or sales promotional use. Online editions are also available for most titles (*http://oreilly.com*). For more information, contact our corporate/institutional sales department: 800-998-9938 or *corporate@oreilly.com*.

Acquisitions Editor: Michelle Smith
Development Editor: Corbin Collins
Production Editor: Caitlin Ghegan
Copyeditor: Sonia Saruba
Proofreader: Rachel Head

Indexer: Sue Klefstad
Interior Designer: David Futato
Cover Designer: Karen Montgomery
Illustrator: Kate Dullea

May 2021:   First Edition

Revision History for the First Edition
2021-05-04:   First Release
2022-12-16:   Second Release

See *http://oreilly.com/catalog/errata.csp?isbn=9781098100889* for release details.

978-1-098-10088-9

[LSI]

Table of Contents

Part II.   Other Programs

Preface

Programming is a force multiplier. We can write computer programs to free ourselves from tedious manual tasks and to accelerate research. Programming in *any* language will likely improve your productivity, but each language has different learning curves and tools that improve or impede the process of coding.

There is an adage in business that says you have three choices:

1. Fast
2. Good
3. Cheap

Pick any two.

When it comes to programming languages, Python hits a sweet spot in that it's *fast* because it's fairly easy to learn and to write a working prototype of an idea—it's pretty much always the first language I'll use to write any program. I find Python to be *cheap* because my programs will usually run well enough on commodity hardware like my laptop or a tiny AWS instance. However, I would contend that it's not necessarily easy to make *good* programs using Python because the language itself is fairly lax. For instance, it allows one to mix characters and numbers in operations that will crash the program.

This book has been written for the aspiring bioinformatics programmer who wants to learn about Python's best practices and tools such as the following:

- Since Python 3.6, you can add type hints to indicate, for instance, that a variable should be a *type* like a number or a list, and you can use the mypy tool to ensure the types are used correctly.

- Testing frameworks like pytest can exercise your code with both good and bad data to ensure that it reacts in some predictable way.

- Tools like `pylint` and `flake8` can find potential errors and stylistic problems that would make your programs more difficult to understand.

- The `argparse` module can document and validate the arguments to your programs.

- The Python ecosystem allows you to leverage hundreds of existing modules like Biopython to shorten programs and make them more reliable.

Using these tools practices individually will improve your programs, but combining them all will improve your code in compounding ways. This book is not a textbook on bioinformatics per se. The focus is on what Python offers that makes it suitable for writing scientific programs that are *reproducible*. That is, I'll show you how to design and test programs that will always produce the same outputs given the same inputs. Bioinformatics is saturated with poorly written, undocumented programs, and my goal is to reverse this trend, one program at a time.

The criteria for program reproducibility include:

Parameters
> All program parameters can be set as runtime arguments. This means no hard-coded values which would require changing the source code to change the program's behavior.

Documentation
> A program should respond to a `--help` argument by printing the parameters and usage.

Testing
> You should be able to run a test suite that proves the code meets some specifications

You might expect that this would logically lead to programs that are perhaps correct, but alas, as Edsger Dijkstra famously said, "Program testing can be used to show the presence of bugs, but never to show their absence!"

Most bioinformaticians are either scientists who've learned programming or programmers who've learned biology (or people like me who had to learn both). No matter how you've come to the field of bioinformatics, I want to show you practical programming techniques that will help you write correct programs quickly. I'll start with how to write programs that document and validate their arguments. Then I'll show how to write and run tests to ensure the programs do what they purport.

For instance, the first chapter shows you how to report the tetranucleotide frequency from a string of DNA. Sounds pretty simple, right? It's a trivial idea, but I'll take about 40 pages to show how to structure, document, and test this program. I'll spend a lot of time on how to write and test several different versions of the program so that I can explore many aspects of Python data structures, syntax, modules, and tools.

Who Should Read This?

You should read this book if you care about the craft of programming, and if you want to learn how to write programs that produce documentation, validate their parameters, fail gracefully, and work reliably. Testing is a key skill both for understanding your code and for verifying its correctness. I'll show you how to use the tests I've written as well as how to write tests for your programs.

To get the most out of this book, you should already have a solid understanding of Python. I will build on the skills I taught in *Tiny Python Projects* (Manning, 2020), where I show how to use Python data structures like strings, lists, tuples, dictionaries, sets, and named tuples. You need not be an expert in Python, but I definitely will push you to understand some advanced concepts I introduce in that book, such as types, regular expressions, and ideas about higher-order functions, along with testing and how to use tools like `pylint`, `flake8`, `yapf`, and `pytest` to check style, syntax, and correctness. One notable difference is that I will consistently use type annotations for all code in this book and will use the `mypy` tool to ensure the correct use of types.

Programming Style: Why I Avoid OOP and Exceptions

I tend to avoid object-oriented programming (OOP). If you don't know what OOP means, that's OK. Python itself is an OO language, and almost every element from a string to a set is technically an object with internal state and methods. You will encounter enough objects to get a feel for what OOP means, but the programs I present will mostly avoid using objects to represent ideas.

That said, Chapter 1 shows how to use a `class` to represent a complex data structure. The `class` allows me to define a data structure with type annotations so that I can verify that I'm using the data types correctly. It does help to understand a bit about OOP. For instance, classes define the attributes of an object, and classes can inherit attributes from parent classes, but this essentially describes the limits of how and why I use OOP in Python. If you don't entirely follow that right now, don't worry. You'll understand it once you see it.

Instead of object-oriented code, I demonstrate programs composed almost entirely of *functions*. These functions are also *pure* in that they will only act on the values given to them. That is, pure functions never rely on some hidden, mutable state like global variables, and they will always return the same values given the same arguments. Additionally, every function will have an associated test that I can run to verify it behaves predictably. It's my opinion that this leads to shorter programs that are more transparent and easier to test than solutions written using OOP. You may disagree and are of course welcome to write your solutions using whatever style of programming you prefer, so long as they pass the tests. The Python Functional Programming HOWTO documentation (*https://docs.python.org/3/howto/functional.html*) makes a good case for why Python is suited for functional programming (FP).

Finally, the programs in this book also avoid the use of exceptions, which I think is appropriate for short programs you write for personal use. Managing exceptions so that they don't interrupt the flow of a program adds another level of complexity that I feel detracts from one's ability to understand a program. I'm generally unhappy with how to write functions in Python that return errors. Many people would raise an exception and let a try/catch block handle the mistakes. If I feel an exception is warranted, I will often choose to *not* catch it, instead letting the program crash. In this respect, I'm following an idea from Joe Armstrong, the creator of the Erlang language, who said, "The Erlang *way* is to write the happy path, and not write twisty little passages full of error correcting code."

If you choose to write programs and modules for public release, you will need to learn much more about exceptions and error handling, but that's beyond the scope of this book.

Structure

The book is divided into two main parts. The first part tackles 14 of the programming challenges found at the Rosalind.info website. (*http://rosalind.info/about*)[1] The second part shows more complicated programs that demonstrate other patterns or concepts I feel are important in bioinformatics. Every chapter of the book describes a coding challenge for you to write and provides a test suite for you to determine when you've written a working program.

Although the "Zen of Python" (*https://oreil.ly/20PSy*) says "There should be one—and preferably only one—obvious way to do it," I believe you can learn quite a bit by attempting many different approaches to a problem. Perl was my gateway into bioinformatics, and the Perl community's spirit of "There's More Than One Way To Do It" (TMTOWTDI) still resonates with me. I generally follow a theme-and-variations approach to each chapter, showing many solutions to explore different aspects of Python syntax and data structures.

---

1 Named for Rosalind Franklin, who should have received a Nobel Prize for her contributions to discovering the structure of DNA.

Test-Driven Development

> More than the act of testing, the act of designing tests is one of the best bug preventers known. The thinking that must be done to create a useful test can discover and eliminate bugs before they are coded—indeed, test-design thinking can discover and eliminate bugs at every stage in the creation of software, from conception to specification, to design, coding, and the rest.
>
> —Boris Beizer, Software Testing Techniques (Thompson Computer Press)

Underlying all my experimentation will be test suites that I'll constantly run to ensure the programs continue to work correctly. Whenever I have the opportunity, I try to teach *test-driven development* (TDD), an idea explained in a book by that title written by Kent Beck (Addison-Wesley, 2002). TDD advocates writing tests for code *before* writing the code. The typical cycle involves the following:

1. Add a test.
2. Run all tests and see if the new test fails.
3. Write the code.
4. Run tests.
5. Refactor code.
6. Repeat.

In the book's GitHub repository (*https://oreil.ly/yrTZZ*), you'll find the tests for each program you'll write. I'll explain how to run and write tests, and I hope by the end of the material you'll believe in the common sense and basic decency of using TDD. I hope that thinking about tests first will start to change the way you understand and explore coding.

Using the Command Line and Installing Python

My experience in bioinformatics has always been centered around the Unix command line. Much of my day-to-day work has been on some flavor of Linux server, stitching together existing command-line programs using shell scripts, Perl, and Python. While I might write and debug a program or a pipeline on my laptop, I will often deploy my tools to a high-performance compute (HPC) cluster where a scheduler will run my programs asynchronously, often in the middle of the night or over a weekend and without any supervision or intervention by me. Additionally, all my work building databases and websites and administering servers is done entirely from the command line, so I feel strongly that you need to master this environment to be successful in bioinformatics.

I used a Macintosh to write and test all the material for this book, and macOS has the Terminal app you can use for a command line. I have also tested all the programs using various Linux distributions, and the GitHub repository includes instructions on how to use a Linux virtual machine with Docker. Additionally, I tested all the programs on Windows 10 using the Ubuntu distribution Windows Subsystem for Linux (WSL) version 1. I *highly* recommend WSL for Windows users to have a true Unix command line, but Windows shells like cmd.exe, PowerShell, and Git Bash can sometimes work sufficiently well for some programs.

I would encourage you to explore integrated development environments (IDEs) like VS Code, PyCharm, or Spyder to help you write, run, and test your programs. These tools integrate text editors, help documentation, and terminals. Although I wrote all the programs, tests, and even this book using the vim editor in a terminal, most people would probably prefer to use at least a more modern text editor like Sublime, TextMate, or Notepad++.

I wrote and tested all the examples using Python versions 3.8.6 and 3.9.1. Some examples use Python syntax that was not present in version 3.6, so I would recommend you not use that version. Python version 2.x is no longer supported and should not be used. I tend to get the latest version of Python 3 from the Python download page (*https://www.python.org/downloads*), but I've also had success using the Anaconda Python distribution (*https://www.anaconda.com*). You may have a package manager like apt on Ubuntu or brew on Mac that can install a recent version, or you may choose to build from source. Whatever your platform and installation method, I would recommend you try to use the most recent version as the language continues to change, mostly for the better.

Note that I've chosen to present the programs as command-line programs and not as Jupyter Notebooks for several reasons. I like Notebooks for data exploration, but the source code for Notebooks is stored in JavaScript Object Notation (JSON) and not as line-oriented text. This makes it very difficult to use tools like diff to find the differences between two Notebooks. Also, Notebooks cannot be parameterized, meaning I cannot pass in arguments from outside the program to change the behavior but instead have to change the source code itself. This makes the programs inflexible and automated testing impossible. While I encourage you to explore Notebooks, especially as an interactive way to run Python, I will focus on how to write command-line programs.

Getting the Code and Tests

All the code and tests are available from the book's GitHub repository. You can use the program Git (which you may need to install) to copy the code to your computer with the following command. This will create a new directory called *biofx_python* on your computer with the contents of the repository:

```
$ git clone https://github.com/kyclark/biofx_python
```

If you enjoy using an IDE, it may be possible to clone the repository through that interface, as shown in Figure P-1. Many IDEs can help you manage projects and write code, but they all work differently. To keep things simple, I will show how to use the command line to accomplish most tasks.

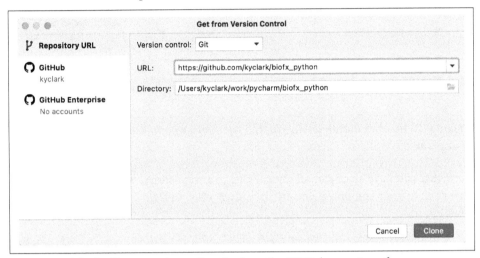

Figure P-1. *The PyCharm tool can directly clone the GitHub repository for you*

 Some tools, like PyCharm, may automatically try to create a *virtual environment* inside the project directory. This is a way to insulate the version of Python and modules from other projects on your computer. Whether or not you use virtual environments is a personal preference. It is not a requirement to use them.

You may prefer to make a copy of the code in your own account so that you can track your changes and share your solutions with others. This is called *forking* because you're breaking off from my code and adding your programs to the repository.

To fork my GitHub repository, do the following:

1. Create an account on GitHub.com.
2. Go to *https://github.com/kyclark/biofx_python*.
3. Click the Fork button in the upper-right corner (see Figure P-2) to make a copy of the repository in your account.

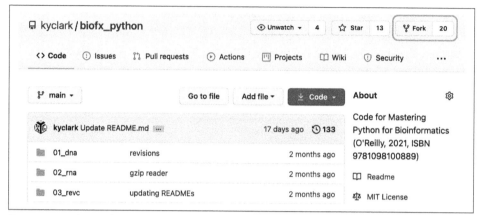

Figure P-2. The Fork button on my GitHub repository will make a copy of the code in your account

Now that you have a copy of all my code in your repository, you can use Git to copy that code to your computer. Be sure to replace *YOUR_GITHUB_ID* with your actual GitHub ID:

```
$ git clone https://github.com/YOUR_GITHUB_ID/biofx_python
```

I may update the repo after you make your copy. If you would like to be able to get those updates, you will need to configure Git to set my repository as an *upstream* source. To do so, after you have cloned your repository to your computer, go into your *biofx_python* directory:

```
$ cd biofx_python
```

Then execute this command:

```
$ git remote add upstream https://github.com/kyclark/biofx_python.git
```

Whenever you would like to update your repository from mine, you can execute this command:

```
$ git pull upstream main
```

Installing Modules

You will need to install several Python modules and tools. I've included a *requirements.txt* file in the top level of the repository. This file lists all the modules needed to run the programs in the book. Some IDEs may detect this file and offer to install these for you, or you can use the following command:

```
$ python3 -m pip install -r requirements.txt
```

Or use the `pip3` tool:

```
$ pip3 install -r requirements.txt
```

Sometimes `pylint` may complain about some of the variable names in the programs, and `mypy` will raise some issues when you import modules that do not have type annotations. To silence these errors, you can create initialization files in your home directory that these programs will use to customize their behavior. In the root of the source repository, there are files called *pylintrc* and *mypy.ini* that you should copy to your home directory like so:

```
$ cp pylintrc ~/.pylintrc
$ cp mypy.ini ~/.mypy.ini
```

Alternatively, you can generate a new *pylintrc* with the following command:

```
$ cd ~
$ pylint --generate-rcfile > .pylintrc
```

Feel free to customize these files to suit your tastes.

Installing the new.py Program

I wrote a Python program called `new.py` that creates Python programs. So meta, I know. I wrote this for myself and then gave it to my students because I think it's quite difficult to start writing a program from an empty screen. The `new.py` program will create a new, well-structured Python program that uses the `argparse` module to interpret command-line arguments. It should have been installed in the preceding section with the module dependencies. If not, you can use the `pip` module to install it, like so:

```
$ python3 -m pip install new-py
```

You should now be able to execute **new.py** and see something like this:

```
$ new.py
usage: new.py [-h] [-n NAME] [-e EMAIL] [-p PURPOSE] [-t] [-f] [--version]
              program
new.py: error: the following arguments are required: program
```

Each exercise will suggest that you use `new.py` to start writing your new programs. For instance, in Chapter 1 you will create a program called `dna.py` in the *01_dna* directory, like so:

```
$ cd 01_dna/
$ new.py dna.py
Done, see new script "dna.py".
```

If you then execute **./dna.py --help**, you will see that it generates help documentation on how to use the program. You should open the `dna.py` program in your editor, modify the arguments, and add your code to satisfy the requirements of the program and the tests.

Note that it's never a requirement that you use new.py. I only offer this as an aid to getting started. This is how I start every one of my own programs, but, while I find it useful, you may prefer to go a different route. As long as your programs pass the test suites, you are welcome to write them however you please.

Why Did I Write This Book?

Richard Hamming spent decades as a mathematician and researcher at Bell Labs. He was known for seeking out people he didn't know and asking them about their research. Then he would ask them what they thought were the biggest, most pressing unanswered questions in their field. If their answers for both of these weren't the same, he'd ask, "So why aren't you working on that?"

I feel that one of the most pressing problems in bioinformatics is that much of the software is poorly written and lacks proper documentation and testing, if it has any at all. I want to show you that it's *less* difficult to use types and tests and linters and formatters because it will prove easier over time to add new features and release more and better software. You will have the confidence to know for certain when your program is correct, for at least some measure of correctness.

To that end, I will demonstrate best practices in software development. Though I'm using Python as the medium, the principles apply to any language from C to R to JavaScript. The most important thing you can learn from this book is the craft of developing, testing, documenting, releasing, and supporting software, so that together we can all advance scientific research computing.

My career in bioinformatics was a product of wandering and happy accidents. I studied English literature and music in college, and then started playing with databases, HTML, and eventually learned programming on the job in the mid-1990s. By 2001, I'd become a decent Perl hacker, and I managed to get a job as a web developer for Dr. Lincoln Stein, an author of several Perl modules and books, at Cold Spring Harbor Laboratory (CSHL). He and my boss, Dr. Doreen Ware, patiently spoon-fed me enough biology to understand the programs they wanted to be written. I spent 13 years working on a comparative plant genomics database called Gramene.org, learning a decent amount of science while continuing to explore programming languages and computer science.

Lincoln was passionate about sharing everything from data and code to education. He started the Programming for Biology course at CSHL, a two-week intensive crash course to teach Unix command-line, Perl programming, and bioinformatics skills. The course is still being taught, although using Python nowadays, and I've had several opportunities to act as a teaching assistant. I've always found it rewarding to help someone learn a skill they will use to further their research.

It was during my tenure at CSHL that I met Bonnie Hurwitz, who eventually left to pursue her PhD at the University of Arizona (UA). When she started her new lab at UA, I was her first hire. Bonnie and I worked together for several years, and teaching became one of my favorite parts of the job. As with Lincoln's course, we introduced basic programming skills to scientists who wanted to branch out into more computational approaches.

Some of the materials I wrote for these classes became the foundation for my first book, *Tiny Python Projects*, where I try to teach the essential elements of Python language syntax as well as how to use tests to ensure that programs are correct and reproducible—elements crucial to scientific programming. This book picks up from there and focuses on the elements of Python that will help you write programs for biology.

Conventions Used in This Book

The following typographical conventions are used in this book:

Italic
> Indicates new terms, URLs, email addresses, filenames, and file extensions, as well as codons and DNA bases.

`Constant width`
> Used for program listings, as well as within paragraphs to refer to program elements such as variable or function names, databases, data types, environment variables, statements, and keywords.

`Constant width bold`
> Shows commands or other text that should be typed literally by the user.

`Constant width italic`
> Shows text that should be replaced with user-supplied values or by values determined by context.

 This element signifies a tip or suggestion.

This element signifies a general note.

This element indicates a warning or caution.

Using Code Examples

Supplemental material (code examples, exercises, etc.) is available for download at *https://github.com/kyclark/biofx_python*.

If you have a technical question or a problem using the code examples, please send email to *bookquestions@oreilly.com*.

This book is here to help you get your job done. In general, if example code is offered with this book, you may use it in your programs and documentation. You do not need to contact us for permission unless you're reproducing a significant portion of the code. For example, writing a program that uses several chunks of code from this book does not require permission. Selling or distributing examples from O'Reilly books does require permission. Answering a question by citing this book and quoting example code does not require permission. Incorporating a significant amount of example code from this book into your product's documentation does require permission.

We appreciate, but generally do not require, attribution. An attribution usually includes the title, author, publisher, and ISBN. For example: "*Mastering Python for Bioinformatics* by Ken Youens-Clark (O'Reilly). Copyright 2021 Charles Kenneth Youens-Clark, 978-1-098-10088-9."

If you feel your use of code examples falls outside fair use or the permission given above, feel free to contact us at *permissions@oreilly.com*.

O'Reilly Online Learning

For more than 40 years, *O'Reilly Media* has provided technology and business training, knowledge, and insight to help companies succeed.

Our unique network of experts and innovators share their knowledge and expertise through books, articles, and our online learning platform. O'Reilly's online learning platform gives you on-demand access to live training courses, in-depth learning paths, interactive coding environments, and a vast collection of text and video from O'Reilly and 200+ other publishers. For more information, visit *http://oreilly.com*.

How to Contact Us

Please address comments and questions concerning this book to the publisher:

O'Reilly Media, Inc.
1005 Gravenstein Highway North
Sebastopol, CA 95472
800-998-9938 (in the United States or Canada)
707-829-0515 (international or local)
707-829-0104 (fax)

We have a web page for this book, where we list errata, examples, and any additional information. You can access this page at *https://oreil.ly/mastering-bioinformatics-python*.

Email *bookquestions@oreilly.com* to comment or ask technical questions about this book.

For news and information about our books and courses, visit *http://oreilly.com*.

Find us on Facebook: *http://facebook.com/oreilly*

Follow us on Twitter: *http://twitter.com/oreillymedia*

Watch us on YouTube: *http://www.youtube.com/oreillymedia*

Acknowledgments

I want to thank the many people who have reviewed this book, including my editor, Corbin Collins; the entire production team but especially my production editor, Caitlin Ghegan; my technical reviewers, Al Scherer, Brad Fulton, Bill Lubanovic, Rangarajan Janani, and Joshua Orvis; and the many other people who provided much-appreciated feedback, including Mark Henderson, Marc Bañuls Tornero, and Dr. Scott Cain.

In my professional career, I've been extremely fortunate to have had many wonderful bosses, supervisors, and colleagues who've helped me grow and pushed me to be better. Eric Thorsen was the first person to see I had the potential to learn how to code, and he helped me learn various languages and databases as well as important lessons about sales and support. Steve Reppucci was my boss at boston.com, and he provided a much deeper understanding of Perl and Unix and how to be an honest and thoughtful team leader. Dr. Lincoln Stein at CSHL took a chance to hire someone who had no knowledge of biology to work in his lab, and he pushed me to create programs I didn't imagine I could. Dr. Doreen Ware patiently taught me biology and pushed me to assume leadership roles and publish. Dr. Bonnie Hurwitz supported me through many years of learning about high-performance computing, more programming languages, mentoring, teaching, and writing. In every position, I also had many colleagues who taught me as much about programming as about being human, and I thank everyone who has helped me along the way.

In my personal life, I would be nowhere without my family, who have loved and supported me. My parents have shown great support throughout my life, and I surely wouldn't be the person I am without them. Lori Kindler and I have been married 25 years, and I can't imagine a life without her. Together we generated three offspring who have been an incredible source of delight and challenge.

## PART I

The Rosalind.info Challenges

The chapters in this part explore the elements of Python's syntax and tooling that will enable you to write well-structured, documented, tested, and reproducible programs. I'll show you how to solve 14 challenges from Rosalind.info. These problems are short and focused and allow for many different solutions that will help you explore Python. I'll also teach you how to write a program, step-by-step, using tests to guide you and to let you know when you're done. I encourage you to read the Rosalind page for each problem, as I do not have space to recapitulate all the background and information there.

## CHAPTER 1

Tetranucleotide Frequency: Counting Things

Counting the bases in DNA is perhaps the "Hello, World!" of bioinformatics. The Rosalind DNA challenge (*https://oreil.ly/maR31*) describes a program that will take a sequence of DNA and print a count of how many *A*s, *C*s, *G*s, and *T*s are found. There are surprisingly many ways to count things in Python, and I'll explore what the language has to offer. I'll also demonstrate how to write a well-structured, documented program that validates its arguments as well as how to write and run tests to ensure the program works correctly.

In this chapter, you'll learn:

- How to start a new program using `new.py`
- How to define and validate command-line arguments using `argparse`
- How to run a test suite using `pytest`
- How to iterate the characters of a string
- Ways to count elements in a collection
- How to create a decision tree using `if`/`elif` statements
- How to format strings

Getting Started

Before you start, be sure you have read "Getting the Code and Tests" on page xvi in the Preface. Once you have a local copy of the code repository, change into the *01_dna* directory:

```
$ cd 01_dna
```

Here you'll find several solution*.py programs along with tests and input data you can use to see if the programs work correctly. To get an idea of how your program should work, start by copying the first solution to a program called dna.py:

```
$ cp solution1_iter.py dna.py
```

Now run the program with no arguments, or with the -h or --help flags. It will print usage documentation (note that *usage* is the first word of the output):

```
$ ./dna.py
usage: dna.py [-h] DNA
dna.py: error: the following arguments are required: DNA
```

 If you get an error like "permission denied," you may need to run **chmod +x dna.py** to change the mode of the program by adding the executable bit.

This is one of the first elements of reproducibility. *Programs should provide documentation on how they work.* While it's common to have something like a *README* file or even a paper to describe a program, the program itself must provide documentation on its parameters and outputs. I'll show you how to use the argparse module to define and validate the arguments as well as to generate the documentation, meaning that there is no possibility that the usage statement generated by the program could be incorrect. Contrast this with how *README* files and change logs and the like can quickly fall out of sync with a program's development, and I hope you'll appreciate that this sort of documentation is quite effective.

You can see from the usage line that the program expects something like DNA as an argument, so let's give it a sequence. As described on the Rosalind page, the program prints the counts for each of the bases *A*, *C*, *G*, and *T*, in that order and separated by a single space each:

```
$ ./dna.py ACCGGGTTTT
1 2 3 4
```

When you go to solve a challenge on the Rosalind.info website, the input for your program will be provided as a downloaded file; therefore, I'll write the program so that it will also read the contents of a file. I can use the command **cat** (for *concatenate*) to print the contents of one of the files in the *tests/inputs* directory:

```
$ cat tests/inputs/input2.txt
AGCTTTTCATTCTGACTGCAACGGGCAATATGTCTCTGTGTGGATTAAAAAAAGAGTGTCTGATAGCAGC
```

This is the same sequence shown in the example on the website. Accordingly, I know that the output of the program should be this:

```
$ ./dna.py tests/inputs/input2.txt
20 12 17 21
```

Throughout the book, I'll use the `pytest` tool to run the tests that ensure programs work as expected. When I run the command **pytest**, it will recursively search the current directory for tests and functions that look like tests. Note that you may need to run **python3 -m pytest** or **pytest.exe** if you are on Windows. Run this now, and you should see something like the following to indicate that the program passes all four of the tests found in the *tests/dna_test.py* file:

```
$ pytest
=========================== test session starts ===========================
...
collected 4 items

tests/dna_test.py ....                                              [100%]

=========================== 4 passed in 0.41s ===========================
```

 A key element to testing software is that you *run your program with known inputs and verify that it produces the correct output*. While that may seem like an obvious idea, I've had to object to "testing" schemes that simply ran programs but never verified that they behaved correctly.

Creating the Program Using new.py

If you copied one of the solutions, as shown in the preceding section, then delete that program so you can start from scratch:

```
$ rm dna.py
```

Without looking at my solutions yet, I want you to try to solve this problem. If you think you have all the information you need, feel free to jump ahead and write your own version of dna.py, using `pytest` to run the provided tests. Keep reading if you want to go step-by-step with me to learn how to write the program and run the tests.

Every program in this book will accept some command-line argument(s) and create some output, like text on the command line or new files. I'll always use the new.py program described in the Preface to start, but this is not a requirement. You can write your programs however you like, starting from whatever point you want, but your programs are expected to have the same features, such as generating usage statements and properly validating arguments.

Create your dna.py program in the *01_dna* directory, as this contains the test files for the program. Here is how I will start the dna.py program. The --purpose argument will be used in the program's documentation:

```
$ new.py --purpose 'Tetranucleotide frequency' dna.py
Done, see new script "dna.py."
```

If you run the new dna.py program, you will see that it defines many different types of arguments common to command-line programs:

```
$ ./dna.py --help
usage: dna.py [-h] [-a str] [-i int] [-f FILE] [-o] str

Tetranucleotide frequency ❶

positional arguments:
  str                   A positional argument ❷

optional arguments:
  -h, --help            show this help message and exit ❸
  -a str, --arg str     A named string argument (default: ) ❹
  -i int, --int int     A named integer argument (default: 0) ❺
  -f FILE, --file FILE  A readable file (default: None) ❻
  -o, --on              A boolean flag (default: False) ❼
```

❶ The --purpose from new.py is used here to describe the program.

❷ The program accepts a single positional string argument.

❸ The -h and --help flags are automatically added by argparse and will trigger the usage.

❹ This is a named option with short (-a) and long (--arg) names for a string value.

❺ This is a named option with short (-i) and long (--int) names for an integer value.

❻ This is a named option with short (-f) and long (--file) names for a file argument.

❼ This is a Boolean flag that will be True when either -o or --on is present and False when they are absent.

This program only needs the `str` positional argument, and you can use `DNA` for the `metavar` value to give some indication to the user as to the meaning of the argument. Delete all the other parameters. Note that you never define the `-h` and `--help` flags, as `argparse` uses those internally to respond to usage requests. See if you can modify your program until it will produce the usage that follows (if you can't produce the usage just yet, don't worry, I'll show this in the next section):

```
$ ./dna.py -h
usage: dna.py [-h] DNA

Tetranucleotide frequency

positional arguments:
  DNA          Input DNA sequence

optional arguments:
  -h, --help  show this help message and exit
```

If you can manage to get this working, I'd like to point out that this program will accept exactly one positional argument. If you try running it with any other number of arguments, the program will immediately halt and print an error message:

```
$ ./dna.py AACC GGTT
usage: dna.py [-h] DNA
dna.py: error: unrecognized arguments: GGTT
```

Likewise, the program will reject any unknown flags or options. With very few lines of code, you have built a documented program that validates the arguments to the program. That's a very basic and important step toward reproducibility.

Using argparse

The program created by `new.py` uses the `argparse` module to define the program's parameters, validate that the arguments are correct, and create the usage documentation for the user. The `argparse` module is a *standard* Python module, which means it's always present. Other modules can also do these things, and you are free to use any method you like to handle this aspect of your program. Just be sure your programs can pass the tests.

I wrote a version of `new.py` for *Tiny Python Projects* that you can find in the *bin* directory of that book's GitHub repo (*https://oreil.ly/7romb*). That version is somewhat simpler than the version I want you to use. I'll start by showing you a version of `dna.py` created using this earlier version of `new.py`:

```
#!/usr/bin/env python3 ❶
""" Tetranucleotide frequency """ ❷

import argparse ❸
```

```
#  --------------------------------------------------
def get_args():  ❹
    """ Get command-line arguments """  ❺

    parser = argparse.ArgumentParser(  ❻
        description='Tetranucleotide frequency',
        formatter_class=argparse.ArgumentDefaultsHelpFormatter)

    parser.add_argument('dna', metavar='DNA', help='Input DNA sequence')  ❼

    return parser.parse_args()  ❽


#  --------------------------------------------------
def main():  ❾
    """ Make a jazz noise here """

    args = get_args()  ❿
    print(args.dna)     ⓫


#  --------------------------------------------------
if __name__ == '__main__':  ⓬
    main()
```

❶ The colloquial *shebang* (#!) tells the operating system to use the env command (*environment*) to find python3 to execute the rest of the program.

❷ This is a *docstring* (documentation string) for the program or module as a whole.

❸ I import the argparse module to handle command-line arguments.

❹ I always define a get_args() function to handle the argparse code.

❺ This is a docstring for a function.

❻ The parser object is used to define the program's parameters.

❼ I define a dna argument, which will be positional because the name dna *does not* start with a dash. The metavar is a short description of the argument that will appear in the short usage. No other arguments are needed.

❽ The function returns the results of parsing the arguments. The help flags or any problems with the arguments will cause argparse to print a usage statement/ error messages and exit the program.

❾ All programs in the book will always start in the main() function.

⑩ The first step in main() will always be to call get_args(). If this call succeeds, then the arguments must have been valid.

⑪ The DNA value is available in the args.dna attribute, as this is the name of the argument.

⑫ This is a common idiom in Python programs to detect when the program is being executed (as opposed to being imported) and to execute the main() function.

The shebang line is used by the Unix shell when the program is invoked as a program, like ./dna.py. It does not work on Windows, where you are required to run **python.exe dna.py** to execute the program.

While this code works completely adequately, the value returned from get_args() is an argparse.Namespace object that is *dynamically generated* when the program runs. That is, I am using code like parser.add_argument() to modify the structure of this object *at runtime*, so Python is unable to know positively *at compile time* what attributes will be available in the parsed arguments or what their types would be. While it may be obvious to you that there can only be a single, required string argument, there is not enough information in the code for Python to discern this.

To *compile* a program is to turn it into the machine code that a computer can execute. Some languages, like C, must be compiled separately before they can be run. Python programs are often compiled and run in one step, but there is still a compilation phase. Some errors can be caught at compilation, and others don't turn up until runtime. For instance, syntax errors will prevent compilation. It is preferable to have compile-time errors over runtime errors.

To see why this could be a problem, I'll alter the main() function to introduce a type error. That is, I'll intentionally misuse the *type* of the args.dna value. Unless otherwise stated, all argument values returned from the command line by argparse are strings. If I try to divide the string args.dna by the integer value 2, Python will raise an exception and crash the program at runtime:

```
def main():
    args = get_args()
    print(args.dna / 2) ❶
```

❶ Dividing a string by an integer will produce an exception.

If I run the program, it crashes as expected:

```
$ ./dna.py ACGT
Traceback (most recent call last):
  File "./dna.py", line 30, in <module>
    main()
  File "./dna.py", line 25, in main
    print(args.dna / 2)
TypeError: unsupported operand type(s) for /: 'str' and 'int'
```

Our big squishy brains know that this is an inevitable error waiting to happen, but Python can't see the problem. What I need is a *static* definition of the arguments that cannot be modified when the program is run. Read on to see how type annotations and other tools can detect these sorts of bugs.

Tools for Finding Errors in the Code

The goal here is to write correct, reproducible programs in Python. Are there ways to spot and avoid problems like misusing a string in a numeric operation? The python3 interpreter found no problems that prevented me from running the code. That is, the program is syntactically correct, so the code in the preceding section produces a *runtime error* because the error happens only when I execute the program. Years back I worked in a group where we joked, "If it compiles, ship it!" This is clearly a myopic approach when coding in Python.

I can use tools like linters and type checkers to find some kinds of problems in code. *Linters* are tools that check for program style and many kinds of errors beyond bad syntax. The pylint tool (*https://www.pylint.org*) is a popular Python linter that I use almost every day. Can it find this problem? Apparently not, as it gives the biggest of thumbs-ups:

```
$ pylint dna.py

--------------------------------------------------------------------
Your code has been rated at 10.00/10 (previous run: 9.78/10, +0.22)
```

The flake8 (*https://oreil.ly/b3Qtj*) tool is another linter that I often use in combination with pylint, as it will report different kinds of errors. When I run flake8 dna.py, I get no output, which means it found no errors to report.

The mypy (*http://mypy-lang.org*) tool is a static *type checker* for Python, meaning it is designed to find misused types such as trying to divide a string by a number. Neither pylint nor flake8 is designed to catch type errors, so I cannot be legitimately surprised they missed the bug. So what does mypy have to say?

```
$ mypy dna.py
Success: no issues found in 1 source file
```

Well, that's just a little disappointing; however, you must understand that mypy is failing to report a problem *because there is no type information.* That is, mypy has no information to say that dividing args.dna by 2 is wrong. I'll fix that shortly.

Using Python's Interactive Interpreter

In the next section, I want to show you how to use Python's interactive interpreter python3 to run short pieces of code. This kind of interface is sometimes called a *REPL*, which stands for *Read-Evaluate-Print-Loop*, and I pronounce this so that it sort of rhymes with *pebble*. Each time you enter code in the REPL, Python will immediately *read* and *evaluate* the expressions, *print* the results, and then *loop* back to wait for more input.

Using a REPL may be new to you, so let's take a moment to introduce the idea. While I'll demonstrate using python3, you may prefer to use idle3, ipython, a Jupyter Notebook, or a Python console inside an integrated development environment (IDE) like VS Code, as in Figure 1-1.

Figure 1-1. You can run an interactive Python interpreter inside VS Code

Regardless of your choice, I *highly* encourage you to type all the examples yourself. You will learn so much from interacting with the Python interpreter. To start the

REPL, type **python3**, or possibly just **python** on your computer if that points to the latest version. Here is what it looks like on my computer:

```
$ python3
Python 3.9.1 (v3.9.1:1e5d33e9b9, Dec  7 2020, 12:10:52)
[Clang 6.0 (clang-600.0.57)] on darwin
Type "help", "copyright", "credits" or "license" for more information.
>>>
```

The standard python3 REPL uses >>> as a prompt. Be sure you don't type the >>> prompt, only the text that it follows. For example, if I demonstrate adding two numbers like this:

```
>>> 3 + 5
8
```

You should only type **3 + 5<Enter>**. If you include the leading prompt, you will get an error:

```
>>> >>> 3 + 5
  File "<stdin>", line 1
    >>> 3 + 5
    ^
SyntaxError: invalid syntax
```

I especially like to use the REPL to read documentation. For instance, type **help(str)** to read about Python's string class. Inside the help documentation, you can move forward with the F key, Ctrl-F, or the space bar, and you can move backward with the B key or Ctrl-B. Search by pressing the / key followed by a string and then Enter. To leave the help, press Q. To quit the REPL, type **exit()** or press Ctrl-D.

Introducing Named Tuples

To avoid the problems with dynamically generated objects, all of the programs in this book will use a named tuple data structure to statically define the arguments from get_args(). *Tuples* are essentially immutable lists, and they are often used to represent record-type data structures in Python. There's quite a bit to unpack with all that, so let's step back to lists.

To start, *lists* are ordered sequences of items. The items can be heterogeneous; in theory, this means all the items can be of different types, but in practice, mixing types is often a bad idea. I'll use the python3 REPL to demonstrate some aspects of lists. I recommend you use **help(list)** to read the documentation.

Use empty square brackets ([]) to create an empty list that will hold some sequences:

```
>>> seqs = []
```

The `list()` function will also create a new, empty list:

```
>>> seqs = list()
```

Verify that this is a list by using the `type()` function to return the variable's type:

```
>>> type(seqs)
<class 'list'>
```

Lists have methods that will add values to the end of the list, like `list.append()` to add one value:

```
>>> seqs.append('ACT')
>>> seqs
['ACT']
```

and `list.extend()` to add multiple values:

```
>>> seqs.extend(['GCA', 'TTT'])
>>> seqs
['ACT', 'GCA', 'TTT']
```

If you type the variable by itself in the REPL, it will be evaluated and stringified into a textual representation:

```
>>> seqs
['ACT', 'GCA', 'TTT']
```

This is basically the same thing that happens when you `print()` a variable:

```
>>> print(seqs)
['ACT', 'GCA', 'TTT']
```

You can modify any of the values *in-place* using the index. Remember that all indexing in Python is 0-based, so 0 is the first element. Change the first sequence to be TCA:

```
>>> seqs[0] = 'TCA'
```

Verify that it was changed:

```
>>> seqs
['TCA', 'GCA', 'TTT']
```

Like lists, tuples are ordered sequences of possibly heterogeneous objects. Whenever you put commas between items in a series, you are creating a tuple:

```
>>> seqs = 'TCA', 'GCA', 'TTT'
>>> type(seqs)
<class 'tuple'>
```

It's typical to place parentheses around tuple values to make this more explicit:

```
>>> seqs = ('TCA', 'GCA', 'TTT')
>>> type(seqs)
<class 'tuple'>
```

Unlike lists, tuples cannot be changed once they are created. If you read help(tuple), you will see that a tuple is a *built-in immutable sequence*, so I cannot add values:

```
>>> seqs.append('GGT')
Traceback (most recent call last):
  File "<stdin>", line 1, in <module>
AttributeError: 'tuple' object has no attribute 'append'
```

or modify existing values:

```
>>> seqs[0] = 'TCA'
Traceback (most recent call last):
  File "<stdin>", line 1, in <module>
TypeError: 'tuple' object does not support item assignment
```

It's fairly common in Python to use tuples to represent records. For instance, I might represent a Sequence having a unique ID and a string of bases:

```
>>> seq = ('CAM_0231669729', 'GTGTTTATTCAATGCTAG')
```

While it's possible to use indexing to get the values from a tuple just as with lists, that's awkward and error-prone. *Named tuples* allow me to assign names to the fields, which makes them more ergonomic to use. To use named tuples, I can import the namedtuple() function from the collections module:

```
>>> from collections import namedtuple
```

As shown in Figure 1-2, I use the namedtuple() function to create the idea of a Sequence that has fields for the id and the seq:

```
>>> Sequence = namedtuple('Sequence', ['id', 'seq'])
```

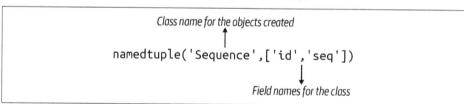

Figure 1-2. The namedtuple() function generates a way to make objects of the class Sequence that have the fields id and seq

What exactly is Sequence here?

```
>>> type(Sequence)
<class 'type'>
```

I've just created a new type. You might call the Sequence() function a *factory* because it's a function used to generate new objects of the class Sequence. It's a common naming convention for these factory functions and class names to be TitleCased to set them apart.

Just as I can use the `list()` function to create a new list, I can use the `Sequence()` function to create a new `Sequence` object. I can pass the `id` and `seq` values *positionally* to match the order they are defined in the class:

```
>>> seq1 = Sequence('CAM_0231669729', 'GTGTTTATTCAATGCTAG')
>>> type(seq1)
<class '__main__.Sequence'>
```

Or I can use the field names and pass them as key/value pairs in any order I like:

```
>>> seq2 = Sequence(seq='GTGTTTATTCAATGCTAG', id='CAM_0231669729')
>>> seq2
Sequence(id='CAM_0231669729', seq='GTGTTTATTCAATGCTAG')
```

While it's possible to use indexes to access the ID and sequence:

```
>>> 'ID = ' + seq1[0]
'ID = CAM_0231669729'
>>> 'seq = ' + seq1[1]
'seq = GTGTTTATTCAATGCTAG'
```

…the whole point of named tuples is to use the field names:

```
>>> 'ID = ' + seq1.id
'ID = CAM_0231669729'
>>> 'seq = ' + seq1.seq
'seq = GTGTTTATTCAATGCTAG'
```

The record's values remain immutable:

```
>>> seq1.id = 'XXX'
Traceback (most recent call last):
  File "<stdin>", line 1, in <module>
AttributeError: can't set attribute
```

I often want a guarantee that a value cannot be accidentally changed in my code. Python doesn't have a way to declare that a variable is *constant* or immutable. Tuples are by default immutable, and I think it makes sense to represent the arguments to a program using a data structure that cannot be altered. The inputs are sacrosanct and should (almost) never be modified.

Adding Types to Named Tuples

As nice as `namedtuple()` is, I can make it even better by importing the `NamedTuple` class from the `typing` module to use as the base class for the `Sequence`. Additionally, I can assign *types* to the fields using this syntax. Note the need to use an empty line in the REPL to indicate that the block is complete:

```
>>> from typing import NamedTuple
>>> class Sequence(NamedTuple):
...     id: str
...     seq: str
...
```

The ... you see are line continuations. The REPL is showing that what's been entered so far is not a complete expression. You need to enter a blank line to let the REPL know that you're done with the code block.

As with the `namedtuple()` method, `Sequence` is a new type:

```
>>> type(Sequence)
<class 'type'>
```

The code to instantiate a new `Sequence` object is the same:

```
>>> seq3 = Sequence('CAM_0231669729', 'GTGTTTATTCAATGCTAG')
>>> type(seq3)
<class '__main__.Sequence'>
```

I can still access the fields by names:

```
>>> seq3.id, seq3.seq
('CAM_0231669729', 'GTGTTTATTCAATGCTAG')
```

Since I defined that both fields have `str` types, you might assume this would *not* work:

```
>>> seq4 = Sequence(id='CAM_0231669729', seq=3.14)
```

I'm sorry to tell you that Python itself ignores the type information. You can see the `seq` field that I hoped would be a `str` is actually a `float`:

```
>>> seq4
Sequence(id='CAM_0231669729', seq=3.14)
>>> type(seq4.seq)
<class 'float'>
```

So how does this help us? It doesn't help me in the REPL, but adding types to my source code will allow type-checking tools like `mypy` to find such errors.

Representing the Arguments with a NamedTuple

I want the data structure that represents the program's arguments to include type information. As with the `Sequence` class, I can define a class that is derived from the `NamedTuple` type where I can *statically define the data structure with types*. I like to call this class `Args`, but you can call it whatever you like. I know this probably seems like driving a finishing nail with a sledgehammer, but trust me, this kind of detail will pay off in the future.

The latest `new.py` uses the `NamedTuple` class from the `typing` module. Here is how I suggest you define and represent the arguments:

```
#!/usr/bin/env python3
"""Tetranucleotide frequency"""
```

```
import argparse
from typing import NamedTuple ❶


class Args(NamedTuple): ❷
    """ Command-line arguments """
    dna: str ❸


# --------------------------------------------------
def get_args() -> Args: ❹
    """ Get command-line arguments """

    parser = argparse.ArgumentParser(
        description='Tetranucleotide frequency',
        formatter_class=argparse.ArgumentDefaultsHelpFormatter)

    parser.add_argument('dna', metavar='DNA', help='Input DNA sequence')

    args = parser.parse_args() ❺

    return Args(args.dna) ❻


# --------------------------------------------------
def main() -> None: ❼
    """ Make a jazz noise here """

    args = get_args()
    print(args.dna / 2) ❽


# --------------------------------------------------
if __name__ == '__main__':
    main()
```

❶ Import the NamedTuple class from the typing module.

❷ Define a class for the arguments which is based on the NamedTuple class. See the following note.

❸ The class has a single field called dna that has the type str.

❹ The type annotation on the get_args() function shows that it returns an object of the type Args.

❺ Parse the arguments as before.

❻ Return a new `Args` object that contains the single value from `args.dna`.

❼ The `main()` function has no `return` statement, so it returns the default `None` value.

❽ This is the type error from the earlier program.

 If you run `pylint` on this program, you may encounter the errors "Inheriting *NamedTuple*, which is not a class. (inherit-non-class)" and "Too few public methods (0/2) (too-few-public-methods)." You can disable these warnings by adding "inherit-non-class" and "too-few-public-methods" to the "disable" section of your *pylintrc* file, or use the *pylintrc* file included in the root of the GitHub repository.

If you run this program, you'll see it still creates the same uncaught exception. Both `flake8` and `pylint` will continue to report that the program looks fine, but see what `mypy` tells me now:

```
$ mypy dna.py
dna.py:32: error: Unsupported operand types for / ("str" and "int")
Found 1 error in 1 file (checked 1 source file)
```

The error message shows that there is a problem on line 32 with the operands, which are the arguments to the division (/) operator. I'm mixing string and integer values. Without the type annotations, `mypy` would be unable to find a bug. Without this warning from `mypy`, I'd have to run my program to find it, being sure to exercise the branch of code that contains the error. In this case, it's all rather obvious and trivial, but in a much larger program with hundreds or thousands of lines of code (LOC) with many functions and logical branches (like `if`/`else`), I might not stumble upon this error. I rely on types and programs like `mypy` (and `pylint` and `flake8` and so on) to correct these kinds of errors rather than relying solely on tests, or worse, waiting for users to report bugs.

Reading Input from the Command Line or a File

When you attempt to prove that your program works on the Rosalind.info website, you will download a data file containing the input to your program. Usually, this data will be much larger than the sample data described in the problem. For instance, the example DNA string for this problem is 70 bases long, but the one I downloaded for one of my attempts was 910 bases.

Let's make the program read input both from the command line and from a text file so that you don't have to copy and paste the contents from a downloaded file. This is a common pattern I use, and I prefer to handle this option inside the get_args() function since this pertains to processing the command-line arguments.

First, correct the program so that it prints the args.dna value without the division:

```
def main() -> None:
    args = get_args()
    print(args.dna) ❶
```

❶   Remove the division type error.

Check that it works:

```
$ ./dna.py ACGT
ACGT
```

For this next part, you need to bring in the os module to interact with your operating system. Add import os to the other import statements at the top, then add these two lines to your get_args() function:

```
def get_args() -> Args:
    """ Get command-line arguments """

    parser = argparse.ArgumentParser(
        description='Tetranucleotide frequency',
        formatter_class=argparse.ArgumentDefaultsHelpFormatter)

    parser.add_argument('dna', metavar='DNA', help='Input DNA sequence')

    args = parser.parse_args()

    if os.path.isfile(args.dna):  ❶
        args.dna = open(args.dna).read().rstrip()   ❷

    return Args(args.dna)
```

❶   Check if the dna value is a file.

❷   Call open() to open a filehandle, then chain the fh.read() method to return a string, then chain the str.rstrip() method to remove trailing whitespace.

 The fh.read() function will read an *entire* file into a variable. In this case, the input file is small and so this should be fine, but it's very common in bioinformatics to process files that are gigabytes in size. Using read() on a large file could crash your program or even your entire computer. Later I will show you how to read a file line-by-line to avoid this.

Now run your program with a string value to ensure it works:

```
$ ./dna.py ACGT
ACGT
```

and then use a text file as the argument:

```
$ ./dna.py tests/inputs/input2.txt
AGCTTTTCATTCTGACTGCAACGGGCAATATGTCTCTGTGTGGATTAAAAAAAGAGTGTCTGATAGCAGC
```

Now you have a flexible program that reads input from two sources. Run **mypy dna.py** to make sure there are no problems.

Testing Your Program

You know from the Rosalind description that given the input ACGT, the program should print 1 1 1 1 since that is the number of *A*s, *C*s, *G*s, and *T*s, respectively. In the *01_dna/tests* directory, there is a file called *dna_test.py* that contains tests for the dna.py program. I wrote these tests for you so you can see what it's like to develop a program using a method to tell you with some certainty when your program is correct. The tests are really basic—given an input string, the program should print the correct counts for the four nucleotides. When the program reports the correct numbers, then it works.

Inside the *01_dna* directory, I'd like you to run **pytest** (or **python3 -m pytest** or **pytest.exe** on Windows). The program will recursively search for all files with names that start with *test_* or end with *_test.py*. It will then run for any functions in these files that have names starting with *test_*.

When you run **pytest**, you will see a lot of output, most of which is failing tests. To understand why these tests are failing, let's look at the *tests/dna_test.py* module:

```
""" Tests for dna.py """ ❶

import os ❷
import platform ❸
from subprocess import getstatusoutput ❹

PRG = './dna.py' ❺
RUN = f'python {PRG}' if platform.system() == 'Windows' else PRG ❻
TEST1 = ('./tests/inputs/input1.txt', '1 2 3 4') ❼
TEST2 = ('./tests/inputs/input2.txt', '20 12 17 21')
TEST3 = ('./tests/inputs/input3.txt', '196 231 237 246')
```

❶ This is the docstring for the module.

❷ The standard os module will interact with the operating system.

❸ The platform module is used to determine if this is being run on Windows.

❹ From the `subprocess` module I import a function to run the `dna.py` program and capture the output and status.

❺ These following lines are global variables for the program. I tend to avoid globals except in my tests. Here I want to define some values that I'll use in the functions. I like to use UPPERCASE_NAMES to highlight the global visibility.

❻ The `RUN` variable determines how to run the `dna.py` program. On Windows, the `python` command must be used to run a Python program, but on Unix platforms, the `dna.py` program can be directly executed.

❼ The `TEST*` variables are tuples that define a file containing a string of DNA and the expected output from the program for that string.

The `pytest` module will run the test functions in the order in which they are defined in the test file. I often structure my tests so that they progress from the simplest cases to more complex, so there's usually no point in continuing after a failure. For instance, the first test is always that the program to test exists. If it doesn't, then there's no point in running more tests. I recommend you run `pytest` with the `-x` flag to stop on the first failing test along with the `-v` flag for verbose output.

Let's look at the first test. The function is called `test_exists()` so that `pytest` will find it. In the body of the function, I use one or more `assert` statements to check if some condition is *truthy*.[1] Here I assert that the program `dna.py` exists. This is why your program must exist in this directory—otherwise it wouldn't be found by the test:

```
def test_exists():  ❶
    """ Program exists """

    assert os.path.exists(PRG)  ❷
```

❶ The function name must start with `test_` to be found by `pytest`.

❷ The `os.path.exists()` function returns `True` if the given argument is a file. If it returns `False`, then the assertion fails and this test will fail.

The next test I write is always to check that the program will produce a usage statement for the `-h` and `--help` flags. The `subprocess.getstatusoutput()` function will run the `dna.py` program with the short and long help flags. In each case, I want to see that the program prints text starting with the word *usage:*. It's not a perfect test. It

---

1 Boolean types are `True` or `False`, but many other data types are *truthy* or conversely *falsey*. The empty `str` (`""`) is falsey, so any nonempty string is truthy. The number 0 is falsey, so any nonzero value is truthy. An empty `list`, `set`, or `dict` is falsey, so any nonempty one of those is truthy.

doesn't check that the documentation is accurate, only that it appears to be something that might be a usage statement. I don't feel that every test needs to be completely exhaustive. Here's the test:

```
def test_usage() -> None:
    """ Prints usage """

    for arg in ['-h', '--help']:  ❶
        rv, out = getstatusoutput(f'{RUN} {arg}')  ❷
        assert rv == 0  ❸
        assert out.lower().startswith('usage:')  ❹
```

❶ Iterate over the short and long help flags.

❷ Run the program with the argument and capture the return value and output.

❸ Verify that the program reports a successful exit value of 0.

❹ Assert that the lowercased output of the program starts with the text *usage:*.

 Command-line programs usually indicate an error to the operating system by returning a nonzero value. If the program runs successfully, it ought to return a 0. Sometimes that nonzero value may correlate to some internal error code, but often it just means that something went wrong. The programs I write will, likewise, always strive to report 0 for successful runs and some nonzero value when there are errors.

Next, I want to ensure that the program will die when given no arguments:

```
def test_dies_no_args() -> None:
    """ Dies with no arguments """

    rv, out = getstatusoutput(RUN)  ❶
    assert rv != 0  ❷
    assert out.lower().startswith('usage:')  ❸
```

❶ Capture the return value and output from running the program with no arguments.

❷ Verify that the return value is a nonzero failure code.

❸ Check that the output looks like a usage statement.

At this point in testing, I know that I have a program with the correct name that can be run to produce documentation. This means that the program is at least syntactically correct, which is a decent place to start testing. If your program has typographical errors, then you'll be forced to correct those to even get to this point.

Running the Program to Test the Output

Now I need to see if the program does what it's supposed to do. There are many ways to test programs, and I like to use two basic approaches I call *inside-out* and *outside-in*. The inside-out approach starts at the level of testing individual functions inside a program. This is often called *unit* testing, as functions might be considered a basic unit of computing, and I'll get to this in the solutions section. I'll start with the outside-in approach. This means that I will run the program from the command line just as the user will run it. This is a holistic approach to check if the pieces of the code can work together to create the correct output, and so it's sometimes called an *integration* test.

The first such test will pass the DNA string as a command-line argument and check if the program produces the right counts formatted in the correct string:

```python
def test_arg():
    """ Uses command-line arg """

    for file, expected in [TEST1, TEST2, TEST3]:   ❶
        dna = open(file).read()   ❷
        retval, out = getstatusoutput(f'{RUN} {dna}')   ❸
        assert retval == 0   ❹
        assert out == expected   ❺
```

❶   Unpack the tuples into the `file` containing a string of DNA and the `expected` value from the program when run with this input.

❷   Open the file and read the dna from the contents.

❸   Run the program with the given DNA string using the function `subprocess.getstatusoutput()`, which gives me both the return value from the program and the text output (also called STDOUT, which is pronounced *standard out*).

❹   Assert that the return value is 0, which indicates success (or 0 errors).

❺   Assert that the output from the program is the string of numbers expected.

The next test is almost identical, but this time I'll pass the filename as the argument to the program to verify that it correctly reads the DNA from a file:

```python
def test_file():
    """ Uses file arg """
```

```
    for file, expected in [TEST1, TEST2, TEST3]:
        retval, out = getstatusoutput(f'{RUN} {file}')  ❶
        assert retval == 0
        assert out == expected
```

❶ The only difference from the first test is that I pass the filename instead of the
contents of the file.

Now that you've looked at the tests, go back and run the tests again. This time, use
pytest -xv, where the -v flag is for verbose output. Since both -x and -v are short
flags, you can combine them like -xv or -vx. Read the output closely and notice that
it's trying to tell you that the program is printing the DNA sequence but that the test
is expecting a sequence of numbers:

```
$ pytest -xv
============================ test session starts ============================
...

tests/dna_test.py::test_exists PASSED                                  [ 25%]
tests/dna_test.py::test_usage PASSED                                   [ 50%]
tests/dna_test.py::test_arg FAILED                                     [ 75%]


================================== FAILURES ==================================
_____ test_arg _____

    def test_arg():
        """ Uses command-line arg """

        for file, expected in [TEST1, TEST2, TEST3]:
            dna = open(file).read()
            retval, out = getstatusoutput(f'{RUN} {dna}')
            assert retval == 0
>           assert out == expected      ❶
E           AssertionError: assert 'ACCGGGTTTT' == '1 2 3 4'  ❷
E             - 1 2 3 4
E             + ACCGGGTTTT

tests/dna_test.py:36: AssertionError
=========================== short test summary info ===========================
FAILED tests/dna_test.py::test_arg - AssertionError: assert 'ACCGGGTTTT' == '...
!!!!!!!!!!!!!!!!!!!!!!!!!!!! stopping after 1 failures !!!!!!!!!!!!!!!!!!!!!!!!!!!!
=========================== 1 failed, 2 passed in 0.35s ===========================
```

❶ The > at the beginning of this line shows that this is the source of the error.

❷ The output from the program was the string ACCGGGTTTT but the expected value
was 1 2 3 4. Since these are not equal, an AssertionError exception is raised.

Let's fix that. If you think you know how to finish the program, please jump right into your solution. First, perhaps try running your program to verify that it will report the correct number of *A*s:

```
$ ./dna.py A
1 0 0 0
```

And then *C*s:

```
$ ./dna.py C
0 1 0 0
```

and so forth with *G*s and *T*s. Then run **pytest** to see if it passes all the tests.

After you have a working version, consider trying to find as many different ways as you can to get the same answer. This is called *refactoring* a program. You need to start with something that works correctly, and then you try to improve it. The improvements can be measured in many ways. Perhaps you find a way to write the same idea using less code, or maybe you find a solution that runs faster. No matter what metric you're using, keep running **pytest** to ensure the program is correct.

Solution 1: Iterating and Counting the Characters in a String

If you don't know where to start, I'll work through the first solution with you. The goal is to travel through all the bases in the DNA string. So, first I need to create a variable called dna by assigning it some value in the REPL:

```
>>> dna = 'ACGT'
```

Note that any value enclosed in quotes, whether single or double, is a string. Even a single character in Python is considered a string. I will often use the type() function to verify the type of a variable, and here I see that dna is of the class str (string):

```
>>> type(dna)
<class 'str'>
```

 Type help(str) in the REPL to see all the wonderful things you can do with strings. This data type is especially important in genomics, where strings comprise so much of the data.

In the parlance of Python, I want to *iterate* through the characters of a string, which in this case are the nucleotides of DNA. A for loop will do that. Python sees a string as an ordered sequence of characters, and a for loop will visit each character from beginning to end:

```
>>> for base in dna: ❶
...     print(base) ❷
...
A
C
G
T
```

❶ Each character in the dna string will be copied into the base variable. You could call this char, or c for *character*, or whatever else you like.

❷ Each call to print() will end with a newline, so you'll see each base on a separate line.

Later you will see that for loops can be used with lists and dictionaries and sets and lines in a file—basically any iterable data structure.

Counting the Nucleotides

Now that I know how to visit each base in the sequence, I need to count each base rather than printing it. That means I'll need some variables to keep track of the numbers for each of the four nucleotides. One way to do this is to create four variables that hold integer counts, one for each base. I will *initialize* four variables for counting by setting their initial values to 0:

```
>>> count_a = 0
>>> count_c = 0
>>> count_g = 0
>>> count_t = 0
```

I could write this in one line by using the tuple unpacking syntax that I showed earlier:

```
>>> count_a, count_c, count_g, count_t = 0, 0, 0, 0
```

Variable Naming Conventions

I could have named my variables countA or CountA or COUNTA or count_A or any number of ways, but I always stick to the suggested naming conventions in the "Style Guide for Python Code" (*https://oreil.ly/UmUYt*), also known as *PEP8*, which says that function and variable names "should be lowercase, with words separated by underscores as necessary to improve readability."

I need to look at each base and determine which variable to *increment*, making its value increase by 1. For instance, if the current base is a C, then I should increment the count_c variable. I could write this:

```
for base in dna:
    if base == 'C':
        count_c = count_c + 1  ❷
```

❶ The == operator is used to compare two values for equality. Here I want to know if the current **base** is equal to the string C.

❷ Set **count_c** equal to 1 greater than the current value.

> The == operator is used to compare two values for equality. It works to compare two strings or two numbers. I showed earlier that division with / will raise an exception if you mix strings and numbers. What happens if you mix types with this operator, for example '3' == 3? Is this a safe operator to use without first comparing the types?

As shown in Figure 1-3, a shorter way to increment a variable uses the += operator to add whatever is on the righthand side (often noted as RHS) of the expression to whatever is on the lefthand side (or LHS):

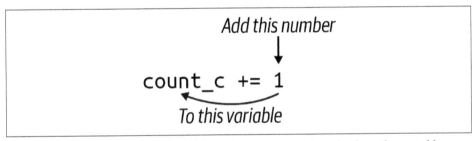

Figure 1-3. The += operator will add the value on the righthand side to the variable on the lefthand side

Since I have four nucleotides to check, I need a way to combine three more **if** expressions. The syntax in Python for this is to use **elif** for *else if* and **else** for any final or default case. Here is a block of code I can enter in the program or the REPL that implements a simple decision tree:

```
dna = 'ACCGGGTTTT'
count_a, count_c, count_g, count_t = 0, 0, 0, 0
for base in dna:
    if base == 'A':
        count_a += 1
    elif base == 'C':
        count_c += 1
    elif base == 'G':
        count_g += 1
```

```
    elif base == 'T':
        count_t += 1
```

I should end up with counts of 1, 2, 3, and 4 for each of the sorted bases:

```
>>> count_a, count_c, count_g, count_t
(1, 2, 3, 4)
```

Now I need to report the outcome to the user:

```
>>> print(count_a, count_c, count_g, count_t)
1 2 3 4
```

That is the exact output the program expects. Notice that print() will accept multiple values to print, and it inserts a space between each value. If you read help(print) in the REPL, you'll find that you can change this with the sep argument:

```
>>> print(count_a, count_c, count_g, count_t, sep='::')
1::2::3::4
```

The print() function will also put a newline at the end of the output, and this can likewise be changed using the end option:

```
>>> print(count_a, count_c, count_g, count_t, end='\n-30-\n')
1 2 3 4
-30-
```

Writing and Verifying a Solution

Using the preceding code, you should be able to create a program that passes all the tests. As you write, I would encourage you to regularly run pylint, flake8, and mypy to check your source code for potential errors. I would even go further and suggest you install the pytest extensions for these so that you can routinely incorporate such tests:

```
$ python3 -m pip install pytest-pylint pytest-flake8 pytest-mypy
```

Alternatively, I've placed a *requirements.txt* file in the root directory of the GitHub repo that lists various dependencies I'll use throughout the book. You can install all these modules with the following command:

```
$ python3 -m pip install -r requirements.txt
```

With those extensions, you can run the following command to run not only the tests defined in the *tests/dna_test.py* file but also tests for linting and type checking using these tools:

```
$ pytest -xv --pylint --flake8 --mypy tests/dna_test.py
=========================== test session starts ============================
...
collected 7 items

tests/dna_test.py::FLAKE8 SKIPPED                               [ 12%]
```

```
tests/dna_test.py::mypy PASSED                                    [ 25%]
tests/dna_test.py::test_exists PASSED                             [ 37%]
tests/dna_test.py::test_usage PASSED                              [ 50%]
tests/dna_test.py::test_dies_no_args PASSED                       [ 62%]
tests/dna_test.py::test_arg PASSED                                [ 75%]
tests/dna_test.py::test_file PASSED                               [ 87%]
::mypy PASSED                                                     [100%]
================================ mypy =================================

Success: no issues found in 1 source file
===================== 7 passed, 1 skipped in 0.58s ===================
```

Some tests are skipped when a cached version indicates nothing has changed since the last test. Run `pytest` with the `---cache-clear` option to force the tests to run. Also, you may find you fail linting tests if your code is not properly formatted or indented. You can automatically format your code using `yapf` or `black`. Most IDEs and editors will provide an auto-format option.

That's a lot to type, so I've created a shortcut for you in the form of a *Makefile* in the directory:

```
$ cat Makefile
.PHONY: test

test:
        python3 -m pytest -xv --flake8 --pylint --pylint-rcfile=../pylintrc \
     --mypy dna.py tests/dna_test.py

all:
        ../bin/all_test.py dna.py
```

You can learn more about these files by reading Appendix A. For now, it's enough to understand that if you have `make` installed on your system, you can use the command **make test** to run the command in the `test` target of the *Makefile*. If you don't have `make` installed or you don't want to use it, that's fine too, but I suggest you explore how a *Makefile* can be used to document and automate processes.

There are many ways to write a passing version of `dna.py`, and I'd like to encourage you to keep exploring before you read the solutions. More than anything, I want to get you used to the idea of changing your program and then running the tests to see if it works. This is the cycle of *test-driven development*, where I first create some metric to decide when the program works correctly. In this instance, that is the *dna_test.py* program that is run by `pytest`.

The tests ensure I don't stray from the goal, and they also let me know when I've met the requirements of the program. They are the specifications (also called *specs*) made incarnate as a program that I can execute. How else would I ever know when a program worked or was finished? Or, as Louis Srygley puts it, "Without requirements or design, programming is the art of adding bugs to an empty text file."

Testing is essential to creating reproducible programs. Unless you can absolutely and automatically prove the correctness and predictability of your program when run with both good and bad data, then you're not writing good software.

Additional Solutions

The program I wrote earlier in this chapter is the *solution1_iter.py* version in the GitHub repo, so I won't bother reviewing that version. I would like to show you several alternate solutions that progress from simpler to more complex ideas. Please do not take this to mean they progress from worse to better. All versions pass the tests, so they are all equally valid. The point is to explore what Python has to offer for solving common problems. Note I will omit code they all have in common, such as the get_args() function.

Solution 2: Creating a count() Function and Adding a Unit Test

The first variation I'd like to show will move all the code in the main() function that does the counting into a count() function. You can define this function anywhere in your program, but I generally like get_args() first, main() second, and then other functions after that but *before* the final couplet that calls main().

For the following function, you will also need to import the typing.Tuple value:

```
def count(dna: str) -> Tuple[int, int, int, int]: ❶
    """ Count bases in DNA """

    count_a, count_c, count_g, count_t = 0, 0, 0, 0 ❷
    for base in dna:
        if base == 'A':
            count_a += 1
        elif base == 'C':
            count_c += 1
        elif base == 'G':
            count_g += 1
        elif base == 'T':
            count_t += 1

    return (count_a, count_c, count_g, count_t) ❸
```

❶ The types show that the function takes a string and returns a tuple containing four integer values.

❷ This is the code from main() that did the counting.

❸ Return a tuple of the four counts.

There are many reasons to move this code into a function. To start, this is a *unit* of computation—given a string of DNA, return the tetranucleotide frequency—so it makes sense to encapsulate it. This will make main() shorter and more readable, and it allows me to write a unit test for the function. Since the function is called count(), I like to call the unit test test_count(). I have placed this function inside the dna.py program just after the count() function rather than in the dna_test.py program just as a matter of convenience. For short programs, I tend to put my functions and unit tests together in the source code, but as projects grow larger, I will segregate unit tests into a separate module. Here's the test function:

```
def test_count() -> None: ❶
    """ Test count """

    assert count('') == (0, 0, 0, 0) ❷
    assert count('123XYZ') == (0, 0, 0, 0)
    assert count('A') == (1, 0, 0, 0) ❸
    assert count('C') == (0, 1, 0, 0)
    assert count('G') == (0, 0, 1, 0)
    assert count('T') == (0, 0, 0, 1)
    assert count('ACCGGGTTTT') == (1, 2, 3, 4)
```

❶ The function name must start with test_ to be found by pytest. The types here show that the test accepts no arguments and, because it has no return statement, returns the default None value.

❷ I like to test functions with both expected and unexpected values to ensure they return something reasonable. The empty string should return all zeros.

❸ The rest of the tests ensure that each base is reported in the correct position.

To verify that my function works, I can use pytest on the dna.py program:

```
$ pytest -xv dna.py
========================= test session starts =========================
...

dna.py::test_count PASSED                                        [100%]

========================= 1 passed in 0.01s =========================
```

The first test passes the empty string and expects to get all zeros for the counts. This is a judgment call, honestly. You might decide your program ought to complain to the user that there's no input. That is, it's possible to run the program using the empty string as the input, and this version will report the following:

```
$ ./dna.py ""
0 0 0 0
```

Likewise, if I passed an empty file, I'd get the same answer. Use the **touch** command to create an empty file:

```
$ touch empty
$ ./dna.py empty
0 0 0 0
```

On Unix systems, /dev/null is a special filehandle that returns nothing:

```
$ ./dna.py /dev/null
0 0 0 0
```

You may feel that no input is an error and report it as such. The important thing about the test is that it forces me to think about it. For instance, should the count() function return zeros or raise an exception if it's given an empty string? Should the program crash on empty input and exit with a nonzero status? These are decisions you will have to make for your programs.

Now that I have a unit test in the dna.py code, I can run pytest on that file to see if it passes:

```
$ pytest -v dna.py
============================ test session starts ============================
...
collected 1 item

dna.py::test_count PASSED                                               [100%]

============================ 1 passed in 0.01s ============================
```

When I'm writing code, I like to write functions that do just one limited thing with as few parameters as possible. Then I like to write a test with a name like test_ plus the function name, usually right after the function in the source code. If I find I have many of these kinds of unit tests, I might decide to move them to a separate file and have pytest execute that file.

To use this new function, modify main() like so:

```
def main() -> None:
    args = get_args()
    count_a, count_c, count_g, count_t = count(args.dna) ❶
    print('{} {} {} {}'.format(count_a, count_c, count_g, count_t)) ❷
```

❶ Unpack the four values returned from count() into separate variables.

❷ Use str.format() to create the output string.

Let's focus for a moment on Python's str.format(). As shown in Figure 1-4, the string '{} {} {} {}' is a template for the output I want to generate, and I'm calling the str.format() function *directly on a string literal*. This is a common idiom in

Python that you'll also see with the `str.join()` function. It's important to remember that, in Python, even a literal string (one that literally exists inside your source code in quotes) is an *object* upon which you can call *methods*.

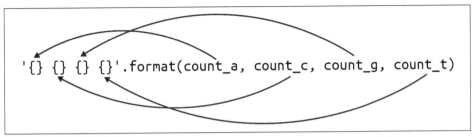

Figure 1-4. The `str.format()` function uses a template containing curly brackets to define placeholders that are filled in with the values of the arguments

Every {} in the string template is a placeholder for some value that is provided as an argument to the function. When using this function, you need to ensure that you have the same number of placeholders as arguments. The arguments are inserted in the order in which they are provided. I'll have much more to say about the `str.format()` function later.

I'm not required to unpack the tuple returned by the `count()` function. I can pass the entire tuple as the argument to the `str.format()` function if I *splat* it by adding an asterisk (*) to the front. This tells Python to expand the tuple into its values:

```
def main() -> None:
    args = get_args()
    counts = count(args.dna) ❶
    print('{} {} {} {}'.format(*counts)) ❷
```

❶ The `counts` variable is a 4-tuple of the integer base counts.

❷ The `*counts` syntax will expand the tuple into the four values needed by the format string; otherwise, the tuple would be interpreted as a single value.

Since I use the `counts` variable only once, I could skip the assignment and shrink this to one line:

```
def main() -> None:
    args = get_args()
    print('{} {} {} {}'.format(*count(args.dna))) ❶
```

❶ Pass the return value from `count()` directly to the `str.format()` method.

The first solution is arguably easier to read and understand, and tools like `flake8` would be able to spot when the number of {} placeholders does not match the number of variables. Simple, verbose, obvious code is often better than compact, clever code. Still, it's good to know about tuple unpacking and splatting variables as I'll use these in ideas in later programs.

Solution 3: Using str.count()

The previous count() function turns out to be quite verbose. I can write the function using a single line of code using the str.count() method. This function will count the number of times one string is found inside another string. Let me show you in the REPL:

```
>>> seq = 'ACCGGGTTTT'
>>> seq.count('A')
1
>>> seq.count('C')
2
```

If the string is not found, it will report 0, making this safe to count all four nucleotides even when the input sequence is missing one or more bases:

```
>>> 'AAA'.count('T')
0
```

Here is a new version of the count() function using this idea:

```
def count(dna: str) -> Tuple[int, int, int, int]: ❶
    """ Count bases in DNA """

    return (dna.count('A'), dna.count('C'), dna.count('G'), dna.count('T')) ❷
```

❶ The signature is the same as before.

❷ Call the dna.count() method for each of the four bases.

This code is much more succinct, and I can use the same unit test to verify that it's correct. This is a key point: *functions should act like black boxes*. That is, I do not know or care what happens inside the box. Something goes in, an answer comes out, and I only really care that the answer is correct. I am free to change what happens inside the box so long as the contract to the outside—the parameters and return value—stays the same.

Here's another way to create the output string in the main() function using Python's f-string syntax:

```
def main() -> None:
    args = get_args()
    count_a, count_c, count_g, count_t = count(args.dna) ❶
    print(f'{count_a} {count_c} {count_g} {count_t}') ❷
```

❶ Unpack the tuple into each of the four counts.

❷ Use an f-string to perform variable interpolation.

 It's called an *f-string* because the f precedes the quotes. I use the mnemonic *format* to remind me this is to format a string. Python also has a *raw* string that is preceded with an r, which I'll discuss later. All strings in Python—bare, f-, or r-strings—can be enclosed in single or double quotes. It makes no difference.

With f-strings, the {} placeholders can perform *variable interpolation*, which is a 50-cent word that means turning a variable into its contents. Those curlies can even execute code. For instance, the len() function will return the length of a string and can be run inside the braces:

```
>>> seq = 'ACGT'
>>> f'The sequence "{seq}" has {len(seq)} bases.'
'The sequence "ACGT" has 4 bases.'
```

I usually find f-strings easier to read than the equivalent code using str.format(). Which you choose is mostly a stylistic decision. I would recommend whichever makes your code more readable.

Solution 4: Using a Dictionary to Count All the Characters

So far I've discussed Python's strings, lists, and tuples. This next solution introduces *dictionaries*, which are key/value stores. I'd like to show a version of the count() function that internally uses dictionaries so that I can hit on some important points to understand:

```
def count(dna: str) -> Tuple[int, int, int, int]: ❶
    """ Count bases in DNA """

    counts = {} ❷
    for base in dna: ❸
        if base not in counts: ❹
            counts[base] = 0 ❺
        counts[base] += 1 ❻

    return (counts.get('A', 0), ❼
            counts.get('C', 0),
            counts.get('G', 0),
            counts.get('T', 0))
```

❶ Internally I'll use a dictionary, but nothing changes about the function signature.

❷ Initialize an empty dictionary to hold the counts.

❸ Use a for loop to iterate through the sequence.

❹ Check if the base does not yet exist in the dictionary.

❺ Initialize the value for this base to 0.

❻ Increment the count for this base by 1.

❼ Use the `dict.get()` method to get each base's count or the default of 0.

Again, the contract for this function—the type signature—hasn't changed. It's still a string in and 4-tuple of integers out. Inside the function, I'm going to use a dictionary that I'll initialize using the empty curly brackets:

```
>>> counts = {}
```

I could also use the `dict()` function. Neither is preferable:

```
>>> counts = dict()
```

I can use the `type()` function to check that this is a dictionary:

```
>>> type(counts)
<class 'dict'>
```

The `isinstance()` function is another way to check the type of a variable:

```
>>> isinstance(counts, dict)
True
```

My goal is to create a dictionary that has each base as a *key* and the number of times it occurs as a *value*. For example, given the sequence ACCGGGTTT, I want `counts` to look like this:

```
>>> counts
{'A': 1, 'C': 2, 'G': 3, 'T': 4}
```

I can access any of the values using square brackets and a key name like so:

```
>>> counts['G']
3
```

Python will raise a `KeyError` exception if I attempt to access a dictionary key that doesn't exist:

```
>>> counts['N']
Traceback (most recent call last):
  File "<stdin>", line 1, in <module>
KeyError: 'N'
```

I can use the `in` keyword to see if a key exists in a dictionary:

```
>>> 'N' in counts
False
>>> 'T' in counts
True
```

As I am iterating through each of the bases in the sequence, I need to see if a base exists in the `counts` dictionary. If it does not, I need to initialize it to 0. Then I can safely use the += assignment to increment the count for a base by 1:

```
>>> seq = 'ACCGGGTTTT'
>>> counts = {}
>>> for base in seq:
...     if not base in counts:
...         counts[base] = 0
...     counts[base] += 1
...
>>> counts
{'A': 1, 'C': 2, 'G': 3, 'T': 4}
```

Finally, I want to return a 4-tuple of the counts for each of the bases. You might think this would work:

```
>>> counts['A'], counts['C'], counts['G'], counts['T']
(1, 2, 3, 4)
```

But ask yourself what would happen if one of the bases was missing from the sequence. Would this pass the unit test I wrote? Definitely not. It would fail on the very first test using an empty string because it would generate a `KeyError` exception. The safe way to ask a dictionary for a value is to use the `dict.get()` method. If the key does not exist, then `None` will be returned:

```
>>> counts.get('T')
4
>>> counts.get('N')
```

Python's None Value

That second call looks like it does nothing because the REPL doesn't show `None`. I'll use the `type()` function to check the return. `NoneType` is the type for the `None` value:

```
>>> type(counts.get('N'))
<class 'NoneType'>
```

I can use the == operator to see if the return value is `None`:

```
>>> counts.get('N') == None
True
```

but PEP8 recommends "Comparisons to singletons like `None` should always be done with `is` or `is not`, never the equality operators." The following is the prescribed method to check if a value is `None`:

```
>>> counts.get('N') is None
True
```

The dict.get() method accepts an optional second argument that is the default value to return when the key does not exist, so this is the safest way to return a 4-tuple of the base counts:

```
>>> counts.get('A', 0), counts.get('C', 0), counts.get('G', 0),
    counts.get('T', 0)
(1, 2, 3, 4)
```

 No matter what you write inside your count() function, ensure that it will pass the test_count() unit test.

Solution 5: Counting Only the Desired Bases

The previous solution will count every character in the input sequence, but what if I only want to count the four nucleotides? In this solution, I will initialize a dictionary with values of 0 for the wanted bases. I'll need to also bring in typing.Dict to run this code:

```
def count(dna: str) -> Dict[str, int]: ❶
    """ Count bases in DNA """

    counts = {'A': 0, 'C': 0, 'G': 0, 'T': 0} ❷
    for base in dna: ❸
        if base in counts: ❹
            counts[base] += 1 ❺

    return counts ❻
```

❶ The signature now indicates I'll be returning a dictionary that has strings for the keys and integers for the values.

❷ Initialize the counts dictionary with the four bases as keys and values of 0.

❸ Iterate through the bases.

❹ Check if the base is found as a key in the counts dictionary.

❺ If so, increment the counts for this base by 1.

❻ Return the counts dictionary.

Since the count() function is now returning a dictionary rather than a tuple, the test_count() function needs to change:

```
def test_count() -> None:
    """ Test count """

    assert count('') == {'A': 0, 'C': 0, 'G': 0, 'T': 0} ❶
    assert count('123XYZ') == {'A': 0, 'C': 0, 'G': 0, 'T': 0} ❷
    assert count('A') == {'A': 1, 'C': 0, 'G': 0, 'T': 0}
    assert count('C') == {'A': 0, 'C': 1, 'G': 0, 'T': 0}
    assert count('G') == {'A': 0, 'C': 0, 'G': 1, 'T': 0}
    assert count('T') == {'A': 0, 'C': 0, 'G': 0, 'T': 1}
    assert count('ACCGGGTTTT') == {'A': 1, 'C': 2, 'G': 3, 'T': 4}
```

❶ The returned dictionary will always have the keys A, C, G, and T. Even for the empty string, these keys will be present and set to 0.

❷ All the other tests have the same inputs, but now I check that the answer comes back as a dictionary.

When writing these tests, note that the order of the keys in the dictionaries is not important. The two dictionaries in the following code have the same content even though they were defined differently:

```
>>> counts1 = {'A': 1, 'C': 2, 'G': 3, 'T': 4}
>>> counts2 = {'T': 4, 'G': 3, 'C': 2, 'A': 1}
>>> counts1 == counts2
True
```

 I would point out that the test_count() function tests the function to ensure it's correct and also serves as documentation. Reading these tests helps me see the structure of the possible inputs and expected outputs from the function.

Here's how I need to change the main() function to use the returned dictionary:

```
def main() -> None:
    args = get_args()
    counts = count(args.dna) ❶
    print('{} {} {} {}'.format(counts['A'], counts['C'], counts['G'], ❷
                               counts['T']))
```

❶ counts is now a dictionary.

❷ Use the str.format() method to create the output using the values from the dictionary.

Solution 6: Using collections.defaultdict()

I can rid my code of all the previous efforts to initialize dictionaries and check for keys and such by using the defaultdict() function from the collections module:

```
>>> from collections import defaultdict
```

When I use the defaultdict() function to create a new dictionary, I tell it the default type for the values. I no longer have to check for a key before using it because the defaultdict type will automatically create any key I reference using a representative value of the default type. For the case of counting the nucleotides, I want to use the int type:

```
>>> counts = defaultdict(int)
```

The default int value will be 0. Any reference to a nonexistent key will cause it to be created with a value of 0:

```
>>> counts['A']
0
```

This means I can instantiate and increment any base in one step:

```
>>> counts['C'] += 1
>>> counts
defaultdict(<class 'int'>, {'A': 0, 'C': 1})
```

Here is how I could rewrite the count() function using this idea:

```
def count(dna: str) -> Dict[str, int]:
    """ Count bases in DNA """

    counts: Dict[str, int] = defaultdict(int) ❶

    for base in dna:
        counts[base] += 1 ❷

    return counts
```

❶ The counts will be a defaultdict with integer values. The type annotation here is required by mypy so that it can be sure that the returned value is correct.

❷ I can safely increment the counts for this base.

The test_count() function looks quite different. I can see at a glance that the answers are very different from the previous versions:

```
def test_count() -> None:
    """ Test count """

    assert count('') == {} ❶
    assert count('123XYZ') == {'1': 1, '2': 1, '3': 1, 'X': 1, 'Y': 1, 'Z': 1} ❷
    assert count('A') == {'A': 1} ❸
    assert count('C') == {'C': 1}
    assert count('G') == {'G': 1}
    assert count('T') == {'T': 1}
    assert count('ACCGGGTTTT') == {'A': 1, 'C': 2, 'G': 3, 'T': 4}
```

❶ Given an empty string, an empty dictionary will be returned.

❷ Notice that every character in the string is a key in the dictionary.

❸ Only A is present, with a count of 1.

Given the fact that the returned dictionary may not contain all the bases, the code in main() needs to use the count.get() method to retrieve each base's frequency:

```
def main() -> None:
    args = get_args()
    counts = count(args.dna) ❶
    print(counts.get('A', 0), counts.get('C', 0), counts.get('G', 0), ❷
        counts.get('T', 0))
```

❶ The counts will be a dictionary that may not contain all of the nucleotides.

❷ It's safest to use the dict.get() method with a default value of 0.

Solution 7: Using collections.Counter()

> Perfection is achieved, not when there is nothing more to add, but when there is nothing left to take away.
>
> —Antoine de Saint-Exupéry

I don't actually like the last three solutions all that much, but I needed to step through how to use a dictionary both manually and with defaultdict() so that you can appreciate the simplicity of using collections.Counter():

```
>>> from collections import Counter
>>> Counter('ACCGGGTTT')
Counter({'G': 3, 'T': 3, 'C': 2, 'A': 1})
```

The best code is code you never write, and Counter() is a prepackaged function that will return a dictionary with the frequency of the items contained in the iterable you pass it. You might also hear this called a *bag* or a *multiset*. Here the iterable is a string composed of characters, and so I get back the same dictionary as in the last two solutions, but *having written no code*.

It's so simple that you could pretty much eschew the count() and test_count() functions and integrate it directly into your main():

```
def main() -> None:
    args = get_args()
    counts = Counter(args.dna) ❶
    print(counts.get('A', 0), counts.get('C', 0), counts.get('G', 0), ❷
        counts.get('T', 0))
```

❶ The `counts` will be a dictionary containing the frequencies of the characters in `args.dna`.

❷ It is still safest to use `dict.get()` as I cannot be certain that all the bases are present.

I could argue that this code belongs in a `count()` function and keep the tests, but the `Counter()` function is already tested and has a well-defined interface. I think it makes more sense to use this function inline.

Going Further

The solutions here only handle DNA sequences provided as UPPERCASE TEXT. It's not unusual to see these sequences provided as lowercase letters. For instance, in plant genomics, it's common to use lowercase bases to denote regions of repetitive DNA. Modify your program to handle both uppercase and lowercase input by doing the following:

1. Add a new input file that mixes case.
2. Add a test to *tests/dna_test.py* that uses this new file and specifies the expected counts insensitive to case.
3. Run the new test and ensure your program fails.
4. Alter the program until it will pass the new test and all of the previous tests.

The solutions that used dictionaries to count all available characters would appear to be more flexible. That is, some of the tests only account for the bases *A*, *C*, *G*, and *T*, but if the input sequence were encoded using IUPAC codes (*https://oreil.ly/qGfsO*) to represent possible ambiguity in sequencing, then the program would have to be entirely rewritten. A program hard-coded to look only at the four nucleotides would also be useless for protein sequences that use a different alphabet. Consider writing a version of the program that will print two columns of output with each character that is found in the first column and the character's frequency in the second. Allow the user to sort ascending or descending by either column.

Review

This was kind of a monster chapter. The following chapters will be a bit shorter, as I'll build upon many of the foundational ideas I've covered here:

- You can use the `new.py` program to create the basic structure of a Python program that accepts and validates command-line arguments using `argparse`.

- The `pytest` module will run all functions with names starting with `test_` and will report the results of how many tests pass.
- Unit tests are for functions, and integration tests check if a program works as a whole.
- Programs like `pylint`, `flake8`, and `mypy` can find various kinds of errors in your code. You can also have `pytest` automatically run tests to see if your code passes these checks.
- Complicated commands can be stored as a target in a *Makefile* and executed using the `make` command.
- You can create a decision tree using a series of `if`/`else` statements.
- There are many ways to count all the characters in a string. Using the `collections.Counter()` function is perhaps the simplest method to create a dictionary of letter frequencies.
- You can annotate variables and functions with types, and use `mypy` to ensure the types are used correctly.
- The Python REPL is an interactive tool for executing code examples and reading documentation.
- The Python community generally follows style guidelines such as PEP8. Tools like `yapf` and `black` can automatically format code according to these suggestions, and tools like `pylint` and `flake8` will report deviations from the guidelines.
- Python strings, lists, tuples, and dictionaries are very powerful data structures, each with useful methods and copious documentation.
- You can create a custom, immutable, typed `class` derived from named tuples.

You may be wondering which is the best of the seven solutions. As with many things in life, it depends. Some programs are shorter to write and easier to understand but may fare poorly when confronting large datasets. In Chapter 2, I'll show you how to *benchmark* programs, pitting them against each other in multiple runs using large inputs to determine which performs the best.

Transcribing DNA into mRNA: Mutating Strings, Reading and Writing Files

To express the proteins necessary to sustain life, regions of DNA must be transcribed into a form of RNA called *messenger RNA* (mRNA). While there are many fascinating biochemical differences between DNA and RNA, for our purposes the only difference is that all the characters *T* representing the base thymine in a sequence of DNA need to be changed to the letter *U*, for uracil. As described on the Rosalind RNA page (*https://oreil.ly/9Dddm*), the program I'll show you how to write will accept a string of DNA like ACGT and print the transcribed mRNA ACGU. I can use Python's str.replace() function to accomplish this in one line:

```
>>> 'GATGGAACTTGACTACGTAAATT'.replace('T', 'U')
'GAUGGAACUUGACUACGUAAAUU'
```

You already saw in Chapter 1 how to write a program to accept a DNA sequence from the command line or a file and print a result, so you won't be learning much if you do that again. I'll make this program more interesting by tackling a very common pattern found in bioinformatics. Namely, I'll show how to process one or more input files and place the results in an output directory. For instance, it's pretty common to get the results of a sequencing run back as a directory of files that need to be quality checked and filtered, with the cleaned sequences going into some new directory for your analysis. Here the input files contain DNA sequences, one per line, and I'll write the mRNA sequences into like-named files in an output directory.

In this chapter, you will learn:

- How to write a program to require one or more file inputs
- How to create directories

- How to read and write files
- How to modify strings

Getting Started

It might help to try running one of the solutions first to see how your program should work. Start by changing into the *02_rna* directory and copying the first solution to the program rna.py:

```
$ cd 02_rna
$ cp solution1_str_replace.py rna.py
```

Request the usage for the program using the -h flag:

```
$ ./rna.py -h
usage: rna.py [-h] [-o DIR] FILE [FILE ...] ❶

Transcribe DNA into RNA

positional arguments: ❷
  FILE                  Input DNA file

optional arguments:
  -h, --help            show this help message and exit
  -o DIR, --out_dir DIR
                        Output directory (default: out) ❸
```

❶ The arguments surrounded by square brackets ([ ]) are optional. The [FILE ...] syntax means that this argument can be repeated.

❷ The input FILE argument(s) will be positional.

❸ The optional output directory has the default value of out.

The goal of the program is to process one or more files, each containing sequences of DNA. Here is the first test input file:

```
$ cat tests/inputs/input1.txt
GATGGAACTTGACTACGTAAATT
```

Run the rna.py program with this input file, and note the output:

```
$ ./rna.py tests/inputs/input1.txt
Done, wrote 1 sequence in 1 file to directory "out".
```

Now there should be an *out* directory containing a file called *input1.txt*:

```
$ ls out/
input1.txt
```

The contents of that file should match the input DNA sequence but with all the *T*s changed to *U*s:

```
$ cat out/input1.txt
GAUGGAACUUGACUACGUAAAUU
```

You should run the program with multiple inputs and verify that you get multiple files in the output directory. Here I will use all the test input files with an output directory called *rna*. Notice how the summary text uses the correct singular/plurals for *sequence(s)* and *file(s)*:

```
$ ./rna.py --out_dir rna tests/inputs/*
Done, wrote 5 sequences in 3 files to directory "rna".
```

I can use the wc (word count) program with the -l option to count the *lines* in the output file and verify that five sequences were written to three files in the *rna* directory:

```
$ wc -l rna/*
      1 rna/input1.txt
      2 rna/input2.txt
      2 rna/input3.txt
      5 total
```

Defining the Program's Parameters

As you can see from the preceding usage, your program should accept the following parameters:

- One or more positional arguments, which must be readable text files each containing strings of DNA to transcribe.
- An optional -o or --out_dir argument that names an output directory to write the sequences of RNA into. The default should be out.

You are free to write and structure your programs however you like (so long as they pass the tests), but I will always start a program using **new.py** and the structure I showed in the first chapter. The --force flag indicates that the existing rna.py should be overwritten:

```
$ new.py --force -p "Transcribe DNA to RNA" rna.py
Done, see new script "rna.py".
```

Defining an Optional Parameter

Modify the get_args() function to accept the parameters described in the previous section. To start, define the out_dir parameter. I suggest you change the -a|--arg option generated by new.py to this:

```
parser.add_argument('-o', ❶
                    '--out_dir', ❷
                    help='Output directory', ❸
                    metavar='DIR', ❹
                    type=str, ❺
                    default='out') ❻
```

❶ This is the short flag name. Short flags start with a single dash and are followed by a single character.

❷ This is the long flag name. Long flags start with two dashes and are followed by a more memorable string than the short flag. This will also be the name `argparse` will use to access the value.

❸ This will be incorporated into the usage statement to describe the argument.

❹ The `metavar` is a short description also shown in the usage.

❺ The default type of all arguments is `str` (string), so this is technically superfluous but still not a bad idea to document.

❻ The default value will be the string `out`. If you do not specify a `default` attribute when defining an option, the default value will be `None`.

Defining One or More Required Positional Parameters

For the `FILE` value(s), I can modify the default `-f|--file` parameter to look like this:

```
parser.add_argument('file', ❶
                    help='Input DNA file(s)', ❷
                    metavar='FILE', ❸
                    nargs='+', ❹
                    type=argparse.FileType('rt')) ❺
```

❶ Remove the `-f` short flag and the two dashes from `--file` so that this becomes a *positional* argument called `file`. Optional parameters start with dashes, and positional ones do not.

❷ The `help` string indicates the argument should be one or more files containing DNA sequences.

❸ This string is printed in the short usage to indicate the argument is a file.

❹ This indicates the number of arguments. The + indicates that one or more values are required.

❺ This is the actual type that `argparse` will enforce. I am requiring any value to be a readable text (`rt`) file.

Using nargs to Define the Number of Arguments

I use `nargs` to describe the *number of arguments* to the program. In addition to using an integer value to describe exactly how many values are allowed, I can use the three symbols shown in Table 2-1.

Table 2-1. Possible values for nargs

Symbol	Meaning
?	Zero or one
*	Zero or more
+	One or more

When you use + with `nargs`, `argparse` will provide the arguments as a list. Even if there is just one argument, you will get a list containing one element. You will never have an empty list because at least one argument is required.

Using argparse.FileType() to Validate File Arguments

The `argparse.FileType()` function is incredibly powerful, and using it can save you loads of time in validating file inputs. When you define a parameter with this type, `argparse` will print an error message and halt the execution of the program if any of the arguments is not a file. For instance, I would assume there is no file in your *02_dna* directory called *blargh*. Notice the result when I pass that value:

```
$ ./rna.py blargh
usage: rna.py [-h] [-o DIR] FILE [FILE ...]
rna.py: error: argument FILE: can't open 'blargh': [Errno 2]
No such file or directory: 'blargh'
```

It's not obvious here, but the program never made it out of the `get_args()` function because `argparse` did the following:

1. Detected that *blargh* is not a valid file
2. Printed the short usage statement
3. Printed a useful error message
4. Exited the program with a nonzero value

This is how a well-written program ought to work, detecting and rejecting bad arguments as soon as possible and notifying the user of the problems. All this happened without my writing anything more than a good description of the kind of argument I wanted. Again, the best code is code you never write.

Because I am using the file *type*, the elements of the list will not be strings representing the filenames but will instead be open filehandles. A *filehandle* is a mechanism to read and write the contents of a file. I used a filehandle in the last chapter when the DNA argument was a filename.

 The order in which you define these parameters in your source code does not matter in this instance. You can define options before or after positional parameters. The order only matters when you have multiple positional arguments—the first parameter will be for the first positional argument, the second parameter for the second positional argument, and so forth.

Defining the Args Class

Finally, I need a way to define the `Args` class that will represent the arguments:

```
from typing import NamedTuple, List, TextIO ❶

class Args(NamedTuple):
    """ Command-line arguments """
    files: List[TextIO] ❷
    out_dir: str ❸
```

❶ I'll need two new imports from the `typing` module, `List` to describe a list, and `TextIO` for an open filehandle.

❷ The `files` attribute will be a list of open filehandles.

❸ The `out_dir` attribute will be a string.

I can use this class to create the return value from `get_args()`. The following syntax uses positional notation such that the `file` is the first field and the `out_dir` is the second. When there are one or two fields, I will tend to use the positional notation:

```
return Args(args.file, args.out_dir)
```

Explicitly using the field names is safer and arguably easier to read, and it will become vital when I have more fields:

```
return Args(files=args.file, out_dir=args.out_dir)
```

Now I have all the code to define, document, and validate the inputs. Next, I'll show how the rest of the program should work.

Outlining the Program Using Pseudocode

I'll sketch out the basics of the program's logic in the `main()` function using a mix of code and pseudocode to generally describe how to handle the input and output files. Whenever you get stuck writing a new program, this approach can help you see *what* needs to be done. Then you can figure out *how* to do it:

```python
def main() -> None:
    args = get_args()

    if not os.path.isdir(args.out_dir): ❶
        os.makedirs(args.out_dir) ❷

    num_files, num_seqs = 0, 0 ❸
    for fh in args.files: ❹
        # open an output file in the output directory ❺
        # for each line/sequence from the input file:
            # write the transcribed sequence to the output file
            # update the number of sequences processed
        # update the number of files processed

    print('Done.') ❻
```

❶ The `os.path.isdir()` function will report if the output directory exists.

❷ The `os.makedirs()` function will create a directory path.

❸ Initialize variables for the number of files and sequences written to use in the feedback you provide when the program exits.

❹ Use a `for` loop to iterate the list of filehandles in `args.files`. The iterator variable `fh` helps remind me of the type.

❺ This is pseudocode describing the steps you need to do with each filehandle.

❻ Print a summary for the user to let them know what happened.

 The `os.makedirs()` function will create a directory and all the parent directories, while the `os.mkdir()` function will fail if the parent directories do not exist. I only ever use the first function in my code.

If you think you know how to finish the program, feel free to proceed. Be sure to run **pytest** (or **make test**) to ensure your code is correct. Stick with me if you need a little more guidance on how to read and write files. I'll tackle the pseudocode in the following sections.

Iterating the Input Files

Remember that `args.files` is a `List[TextIO]`, meaning that it is a list of filehandles. I can use a `for` loop to visit each element in any iterable in such a list:

```
for fh in args.files:
```

I'd like to stress here that I chose an iterator variable called `fh` because each value is a filehandle. I sometimes see people who always use an iterator variable name like `i` or `x` with a `for` loop, but those are not descriptive variable names.[1] I'll concede that it's very common to use variable names like `n` (for *number*) or `i` (for *integer*) when iterating numbers like so:

```
for i in range(10):
```

And I will sometimes use `x` and `xs` (pronounced *exes*) to stand for *one* and *many* of some generic value:

```
for x in xs:
```

Otherwise, it's very important to use variable names that accurately describe the thing they represent.

Creating the Output Filenames

Per the pseudocode, the first goal is to open an output file. For that, I need a filename that combines the name of the output directory with the *basename* of the input file. That is, if the input file is *dna/input1.txt* and the output directory is *rna*, then the output file path should be *rna/input1.txt*.

The `os` module is used to interact with the operating system (like Windows, macOS, or Linux), and the `os.path` module has many handy functions I can use, like the `os.path.dirname()` function to get the name of the directory from a file path and `os.path.basename()` to get the file's name (see Figure 2-1):

```
>>> import os
>>> os.path.dirname('./tests/inputs/input1.txt')
'./tests/inputs'
>>> os.path.basename('./tests/inputs/input1.txt')
'input1.txt'
```

---

1 As Phil Karlton says, "There are only two hard things in Computer Science: cache invalidation and naming things."

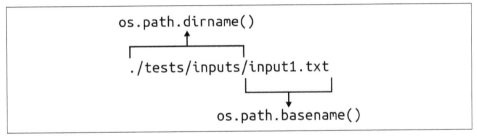

Figure 2-1. The os.path module contains useful functions like dirname() and basename() to extract parts from a file's path

The new sequences will be written to an output file in args.out_dir. I suggest you use the os.path.join() function with the basename of the input file to create the output filename, as shown in Figure 2-2. This will ensure that the output filename works both on Unix and Windows, which use different path dividers—the slash (/) and backslash (\), respectively. You may also want to investigate the pathlib module for similar functionality.

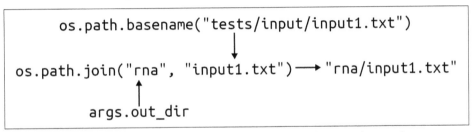

Figure 2-2. The os.path.join() will create the output path by combining the output directory with the basename of the input file

You can get the file's path from the fh.name attribute of the filehandle:

```
for fh in args.files:
    out_file = os.path.join(args.out_dir, os.path.basename(fh.name))
    print(fh.name, '->', out_file)
```

Run your program to verify that it looks like this:

```
$ ./rna.py tests/inputs/*
tests/inputs/input1.txt -> out/input1.txt
tests/inputs/input2.txt -> out/input2.txt
tests/inputs/input3.txt -> out/input3.txt
```

I'm taking baby steps toward what the program is supposed to do. It's very important to write just one or two lines of code and then run your program to see if it's correct. I often see students try to write many lines of code—whole programs, even—before they attempt to run them. That never works out well.

Opening the Output Files

Using this output filename, you need to open() the filehandle. I used this function in the first chapter to read DNA from an input file. By default, open() will only allow me to read a file, but I need to write a file. I can indicate that I want to open the file for writing by passing an optional second argument: the string w for *write*.

 When you open an existing file with a mode of w, the file will be *overwritten*, meaning its previous contents will be immediately and permanently lost. If needed, you can use the os.path.isfile() function to check if you're opening an existing file.

As shown in Table 2-2, you can also use the values r for *read* (the default) and a to *append*, which allows you to open for writing more content at the end of an existing file.

Table 2-2. File writing modes

Mode	Meaning
w	Write
r	Read
a	Append

Table 2-3 shows that you can also read and write either text or raw bytes using the modes t and b, respectively.

Table 2-3. File content modes

Mode	Meaning
t	Text
b	Bytes

You can combine these, for example using rb to *read bytes* and wt to *write text*, which is what I want here:

```
for fh in args.files:
    out_file = os.path.join(args.out_dir, os.path.basename(fh.name))
    out_fh = open(out_file, 'wt') ❶
```

❶ Note that I named my variable out_fh to remind me this is the output filehandle.

Writing the Output Sequences

Looking at the pseudocode again, I have two levels of iterating—one for each filehandle of input, and then one for each line of DNA in the filehandles. To read each line from an open filehandle, I can use another for loop:

```
for fh in args.files:
    for dna in fh:
```

The *input2.txt* file has two sequences, each ending with a newline:

```
$ cat tests/inputs/input2.txt
TTAGCCCAGACTAGGACTTT
AACTAGTCAAAGTACACC
```

To start, I'll show you how to print each sequence to the console, then I'll demonstrate how to use print() to write content to a filehandle. Chapter 1 mentions that the print() function will automatically append a newline (\n on Unix platforms and \r\n on Windows) unless I tell it not to. To avoid having two newlines from the following code, one from the sequence and one from print(), I can either use the str.rstrip() function to remove the newline from the sequence like this:

```
>>> fh = open('./tests/inputs/input2.txt')
>>> for dna in fh:
...     print(dna.rstrip()) ❶
...
TTAGCCCAGACTAGGACTTT
AACTAGTCAAAGTACACC
```

❶ Use dna.rstrip() to remove the trailing newline.

or use the end option to print():

```
>>> fh = open('./tests/inputs/input2.txt')
>>> for dna in fh:
...     print(dna, end='') ❶
...
TTAGCCCAGACTAGGACTTT
AACTAGTCAAAGTACACC
```

❶ Use the empty string at the end instead of a newline.

The goal is to transcribe each DNA sequence to RNA and write the result to out_fh. In the introduction to this chapter, I suggested you could use the str.replace() function. If you read help(str.replace) in the REPL, you'll see that it will "Return a copy with all occurrences of substring old replaced by new":

```
>>> dna = 'ACTG'
>>> dna.replace('T', 'U')
'ACUG'
```

There are other ways to change the *T*s to *U*s that I will explore later. First, I'd like to point out that strings in Python are immutable, meaning they cannot be changed in place. That is, I could check to see if the letter *T* is in the DNA string and then use the `str.index()` function to find the location and try to overwrite it with the letter *U*, but this will raise an exception:

```
>>> dna = 'ACTG'
>>> if 'T' in dna:
...     dna[dna.index('T')] = 'U'
...
Traceback (most recent call last):
  File "<stdin>", line 2, in <module>
TypeError: 'str' object does not support item assignment
```

Instead, I'll use `str.replace()` to create a new string:

```
>>> dna.replace('T', 'U')
'ACUG'
>>> dna
'ACTG'
```

I need to write this new string into the `out_fh` output filehandle. I have two options. First, I can use the `print()` function's `file` option to describe *where* to print the string. Consult the `help(print)` documentation in the REPL:

```
print(...)
    print(value, ..., sep=' ', end='\n', file=sys.stdout, flush=False)

    Prints the values to a stream, or to sys.stdout by default.
    Optional keyword arguments:
    file:  a file-like object (stream); defaults to the current sys.stdout. ❶
    sep:   string inserted between values, default a space.
    end:   string appended after the last value, default a newline.
    flush: whether to forcibly flush the stream.
```

❶   This is the option I need to print the string to the open filehandle.

I need to use the `out_fh` filehandle as the `file` argument. I want to point out that the default `file` value is `sys.stdout`. On the command line, STDOUT (pronounced *standard out*) is the standard place for program output to appear, which is usually the console.

Another option is to use the `out_fh.write()` method of the filehandle itself, but note that this function *does not* append a newline. It's up to you to decide when to add newlines. In the case of reading these sequences that are terminated with newlines, they are not needed.

Printing the Status Report

I almost always like to print something when my programs have finished running so I at least know they got to the end. It may be something as simple as "Done!" Here, though, I'd like to know how many sequences in how many files were processed. I also want to know where I can find the output, something that's especially helpful if I forget the name of the default output directory.

The tests expect that you will use proper grammar[2] to describe the numbers—for example, *1 sequence* and *1 file*:

```
$ ./rna.py tests/inputs/input1.txt
Done, wrote 1 sequence in 1 file to directory "out".
```

or *3 sequences* and *2 files*:

```
$ ./rna.py --out_dir rna tests/inputs/input[12].txt ❶
Done, wrote 3 sequences in 2 files to directory "rna".
```

❶ The syntax input[12].txt is a way to say either 1 or 2 can occur, so *input1.txt* and *input2.txt* will both match.

Using the Test Suite

You can run **pytest -xv** to run *tests/rna_test.py*. A passing test suite looks like this:

```
$ pytest -xv
======================== test session starts ========================
...

tests/rna_test.py::test_exists PASSED                    [ 14%] ❶
tests/rna_test.py::test_usage PASSED                     [ 28%] ❷
tests/rna_test.py::test_no_args PASSED                   [ 42%] ❸
tests/rna_test.py::test_bad_file PASSED                  [ 57%] ❹
tests/rna_test.py::test_good_input1 PASSED               [ 71%] ❺
tests/rna_test.py::test_good_input2 PASSED               [ 85%]
tests/rna_test.py::test_good_multiple_inputs PASSED      [100%]


======================== 7 passed in 0.37s ========================
```

❶ The rna.py program exists.

❷ The program prints a usage statement when requested.

❸ The program exits with an error when given no arguments.

---

2 Sorry, but I can't stop being an English major.

❹ The program prints an error message when given a bad file argument.

❺ The next tests all verify that the program works properly given good inputs.

Generally speaking, I first write tests that try to break a program before giving it good input. For instance, I want the program to fail when given no files or when given nonexistent files. Just as the best detectives can think like criminals, I try to imagine all the ways to break my programs and test that they behave predictably under those circumstances.

The first three tests are exactly as from Chapter 1. For the fourth test, I pass a non-existent file and expect a nonzero exit value along with the usage and the error message. Note that the error specifically mentions the offending value, here the bad filename. You should strive to create feedback that lets the user know exactly what the problem is and how to fix it:

```python
def test_bad_file():
    """ Die on missing input """

    bad = random_filename() ❶
    retval, out = getstatusoutput(f'{RUN} {bad}') ❷
    assert retval != 0 ❸
    assert re.match('usage:', out, re.IGNORECASE) ❹
    assert re.search(f"No such file or directory: '{bad}'", out) ❺
```

❶ This is a function I wrote to generate a string of random characters.

❷ Run the program with this nonexistent file.

❸ Make sure the exit value is not 0.

❹ Use a regular expression (*regex*) to look for the usage in the output.

❺ Use another regex to look for the error message describing the bad input filename.

I haven't introduced regular expressions yet, but they will become central to solutions I write later. To see why they are useful, look at the output from the program when run with a bad file input:

```
$ ./rna.py dKej82
usage: rna.py [-h] [-o DIR] FILE [FILE ...]
rna.py: error: argument FILE: can't open 'dKej82':
[Errno 2] No such file or directory: 'dKej82'
```

Using the re.match() function, I am looking for a pattern of text starting at the beginning of the out text. Using the re.search() function, I am looking for another pattern that occurs somewhere inside the out text. I'll have much more to say about regexes later. For now, it's enough to point out that they are very useful.

I'll show one last test that verifies the program runs correctly when provided good input. There are many ways to write such a test, so don't get the impression this is canon:

```python
def test_good_input1():
    """ Runs on good input """

    out_dir = 'out' ❶
    try: ❷
        if os.path.isdir(out_dir): ❸
            shutil.rmtree(out_dir) ❹

        retval, out = getstatusoutput(f'{RUN} {INPUT1}') ❺
        assert retval == 0
        assert out == 'Done, wrote 1 sequence in 1 file to directory "out".'
        assert os.path.isdir(out_dir) ❻
        out_file = os.path.join(out_dir, 'input1.txt')
        assert os.path.isfile(out_file) ❼
        assert open(out_file).read().rstrip() == 'GAUGGAACUUGACUACGUAAAUU' ❽

    finally: ❾
        if os.path.isdir(out_dir): ❿
            shutil.rmtree(out_dir)
```

❶ This is the default output directory name.

❷ The try/finally blocks help to ensure cleanup when tests fail.

❸ See if the output directory has been left over from a previous run.

❹ Use the shutil.rmtree() function to remove the directory and its contents.

❺ Run the program with a known good input file.

❻ Make sure the expected output directory was created.

❼ Make sure the expected output file was created.

❽ Make sure the contents of the output file are correct.

❾ Even if something fails in the try block, this finally block will be run.

❿ Clean up the testing environment.

I want to stress how important it is to check every aspect of what your program is supposed to do. Here, the program should process some number of input files, create an output directory, and then place the processed data into files in the output directory. I'm testing every one of those requirements using known input to verify that the expected output is created.

There are a couple of other tests I won't cover here as they are similar to what I've already shown, but I would encourage you to read the entire *tests/rna_test.py* program. The first input file has one sequence. The second input file has two sequences, and I use that to test that two sequences are written to the output file. The third input file has two very long sequences. By using these inputs individually and together, I try to test every aspect of my program that I can imagine.

Although you can run the tests in *tests/rna_test.py* using `pytest`, I also urge you to use `pylint`, `flake8`, and `mypy` to check your program. The `make test` shortcut can do this for you as it will execute `pytest` with the additional arguments to run those tools. Your goal should be a completely clean test suite.

 You may find that `pylint` will complain about variable names like `fh` being too short or not being *snake_case*, where lowercase words are joined with underscores. I have included a *pylintrc* configuration file in the top level of the GitHub repository. Copy this to the file *.pylintrc* in your home directory to silence these errors.

You should have enough information and tests now to help you finish this program. You'll get the most benefit from this book if you try to write working programs on your own before you look at my solutions. Once you have one working version, try to find other ways to solve it. If you know about regular expressions, that's a great solution. If you don't, I will demonstrate a version that uses them.

Solutions

The following two solutions differ only in how I substitute the *T*s for *U*s. The first uses the `str.replace()` method, and the second introduces regular expressions and uses the Python `re.sub()` function.

Solution 1: Using str.replace()

Here is the entirety of one solution that uses the `str.replace()` method I discussed in the introduction to this chapter:

```
#!/usr/bin/env python3
""" Transcribe DNA into RNA """

import argparse
```

```
import os
from typing import NamedTuple, List, TextIO


class Args(NamedTuple):
    """ Command-line arguments """
    files: List[TextIO]
    out_dir: str


# --------------------------------------------------
def get_args() -> Args:
    """ Get command-line arguments """

    parser = argparse.ArgumentParser(
        description='Transcribe DNA into RNA',
        formatter_class=argparse.ArgumentDefaultsHelpFormatter)

    parser.add_argument('file',
                        help='Input DNA file',
                        metavar='FILE',
                        type=argparse.FileType('rt'),
                        nargs='+')

    parser.add_argument('-o',
                        '--out_dir',
                        help='Output directory',
                        metavar='DIR',
                        type=str,
                        default='out')

    args = parser.parse_args()

    return Args(args.file, args.out_dir)


# --------------------------------------------------
def main() -> None:
    """ Make a jazz noise here """

    args = get_args()

    if not os.path.isdir(args.out_dir):
        os.makedirs(args.out_dir)

    num_files, num_seqs = 0, 0  ❶
    for fh in args.files:  ❷
        num_files += 1  ❸
        out_file = os.path.join(args.out_dir, os.path.basename(fh.name))
        out_fh = open(out_file, 'wt')  ❹

        for dna in fh:  ❺
```

```
            num_seqs += 1 ❻
            out_fh.write(dna.replace('T', 'U')) ❼

        out_fh.close() ❽

    print(f'Done, wrote {num_seqs} sequence{"" if num_seqs == 1 else "s"} '
          f'in {num_files} file{"" if num_files == 1 else "s"} '
          f'to directory "{args.out_dir}".') ❾


# --------------------------------------------------
if __name__ == '__main__':
    main()
```

❶ Initialize the counters for files and sequences.

❷ Iterate the filehandles.

❸ Increment the counter for files.

❹ Open the output file for this input file.

❺ Iterate the sequences in the input file.

❻ Increment the counter for sequences.

❼ Write the transcribed sequence to the output file.

❽ Close the output filehandle.

❾ Print the status. Note that I'm relying on Python's implicit concatenation of adjacent strings to create one output string.

Solution 2: Using re.sub()

I suggested earlier that you might explore how to use regular expressions to solve this. Regexes are a language for describing patterns of text. They have been around for decades, long before Python was even invented. Though they may seem somewhat daunting at first, regexes are well worth the effort to learn.[3]

To use regular expressions in Python, I must import the re module:

```
>>> import re
```

---

3 *Mastering Regular Expressions* by Jeffrey Friedl (O'Reilly, 2006) is one of the best books I've found.

Previously, I used the re.search() function to look for a pattern of text inside another string. For this program, the pattern I am looking for is the letter *T*, which I can write as a literal string:

```
>>> re.search('T', 'ACGT') ❶
<re.Match object; span=(3, 4), match='T'> ❷
```

❶ Search for the pattern T inside the string ACGT.

❷ Because T was found, the return value is a Re.Match object showing the location of the found pattern. A failed search would return None.

The span=(3, 4) reports the start and stop indexes where the pattern T is found. I can use these positions to extract the substring using a slice:

```
>>> 'ACGT'[3:4]
'T'
```

But instead of just finding the *T*, I want to replace the string T with U. As shown in Figure 2-3, the re.sub() (for *substitute*) function will do this.

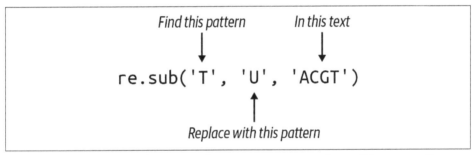

Figure 2-3. The re.sub() function will return a new string where all instances of a pattern have been replaced with a new string

The result is a new string where the *T*s have all been replaced with *U*s:

```
>>> re.sub('T', 'U', 'ACGT') ❶
'ACGU' ❷
```

❶ Replace every T with U in the string ACGT.

❷ The result is a new string with the substitutions.

To use this version, I can modify the inner for loop, as shown. Note that I have chosen to use the str.strip() method to remove the newline terminating the input DNA string because print() will add a newline:

```
for dna in fh:
    num_seqs += 1
    print(re.sub('T', 'U', dna.rstrip()), file=out_fh) ❶
```

❶ Remove the newline from dna, substitute all the Ts with Us, and print the resulting string to the output filehandle.

Benchmarking

You might be curious to know which solution is faster. Comparing the relative runtimes of programs is called *benchmarking*, and I'll show you a simple way to compare these two solutions using some basic bash commands. I'll use the *./tests/inputs/input3.txt* file, as it is the largest test file. I can write a for loop in bash with almost the same syntax as Python. Note that I am using newlines in this command to make it more readable, and bash notes the line continuation with >. You can substitute semicolons (;) to write this on one line:

```
$ for py in ./solution*
> do echo $py && time $py ./tests/inputs/input3.txt
> done
./solution1_str_replace.py
Done, wrote 2 sequences in 1 file to directory "out".

real    0m1.539s
user    0m0.046s
sys        0m0.036s
./solution2_re_sub.py
Done, wrote 2 sequences in 1 file to directory "out".

real    0m0.179s
user    0m0.035s
sys        0m0.013s
```

It would appear the second solution using regular expressions is faster, but I don't have enough data to be sure. I need a more substantial input file. In the *02_rna* directory, you'll find a program called genseq.py I wrote that will generate 1,000 sequences of 1,000,000 bases in a file called *seq.txt*. You can, of course, modify the parameters:

```
$ ./genseq.py --help
usage: genseq.py [-h] [-l int] [-n int] [-o FILE]

Generate long sequence

optional arguments:
  -h, --help           show this help message and exit
  -l int, --len int    Sequence length (default: 1000000)
  -n int, --num int    Number of sequences (default: 100)
```

```
    -o FILE, --outfile FILE
                          Output file (default: seq.txt)
```

The file *seq.txt* that is generated using the defaults is about 95 MB. Here's how the programs do with a more realistic input file:

```
$ for py in ./solution*; do echo $py && time $py seq.txt; done
./solution1_str_replace.py
Done, wrote 100 sequences in 1 file to directory "out".

real    0m0.456s
user    0m0.372s
sys     0m0.064s
./solution2_re_sub.py
Done, wrote 100 sequences in 1 file to directory "out".

real    0m3.100s
user    0m2.700s
sys     0m0.385s
```

It now appears that the first solution is faster. For what it's worth, I came up with several other solutions, all of which fared much worse than these two. I thought I was creating more and more clever solutions that would ultimately lead to the best performance. My pride was sorely wounded when what I thought was my best program turned out to be orders of magnitude slower than these two. When you have assumptions, you should, as the saying goes, "Trust, but verify."

Going Further

Modify your program to print the lengths of the sequences to the output file rather than the transcribed RNA. Have the final status report the maximum, minimum, and average sequence lengths.

Review

Key points from this chapter:

- The `argparse.FileType` option will validate file arguments.
- The `nargs` option to `argparse` allows you to define the number of valid arguments for a parameter.
- The `os.path.isdir()` function can detect if a directory exists.
- The `os.makedirs()` function will create a directory structure.
- The `open()` function by default allows only reading files. The `w` option must be used to write to the filehandle, and the `a` option is for appending values to an existing file.

- File handles can be opened with the t option for *text* (the default) or b for *bytes*, such as when reading image files.
- Strings are immutable, and there are many methods to alter strings into new strings, including `str.replace()` and `re.sub()`.

## CHAPTER 3

Reverse Complement of DNA: String Manipulation

The Rosalind REVC challenge (*https://oreil.ly/ot4z6*) explains that the bases of DNA form pairs of *A-T* and *G-C*. Additionally, DNA has directionality and is usually read from the 5'-end (*five-prime end*) toward the 3'-end (*three-prime end*). As shown in Figure 3-1, the complement of the DNA string *AAAACCCGGT* is *TTTTGGGCCA*. I then reverse this string (reading from the 3'-end) to get *ACCGGGTTTT* as the reverse complement.

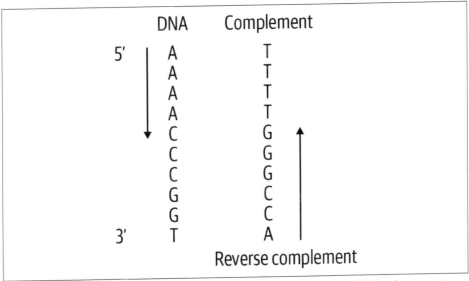

Figure 3-1. The reverse complement of DNA is the complement read from the opposite direction

Although you can find many existing tools to generate the reverse complement of DNA—and I'll drop a spoiler alert that the final solution will use a function from the Biopython library—the point of writing our own algorithm is to explore Python. In this chapter, you will learn:

- How to implement a decision tree using a dictionary as a lookup table
- How to dynamically generate a list or a string
- How to use the reversed() function, which is an example of an iterator
- How Python treats strings and lists similarly
- How to use a list comprehension to generate a list
- How to use str.maketrans() and str.translate() to transform a string
- How to use Biopython's Bio.Seq module
- That the real treasure is the friends you make along the way

Getting Started

The code and tests for this program are in the *03_revc* directory. To get a feel for how the program will work, change into that directory and copy the first solution to a program called revc.py:

```
$ cd 03_revc
$ cp solution1_for_loop.py revc.py
```

Run the program with --help to read the usage:

```
$ ./revc.py --help
usage: revc.py [-h] DNA

Print the reverse complement of DNA

positional arguments:
  DNA          Input sequence or file

optional arguments:
  -h, --help  show this help message and exit
```

The program wants DNA and will print the reverse complement, so I'll give it a string:

```
$ ./revc.py AAAACCCGGT
ACCGGGTTTT
```

As the help indicates, the program will also accept a file as input. The first test input has the same string:

```
$ cat tests/inputs/input1.txt
AAAACCCGGT
```

So the output should be the same:

```
$ ./revc.py tests/inputs/input1.txt
ACCGGGTTTT
```

I want to make the specs for the program just a little harder, so the tests will pass both a mix of uppercase and lowercase. The output should respect the case of the input:

```
$ ./revc.py aaaaCCCGGT
ACCGGGtttt
```

Run **pytest** (or **make test**) to see what kinds of tests the program should pass. When you're satisfied you have a feel for what the program should do, start anew:

```
$ new.py -f -p 'Print the reverse complement of DNA' revc.py
Done, see new script "revc.py".
```

Edit the get_args() function until the program will print the preceding usage. Then modify your program so that it will echo back input either from the command line or from an input file:

```
$ ./revc.py AAAACCCGGT
AAAACCCGGT
$ ./revc.py tests/inputs/input1.txt
AAAACCCGGT
```

If you run the test suite, you should find your program passes the first three tests:

```
$ pytest -xv
============================ test session starts ============================
...

tests/revc_test.py::test_exists PASSED                                  [ 14%]
tests/revc_test.py::test_usage PASSED                                   [ 28%]
tests/revc_test.py::test_no_args PASSED                                 [ 42%]
tests/revc_test.py::test_uppercase FAILED                               [ 57%]


================================= FAILURES =================================
_____ test_uppercase _____

    def test_uppercase():
        """ Runs on uppercase input """

        rv, out = getstatusoutput(f'{RUN} AAAACCCGGT')
        assert rv == 0
>       assert out == 'ACCGGGTTTT'
E       AssertionError: assert 'AAAACCCGGT' == 'ACCGGGTTTT'
E         - ACCGGGTTTT
E         + AAAACCCGGT

tests/revc_test.py:47: AssertionError
========================= short test summary info =========================
FAILED tests/revc_test.py::test_uppercase - AssertionError: assert 'AAAACCCGG...
```

```
!!!!!!!!!!!!!!!!!!!!!!!!!! stopping after 1 failures !!!!!!!!!!!!!!!!!!!!!!!!!!!
========================= 1 failed, 3 passed in 0.33s =========================
```

The program is being passed the input string AAAACCCGGT, and the test expects it to print ACCGGGTTTT. Since the program is echoing the input, this test fails. If you think you know how to write a program to satisfy these tests, have at it. If not, I'll show you how to create the reverse complement of DNA, starting with a simple approach and working up to more elegant solutions.

Iterating Over a Reversed String

When creating the reverse complement of DNA, it doesn't matter if you first reverse the sequence and then complement it or vice versa. You will get the same answer either way, so I'll start with how you can reverse a string. In Chapter 2, I showed how you can use a string slice to get a portion of a string. If you leave out the start position, it will start from the beginning:

```
>>> dna = 'AAAACCCGGT'
>>> dna[:2]
'AA'
```

If you leave out the stop position, it will go to the end:

```
>>> dna[-2:]
'GT'
```

If you leave out both start and stop, it will return a copy of the entire string:

```
>>> dna[:]
'AAAACCCGGT'
```

It also takes an optional third argument to indicate the step size. I can use no arguments for the start and stop, and a step of -1 to reverse the string:

```
>>> dna[::-1]
'TGGCCCAAAA'
```

Python also has a built-in reversed() function, so I'll try that:

```
>>> reversed(dna)
<reversed object at 0x7ffc4c9013a0>
```

Surprise! You were probably expecting to see the string TGGCCCAAAA. If you read help(reversed) in the REPL, however, you'll see that this function will "Return a reverse iterator over the values of the given sequence."

What is an *iterator*? Python's Functional Programming HOWTO (*https://oreil.ly/dIzn3*) describes an iterator as "an object representing a stream of data." I've mentioned that an *iterable* is some collection of items that Python can visit individually; for example, the characters of a string or the elements in a list. An iterator is something that will generate values until it is exhausted. Just as I can start with the first character of a string (or the first element of a list or the first line of a file) and read until the end of the string (or list or file), an iterator can be iterated from the first value it produces until it finishes.

In this case, the `reversed()` function is returning a promise to produce the reversed values as soon as it appears that you need them. This is an example of a *lazy* function because it waits until forced to do any work. One way to coerce the values from `reversed()` is to use a function that will consume the values. For instance, if the only goal is to reverse the string, then I could use the `str.join()` function. I always feel the syntax is backward on this function, but you will often invoke the `str.join()` method on a string literal that is the element used to join the sequence:

```
>>> ''.join(reversed(dna)) ❶
'TGGCCCAAAA'
```

❶ Use the empty string to join the reversed characters of the DNA string.

Another method uses the `list()` function to force `reversed()` to produce the values:

```
>>> list(reversed(dna))
['T', 'G', 'G', 'C', 'C', 'C', 'A', 'A', 'A', 'A']
```

Wait, what happened? The dna variable is a string, but I got back a list—and not just because I used the `list()` function. The documentation for `reversed()` shows that the function takes a *sequence*, which means any data structure or function that returns one thing followed by another. In a list or iterator context, Python treats strings as lists of characters:

```
>>> list(dna)
['A', 'A', 'A', 'A', 'C', 'C', 'C', 'G', 'G', 'T']
```

A longer way to build up the reversed DNA sequence is to use a for loop to iterate over the reversed bases and append them to a string. First I'll declare a rev variable, and I'll append each base in reverse order using the += operator:

```
>>> rev = '' ❶
>>> for base in reversed(dna): ❷
...        rev += base ❸
...
...
>>> rev
'TGGCCCAAAA'
```

❶ Initialize the rev variable with the empty string.

❷ Iterate through the reversed bases of DNA.

❸ Append the current base to the rev variable.

But since I still need to complement the bases, I'm not quite done.

Creating a Decision Tree

There are a total of eight complements: *A* to *T* and *G* to *C*, both upper- and lowercase, and then vice versa. I also need to handle the case of a character *not* being *A, C, G,* or *T.* I can use if/elif statements to create a decision tree. I'll change my variable to revc since it's now the reverse complement, and I'll figure out the correct complement for each base:

```
revc = '' ❶
for base in reversed(dna): ❷
    if base == 'A': ❸
        revc += 'T' ❹
    elif base == 'T':
        revc += 'A'
    elif base == 'G':
        revc += 'C'
    elif base == 'C':
        revc += 'G'
    elif base == 'a':
        revc += 't'
    elif base == 't':
        revc += 'a'
    elif base == 'g':
        revc += 'c'
    elif base == 'c':
        revc += 'g'
    else: ❺
        revc += base
```

❶ Initialize a variable to hold the reverse complement string.

❷ Iterate through the reversed bases in the DNA string.

❸ Test each uppercase and lowercase base.

❹ Append the complementing base to the variable.

❺ If the base doesn't match any of these tests, use the base as is.

If you inspect the revc variable, it appears to be correct:

```
>>> revc
'ACCGGGTTTT'
```

You should be able to incorporate these ideas into a program that will pass the test suite. To understand what exactly is expected of your program, take a look at the *tests/ revc_test.py* file. After you pass the test_uppercase() function, see what is expected by test_lowercase():

```
def test_lowercase():
    """ Runs on lowercase input """

    rv, out = getstatusoutput(f'{RUN} aaaaCCCGGT') ❶
    assert rv == 0 ❷
    assert out == 'ACCGGGtttt' ❸
```

❶ Run the program using lowercase and uppercase DNA strings.

❷ The exit value should be 0.

❸ The output from the program should be the indicated string.

The next tests will pass filenames rather than strings as input:

```
def test_input1():
    """ Runs on file input """

    file, expected = TEST1 ❶
    rv, out = getstatusoutput(f'{RUN} {file}') ❷
    assert rv == 0 ❸
    assert out == open(expected).read().rstrip() ❹
```

❶ The TEST1 tuple is a file of input and a file of expected output.

❷ Run the program with the filename.

❸ Make sure the exit value is 0.

❹ Open and read the expected file and compare that to the output.

It's equally important to read and understand the testing code as it is to learn how to write the solutions. When you write your programs, you may find you can copy many of the ideas from these tests and save yourself time.

Refactoring

While the algorithm in the preceding section will produce the correct answer, it is not an elegant solution. Still, it's a place to start that passes the tests. Now that you perhaps have a better idea of the challenge, it's time to refactor the program. Some of the solutions I present are as short as one or two lines of code. Here are some ideas you might consider:

- Use a dictionary as a lookup table instead of the if/elif chain.

- Rewrite the for loop as a list comprehension.

- Use the `str.translate()` method to complement the bases.
- Create a `Bio.Seq` object and find the method that will do this for you.

There's no hurry to read ahead. Take your time to try other solutions. I haven't introduced all these ideas yet, so I encourage you to research any unknowns and see if you can figure them out on your own.

I remember one of my teachers in music school sharing this quote with me:

> Then said a teacher, Speak to us of Teaching.
>
> And he said:
>
> No man can reveal to you aught but that which already lies half asleep in the dawning of your knowledge.
>
> The teacher who walks in the shadow of the temple, among his followers, gives not of his wisdom but rather of his faith and his lovingness.
>
> If he is indeed wise he does not bid you enter the house of his wisdom, but rather leads you to the threshold of your own mind.
>
> —Kahlil Gibran

Solutions

All of the solutions share the same `get_args()` function, as follows:

```
class Args(NamedTuple): ❶
    """ Command-line arguments """
    dna: str


# -------------------------------------------------
def get_args() -> Args:
    """ Get command-line arguments """

    parser = argparse.ArgumentParser(
        description='Print the reverse complement of DNA',
        formatter_class=argparse.ArgumentDefaultsHelpFormatter)

    parser.add_argument('dna', metavar='DNA', help='Input sequence or file')

    args = parser.parse_args()

    if os.path.isfile(args.dna): ❷
        args.dna = open(args.dna).read().rstrip()

    return Args(args.dna) ❸
```

❶ The only argument to the program is a string of DNA.

❷ Handle the case when reading a file input.

❸ Return an Args object in compliance with the function signature.

Solution 1: Using a for Loop and Decision Tree

Here is my first solution using the if/else decision tree:

```
def main() -> None:
    args = get_args()
    revc = ''  ❶

    for base in reversed(args.dna):  ❷
        if base == 'A':  ❸
            revc += 'T'
        elif base == 'T':
            revc += 'A'
        elif base == 'G':
            revc += 'C'
        elif base == 'C':
            revc += 'G'
        elif base == 'a':
            revc += 't'
        elif base == 't':
            revc += 'a'
        elif base == 'g':
            revc += 'c'
        elif base == 'c':
            revc += 'g'
        else:
            revc += base

    print(revc)  ❹
```

❶ Initialize a variable to hold the reverse complement.

❷ Iterate through the reversed bases of the DNA argument.

❸ Create an if/elif decision tree to determine each base's complement.

❹ Print the result.

Solution 2: Using a Dictionary Lookup

I mentioned that the if/else chain is something you should try to replace. That is 18 lines of code (LOC) that could be represented more easily using a dictionary lookup:

```
>>> trans = {
...     'A': 'T', 'C': 'G', 'G': 'C', 'T': 'A',
```

```
...      'a': 't', 'c': 'g', 'g': 'c', 't': 'a'
... }
```

If I use a `for` loop to iterate through a string of DNA, I can use the `dict.get()` method to safely request each base in a string of DNA to create the complement (see Figure 3-1). Note that I will use the `base` as the optional second argument to `dict.get()`. If the base doesn't exist in the lookup table, then I'll default to using the base as is, just like the `else` case from the first solution:

```
>>> for base in 'AAAACCCGGT':
...        print(base, trans.get(base, base))
...
A T
A T
A T
A T
C G
C G
C G
G C
G C
T A
```

I can create a `complement` variable to hold the new string I generate:

```
>>> complement = ''
>>> for base in 'AAAACCCGGT':
...        complement += trans.get(base, base)
...
>>> complement
'TTTTGGGCCA'
```

You saw before that using the `reversed()` function on a string will return a list of the characters of the string in reverse order:

```
>>> list(reversed(complement))
['A', 'C', 'C', 'G', 'G', 'G', 'T', 'T', 'T', 'T']
```

I can use the `str.join()` function to create a new string from a list:

```
>>> ''.join(reversed(complement))
'ACCGGGTTTT'
```

When I put all these ideas together, the `main()` function becomes significantly shorter. It also becomes easier to expand because adding a new branch to the decision tree only requires adding a new key/value pair to the dictionary:

```
def main() -> None:
    args = get_args()
    trans = { ❶
        'A': 'T', 'C': 'G', 'G': 'C', 'T': 'A',
        'a': 't', 'c': 'g', 'g': 'c', 't': 'a'
    }
```

```
        complement = '' ❷
        for base in args.dna: ❸
            complement += trans.get(base, base) ❹

        print(''.join(reversed(complement))) ❺
```

❶ This is a dictionary showing how to translate one base to its complement.

❷ Initialize a variable to hold the DNA complement.

❸ Iterate through each base in the DNA string.

❹ Append the translation of the base or the base itself to the complement.

❺ Reverse the complement and join the results on an empty string.

Python strings and lists are somewhat interchangeable. I can change the complement variable to a list, and nothing else in the program changes:

```
def main() -> None:
    args = get_args()
    trans = {
        'A': 'T', 'C': 'G', 'G': 'C', 'T': 'A',
        'a': 't', 'c': 'g', 'g': 'c', 't': 'a'
    }

    complement = [] ❶
    for base in args.dna:
        complement += trans.get(base, base)

    print(''.join(reversed(complement)))
```

❶ Initialize the complement to an empty list instead of a string.

I am highlighting here that the += operator works with both strings and lists to append a new value at the end. There is also a list.append() method which does the same:

```
    for base in args.dna:
        complement.append(trans.get(base, base))
```

The reversed() function works just as well on a list as it does a string. It's somewhat remarkable to me that using two different types for the complement results in so few changes to the code.

Solution 3: Using a List Comprehension

I suggested that you use a list comprehension without telling you what that is. If you've never used one before, it's essentially a way to write a for loop inside the square brackets ([]) used to create a new list (see Figure 3-2). When the goal of a for loop is to build up a new string or list, it makes much more sense to use a list comprehension.

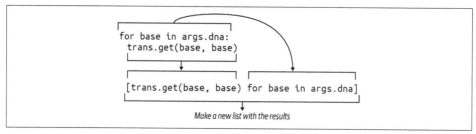

Figure 3-2. A list comprehension uses a for loop to generate a new list

This shortens the three lines to initialize a complement and loop through the string of DNA down to one line:

```
def main() -> None:
    args = get_args()
    trans = {
        'A': 'T', 'C': 'G', 'G': 'C', 'T': 'A',
        'a': 't', 'c': 'g', 'g': 'c', 't': 'a'
    }

    complement = [trans.get(base, base) for base in args.dna]  ❶
    print(''.join(reversed(complement)))
```

❶ Replace the for loop with a list comprehension.

Since the complement variable is only used once, I might even shorten this further by using the list comprehension directly:

```
print(''.join(reversed([trans.get(base, base) for base in args.dna])))
```

This is acceptable because the line is shorter than the maximum of 79 characters recommended by PEP8, but it's not as readable as the longer version. You should use whatever version you feel is most immediately understandable.

Solution 4: Using str.translate()

In Chapter 2, I used the str.replace() method to substitute all the *T*s with *U*s when transcribing DNA to RNA. Could I use that here? Let's try. I'll start by initializing the DNA string and replacing the *A*s with *T*s. Remember that strings are *immutable*, meaning I can't change a string in place, but rather must overwrite the string with a new value:

```
>>> dna = 'AAAACCCGGT'
>>> dna = dna.replace('A', 'T')
```

Now let's look at the DNA string:

```
>>> dna
'TTTTCCCGGT'
```

Can you see where this has started to go wrong? I'll complement the *T*s to *A*s now, and see if you can spot the problem:

```
>>> dna = dna.replace('T', 'A')
>>> dna
'AAAACCCGGA'
```

As shown in Figure 3-3, all the *A*s that turned into *T*s in the first move were just changed back to *A*s. Oh, that way madness lies.

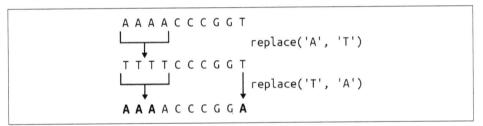

Figure 3-3. Iteratively using str.replace() *leads to double replacements of values and the wrong answer*

Fortunately, Python has the str.translate() function for exactly this purpose. If you read **help(str.translate)**, you will find the function requires a table "which must be a mapping of Unicode ordinals to Unicode ordinals, strings, or None." The trans dictionary table will serve, but first, it must be passed to the str.maketrans() function to transform the complement table into a form that uses the *ordinal* values of the keys:

```
>>> trans = {
...     'A': 'T', 'C': 'G', 'G': 'C', 'T': 'A',
...     'a': 't', 'c': 'g', 'g': 'c', 't': 'a'
... }
>>> str.maketrans(trans)
{65: 'T', 67: 'G', 71: 'C', 84: 'A', 97: 't', 99: 'g', 103: 'c', 116: 'a'}
```

You can see that the string key A was turned into the integer value 65, which is the same value returned by the ord() function:

```
>>> ord('A')
65
```

This value represents the ordinal position of the character *A* in the ASCII (American Standard Code for Information Interchange, pronounced *as-key*) table. That is, *A* is the 65th character in the table. The chr() function will reverse this process, providing the character represented by an ordinal value:

```
>>> chr(65)
'A'
```

The str.translate() function requires the complement table to have ordinal values for the keys, which is what I get from str.maketrans():

```
>>> 'AAAACCCGGT'.translate(str.maketrans(trans))
'TTTTGGGCCA'
```

Finally, I need to reverse the complement. Here is a solution that incorporates all these ideas:

```
def main() -> None:
    args = get_args()

    trans = str.maketrans({ ❶
        'A': 'T', 'C': 'G', 'G': 'C', 'T': 'A',
        'a': 't', 'c': 'g', 'g': 'c', 't': 'a'
    })
    print(''.join(reversed(args.dna.translate(trans)))) ❷
```

❶ Create the translation table needed for the str.translate() function.

❷ Complement the DNA using the trans table. Reverse, and join for a new string.

But, wait—there's more! There's another even shorter way to write this. According to the help(str.translate) documentation:

> If there is only one argument, it must be a dictionary mapping Unicode ordinals (integers) or characters to Unicode ordinals, strings or None. Character keys will be then converted to ordinals. *If there are two arguments, they must be strings of equal length, and in the resulting dictionary, each character in x will be mapped to the character at the same position in y.*

So I can remove the trans dictionary and write the entire solution like this:

```
def main() -> None:
    args = get_args()
    trans = str.maketrans('ACGTacgt', 'TGCAtgca') ❶
    print(''.join(reversed(args.seq.translate(trans)))) ❷
```

❶ Make the translation table using two strings of equal lengths.

❷ Create the reverse complement.

If you wanted to ruin someone's day—and in all likelihood, that person will be future you—you could even condense this into a single line of code.

Solution 5: Using Bio.Seq

I told you at the beginning of this chapter that the final solution would involve an existing function.[1] Many Python programmers working in bioinformatics have contributed to a set of modules under the name of Biopython (*https://biopython.org*). They have written and tested many incredibly useful algorithms, and it rarely makes sense to write your own code when you can use someone else's.

Be sure that you have first installed biopython by running the following:

```
$ python3 -m pip install biopython
```

I could import the entire module using import Bio, but it makes much more sense to only import the code I need. Here I only need the Seq class:

```
>>> from Bio import Seq
```

Now I can use the Seq.reverse_complement() function:

```
>>> Seq.reverse_complement('AAAACCCGGT')
'ACCGGGTTTT'
```

This final solution is the version I would recommend, as it is the shortest and also uses existing, well-tested, documented modules that are almost ubiquitous in bioinformatics with Python:

```
def main() -> None:
    args = get_args()
    print(Seq.reverse_complement(args.dna))  ❶
```

❶ Use the Bio.Seq.reverse_complement() function.

When you run mypy on this solution (you *are* running mypy on every one of your programs, right?), you may get the following error:

```
=================================== FAILURES ===================================
_____ revc.py _____
6: error: Skipping analyzing 'Bio': found module but no type hints or library
   stubs
6: note: See https://mypy.readthedocs.io/en/latest/running_mypy.html#missing
   -imports
=================================== mypy =======================================
Found 1 error in 1 file (checked 2 source files)
```

---

1 This is kind of like how my high school calculus teacher spent a week teaching us how to perform manual derivatives, then showed us how it could be done in 20 seconds by pulling down the exponent and yada yada yada.

```
mypy.ini: No [mypy] section in config file

=========================== short test summary info ===========================
FAILED revc.py::mypy
!!!!!!!!!!!!!!!!!!!!!!!!!! stopping after 1 failures !!!!!!!!!!!!!!!!!!!!!!!!!!!
=========================== 1 failed, 1 skipped in 0.20s =======================
```

To silence this error, you can tell `mypy` to ignore imported files that are missing type annotations. In the root directory of the GitHub repository for this book (*https://oreil.ly/RpMgV*), you will find a file called *mypy.ini* with the following contents:

```
$ cat mypy.ini
[mypy]
ignore_missing_imports = True
```

Adding a *mypy.ini* file to any working directory allows you to make changes to the defaults that `mypy` uses *when you run it in the same directory*. If you would like to make this a global change so that `mypy` will use this no matter what directory you are in, then put this same content into *$HOME/.mypy.ini*.

Review

Manually creating the reverse complement of DNA is something of a rite of passage. Here's what I showed:

- You can write a decision tree using a series of `if`/`else` statements or by using a dictionary as a lookup table.

- Strings and lists are very similar. Both can be iterated using a `for` loop, and the `+=` operator can be used to append to both.

- A list comprehension uses a `for` loop to iterate a sequence and generate a new list.

- The `reversed()` function is a lazy function that will return an iterator of the elements of a sequence in reverse order.

- You can use the `list()` function in the REPL to coerce lazy functions, iterators, and generators to generate their values.

- The `str.maketrans()` and `str.translate()` functions can perform string substitution and generate a new string.

- The `ord()` function returns the ordinal value of a character, and conversely, the `chr()` function returns the character for a given ordinal value.

- Biopython is a collection of modules and functions specific to bioinformatics. The preferred way to create the reverse complement of DNA is to use the `Bio.Seq.reverse_complement()` function.

Creating the Fibonacci Sequence: Writing, Testing, and Benchmarking Algorithms

Writing an implementation of the Fibonacci sequence is another step in the hero's journey to becoming a coder. The Rosalind Fibonacci description (*https://oreil.ly/ 7vkRw*) notes that the genesis for the sequence was a mathematical simulation of breeding rabbits that relies on some important (and unrealistic) assumptions:

- The first month starts with a pair of newborn rabbits.
- Rabbits can reproduce after one month.
- Every month, every rabbit of reproductive age mates with another rabbit of reproductive age.
- Exactly one month after two rabbits mate, they produce a litter of the same size.
- Rabbits are immortal and never stop mating.

The sequence always begins with the numbers 0 and 1. The subsequent numbers can be generated *ad infinitum* by adding the two immediately previous values in the list, as shown in Figure 4-1.

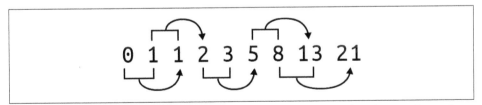

Figure 4-1. The first eight numbers of the Fibonacci sequence—after the initial 0 and 1, subsequent numbers are created by adding the two previous numbers

If you search the internet for solutions, you'll find dozens of different ways to generate the sequence. I want to focus on three fairly different approaches. The first solution uses an *imperative* approach where the algorithm strictly defines every step. The next solution uses a *generator* function, and the last will focus on a *recursive* solution. Recursion, while interesting, slows drastically as I try to generate more of the sequence, but it turns out the performance problems can be solved using caching.

You will learn:

- How to manually validate arguments and throw errors
- How to use a list as a stack
- How to write a generator function
- How to write a recursive function
- Why recursive functions can be slow and how to fix this with memoization
- How to use function decorators

Getting Started

The code and tests for this chapter are found in the *04_fib* directory. Start by copying the first solution to `fib.py`:

```
$ cd 04_fib/
$ cp solution1_list.py fib.py
```

Ask for the usage to see how the parameters are defined. You can use n and k, but I chose to use the names `generations` and `litter`:

```
$ ./fib.py -h
usage: fib.py [-h] generations litter

Calculate Fibonacci

positional arguments:
  generations  Number of generations
  litter       Size of litter per generation

optional arguments:
  -h, --help   show this help message and exit
```

This will be the first program to accept arguments that are not strings. The Rosalind challenge indicates that the program should accept two positive integer values:

- $n \leq 40$ representing the number of generations
- $k \leq 5$ representing the litter size produced by mate pairs

Try to pass noninteger values and notice how the program fails:

```
$ ./fib.py foo
usage: fib.py [-h] generations litter
fib.py: error: argument generations: invalid int value: 'foo'
```

You can't tell, but in addition to printing the brief usage and a helpful error message, the program also generated a nonzero exit value. On the Unix command line, an exit value of 0 indicates success. I think of this as "zero errors." In the bash shell, I can inspect the $? variable to look at the exit status of the most recent process. For instance, the command echo Hello should exit with a value of 0, and indeed it does:

```
$ echo Hello
Hello
$ echo $?
0
```

Try the previously failing command again, and then inspect $?:

```
$ ./fib.py foo
usage: fib.py [-h] generations litter
fib.py: error: argument generations: invalid int value: 'foo'
$ echo $?
2
```

That the exit status is 2 is not as important as the fact that the value is not zero. This is a well-behaved program because it rejects an invalid argument, prints a useful error message, and exits with a nonzero status. If this program were part of a pipeline of data processing steps (such as a *Makefile*, discussed in Appendix A), a nonzero exit value would cause the entire process to stop, which is a good thing. Programs that silently accept invalid values and fail quietly or not at all can lead to unreproducible results. It's vitally important that programs properly validate arguments and fail very convincingly when they cannot proceed.

The program is very strict even about the type of number it accepts. The values must be integers. It will also repel any floating-point values:

```
$ ./fib.py 5 3.2
usage: fib.py [-h] generations litter
fib.py: error: argument litter: invalid int value: '3.2'
```

 All command-line arguments to the program are technically received as strings. Even though 5 on the command line looks like the *number* 5, it's the *character* "5". I am relying on argparse in this situation to attempt to convert the value from a string to an integer. When that fails, argparse generates these useful error messages.

Additionally, the program rejects values for the generations and litter parameters that are not in the allowed ranges. Notice that the error message includes the name of the argument and the offending value to provide sufficient feedback to the user so you can fix it:

```
$ ./fib.py -3 2
usage: fib.py [-h] generations litter
fib.py: error: generations "-3" must be between 1 and 40 ❶
$ ./fib.py 5 10
usage: fib.py [-h] generations litter
fib.py: error: litter "10" must be between 1 and 5 ❷
```

❶ The generations argument of -3 is not in the stated range of values.

❷ The litter argument of 10 is too high.

Look at the first part of the solution to see how to make this work:

```
import argparse
from typing import NamedTuple


class Args(NamedTuple):
    """ Command-line arguments """
    generations: int ❶
    litter: int ❷


def get_args() -> Args:
    """ Get command-line arguments """

    parser = argparse.ArgumentParser(
        description='Calculate Fibonacci',
        formatter_class=argparse.ArgumentDefaultsHelpFormatter)

    parser.add_argument('gen', ❸
                        metavar='generations',
                        type=int, ❹
                        help='Number of generations')

    parser.add_argument('litter', ❺
                        metavar='litter',
                        type=int,
                        help='Size of litter per generation')

    args = parser.parse_args() ❻

    if not 1 <= args.gen <= 40: ❼
        parser.error(f'generations "{args.gen}" must be between 1 and 40') ❽

    if not 1 <= args.litter <= 5: ❾
        parser.error(f'litter "{args.litter}" must be between 1 and 5') ❿

    return Args(generations=args.gen, litter=args.litter) ⓫
```

❶ The generations field must be an int.

❷ The litter field must also be an int.

❸ The gen positional parameter is defined first, so it will receive the first positional value.

❹ The type=int indicates the required class of the value. Notice that int indicates the class itself, not the name of the class.

❺ The litter positional parameter is defined second, so it will receive the second positional value.

❻ Attempt to parse the arguments. Any failure will result in error messages and the program exiting with a nonzero value.

❼ The args.gen value is now an actual int value, so it's possible to perform numeric comparisons on it. Check if it is in the acceptable range.

❽ Use the parser.error() function to generate an error and exit the program.

❾ Likewise check the value of the args.litter argument.

❿ Generate an error that includes information the user needs to fix the problem.

⓫ If the program makes it to this point, then the arguments are valid integer values in the accepted range, so return the Args.

I could check that the generations and litter values are in the correct ranges in the main() function, but I prefer to do as much argument validation as possible inside the get_args() function so that I can use the parser.error() function to generate useful messages and exit the program with a nonzero value.

Remove the fib.py program and start anew with **new.py** or your preferred method for creating a program:

```
$ new.py -fp 'Calculate Fibonacci' fib.py
Done, see new script "fib.py".
```

You can replace the get_args() definition with the preceding code, then modify your main() function like so:

```
def main() -> None:
    args = get_args()
    print(f'generations = {args.generations}')
    print(f'litter = {args.litter}')
```

Run your program with invalid inputs and verify that you see the kinds of error messages shown earlier. Try your program with acceptable values and verify that you see this kind of output:

```
$ ./fib.py 1 2
generations = 1
litter = 2
```

Run **pytest** to see what your program passes and fails. You should pass the first four tests and fail the fifth:

```
$ pytest -xv
========================= test session starts =========================
...
tests/fib_test.py::test_exists PASSED                          [ 14%]
tests/fib_test.py::test_usage PASSED                           [ 28%]
tests/fib_test.py::test_bad_generations PASSED                 [ 42%]
tests/fib_test.py::test_bad_litter PASSED                      [ 57%]
tests/fib_test.py::test_1 FAILED                               [ 71%] ❶

=============================== FAILURES ===============================
_____ test_1 _____

    def test_1():
        """runs on good input"""

        rv, out = getstatusoutput(f'{RUN} 5 3') ❷
        assert rv == 0
>       assert out == '19' ❸
E       AssertionError: assert 'generations = 5\nlitter = 3' == '19' ❹
E         - 19    ❺
E         + generations = 5 ❻
E         + litter = 3

tests/fib_test.py:60: AssertionError
======================= short test summary info =======================
FAILED tests/fib_test.py::test_1 - AssertionError: assert 'generations...
!!!!!!!!!!!!!!!!!!!!!!!!! stopping after 1 failures !!!!!!!!!!!!!!!!!!!!!!!!!
===================== 1 failed, 4 passed in 0.38s =====================
```

❶ The first failing test. Testing halts here because of the -x flag.

❷ The program is run with 5 for the number of generations and 3 for the litter size.

❸ The output should be 19.

❹ This shows the two strings being compared are not equal.

❺ The expected value was 19.

❻ This is the output that was received.

The output from `pytest` is trying very hard to point out exactly what went wrong. It shows how the program was run and what was expected versus what was produced. The program is supposed to print 19, which is the fifth number of the Fibonacci sequence when using a litter size of 3. If you want to finish the program on your own, please jump right in. You should use **pytest** to verify that you are passing all the tests. Also, run **make test** to check your program using `pylint`, `flake8`, and `mypy`. If you want some guidance, I'll cover the first approach I described.

An Imperative Approach

Figure 4-2 depicts the growth of the Fibonacci sequence. The smaller rabbits indicate nonbreeding pairs that must mature into larger, breeding pairs.

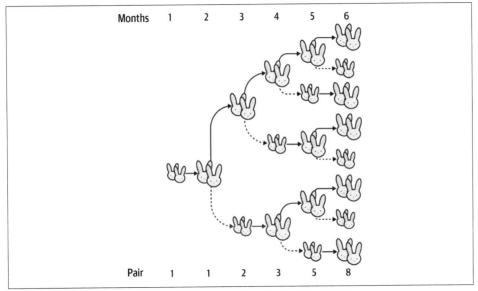

Figure 4-2. A visualization of the growth of the Fibonacci sequence as mating pairs of rabbits using a litter size of 1

You can see that to generate any number after the first two, I need to know the two previous numbers. I can use this formula to describe the value of any position n of the Fibonacci sequence (F):

$$F_n = F_{n-1} + F_{n-2}$$

What kind of a data structure in Python would allow me to keep a sequence of numbers in order and refer to them by their position? A list. I'll start off with $F_1 = 0$ and $F_2 = 1$:

```
>>> fib = [0, 1]
```

The F_3 value is $F_2 + F_1 = 1 + 0 = 1$. When generating the next number, I'll always be referencing the *last two* elements of the sequence. It will be easiest to use negative indexing to indicate a position from the *end* of the list. The last value in a list is always at position -1:

```
>>> fib[-1]
1
```

The penultimate value is at -2:

```
>>> fib[-2]
0
```

I need to multiply this value by the litter size to calculate the number of offspring that generation created. To start, I'll consider a litter size of 1:

```
>>> litter = 1
>>> fib[-2] * litter
0
```

I want to add these two numbers together and append the result to the list:

```
>>> fib.append((fib[-2] * litter) + fib[-1])
>>> fib
[0, 1, 1]
```

If I do this again, I can see that the correct sequence is emerging:

```
>>> fib.append((fib[-2] * litter) + fib[-1])
>>> fib
[0, 1, 1, 2]
```

I need to repeat this action generations times. (Technically it will be generations-1 times because Python uses 0-based indexing.) I can use Python's range() function to generate a list of numbers from 0 up to but not including the end value. I'm calling this function solely for the side effect of iterating a particular number of times and so don't need the values produced by the range() function. It's common to use the underscore (_) variable to indicate one's intent to ignore a value:

```
>>> fib = [0, 1]
>>> litter = 1
>>> generations = 5
>>> for _ in range(generations - 1):
...     fib.append((fib[-2] * litter) + fib[-1])
...
>>> fib
[0, 1, 1, 2, 3, 5]
```

This should be enough for you to create a solution that passes the tests. In the next section, I'll cover two other solutions that highlight some very interesting parts of Python.

Solutions

All the following solutions share the same `get_args()` shown previously.

Solution 1: An Imperative Solution Using a List as a Stack

Here is how I wrote my imperative solution. I'm using a list as a kind of *stack* to keep track of past values. I don't need all the values, just the last two, but it's pretty easy to keep growing the list and referring to the last two values:

```python
def main() -> None:
    args = get_args()

    fib = [0, 1] ❶
    for _ in range(args.generations - 1): ❷
        fib.append((fib[-2] * args.litter) + fib[-1]) ❸

    print(fib[-1]) ❹
```

❶ Start with 0 and 1.

❷ Use the `range()` function to create the right number of loops.

❸ Append the next value to the sequence.

❹ Print the last number of the sequence.

 I used the _ variable name in the for loop to indicate that I don't intend to use the variable. The underscore is a valid Python identifier, and it's also a convention to use this to indicate a *throwaway* value. Linting tools, for instance, might see that I've assigned a variable some value but never used it, which would normally indicate a possible error. The underscore variable shows that I do not intend to use the value. In this case, I'm using the `range()` function purely for the side effect of creating the number of loops needed.

This is considered an *imperative* solution because the code directly encodes every instruction of the algorithm. When you read the recursive solution, you will see that the algorithm can be written in a more declarative manner, which also has unintended consequences that I must handle.

A slight variation on this would be to place this code inside a function I'll call `fib()`. Note that it's possible in Python to declare a function inside another function, as here I'll create `fib()` inside `main()`. I do this so I can reference the `args.litter` parameter, creating a *closure* because the function is capturing the runtime value of the litter size:

```python
def main() -> None:
    args = get_args()

    def fib(n: int) -> int:  ❶
        nums = [0, 1]  ❷
        for _ in range(n - 1):  ❸
            nums.append((nums[-2] * args.litter) + nums[-1])  ❹
        return nums[-1]  ❺

    print(fib(args.generations))  ❻
```

❶ Create a function called `fib()` that accepts an integer parameter `n` and returns an integer.

❷ This is the same code as before. Note this list is called `nums` so it doesn't clash with the function name.

❸ Use the `range()` function to iterate the generations. Use `_` to ignore the values.

❹ The function references the `args.litter` parameter and so creates a closure.

❺ Use `return` to send the final value back to the caller.

❻ Call the `fib()` function with the `args.generations` parameter.

The scope of the `fib()` function in the preceding example is limited to the `main()` function. *Scope* refers to the part of the program where a particular function name or variable is visible or legal.

I don't have to use a closure. Here is how I can express the same idea with a standard function:

```python
def main() -> None:
    args = get_args()

    print(fib(args.generations, args.litter))  ❶

def fib(n: int, litter: int) -> int:  ❷
    nums = [0, 1]
    for _ in range(n - 1):
        nums.append((nums[-2] * litter) + nums[-1])

    return nums[-1]
```

❶ The `fib()` function must be called with two arguments.

❷ The function requires both the number of generations and the litter size. The function body is essentially the same.

In the preceding code, you see that I must pass two arguments to `fib()`, whereas the closure required only one argument because the `litter` was captured. Binding values and reducing the number of parameters is a valid reason for creating a closure. Another reason to write a closure is to limit the scope of a function. The closure definition of `fib()` is valid only inside the `main()` function, but the preceding version is visible throughout the program. Hiding a function inside another function makes it harder to test. In this case, the `fib()` function is almost the entire program, so the tests have already been written in *tests/fib_test.py*.

Solution 2: Creating a Generator Function

In the previous solution, I generated the Fibonacci sequence up to the value requested and then stopped; however, the sequence is infinite. Could I create a function that could generate *all* the numbers of the sequence? Technically, yes, but it would never finish, what with being infinite and all.

Python has a way to suspend a function that generates a possibly infinite sequence. I can use `yield` to return a value from a function, temporarily leaving the function later to resume at the same state when the next value is requested. This kind of function is called a *generator*, and here is how I can use it to generate the sequence:

```
def fib(k: int) -> Generator[int, None, None]: ❶
    x, y = 0, 1 ❷
    yield x ❸

    while True: ❹
        yield y ❺
        x, y = y * k, x + y ❻
```

❶ The type signature indicates the function takes the parameter k (litter size), which must be an `int`. It returns a special function of the type `Generator` which yields `int` values and has no send or return types.

❷ I only ever need to track the last two generations, which I initialize to 0 and 1.

❸ Yield the 0.

❹ Create an infinite loop.

❺ Yield the last generation.

❻ Set x (two generations back) to the current generation times the litter size. Set y (one generation back) to the sum of the two current generations.

A generator acts like an iterator, producing values as requested by the code until it is exhausted. Since this generator will only generate yield values, the send and return types are None. Otherwise, this code does exactly what the first version of the program did, only inside a fancy-pants generator function. See Figure 4-3 to consider how the function works for two different litter sizes.

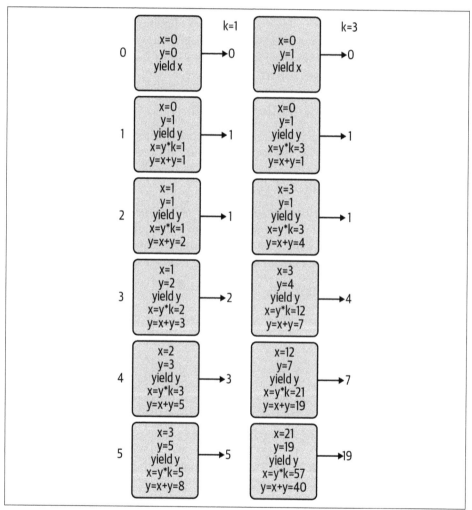

Figure 4-3. A depiction of how the fib() generator's state changes over time (n=5) for two litter sizes (k=1 and k=3)

The type signature for `Generator` looks a little complicated since it defines types for yield, send, and return. I don't need to dive into it further here, but I recommend you read the docs on the `typing` module (*https://oreil.ly/Oir3d*).

Here's how to use this:

```
def main() -> None:
    args = get_args()
    gen = fib(args.litter) ❶
    seq = [next(gen) for _ in range(args.generations + 1)] ❷
    print(seq[-1]) ❸
```

❶ The `fib()` function takes the litter size as an argument and returns a generator.

❷ Use the `next()` function to retrieve the next value from a generator. Use a list comprehension to do this the correct number of times to generate the sequence up to the requested value.

❸ Print the last number in the sequence.

 The `range()` is function different because the first version already had the 0 and 1 in place. Here I have to call the generator two extra times to produce those values.

Although I prefer the list comprehension, I don't need the entire list. I only care about the final value, so I could have written it like so:

```
def main() -> None:
    args = get_args()
    gen = fib(args.litter)
    answer = 0 ❶
    for _ in range(args.generations + 1): ❷
        answer = next(gen) ❸
    print(answer) ❹
```

❶ Initialize the answer to 0.

❷ Create the correct number of loops.

❸ Get the value for the current generation.

❹ Print the answer.

As it happens, it's quite common to call a function repeatedly to generate a list, so there is a function to do this for us. The `itertools.islice()` function will "Make an iterator that returns selected elements from the iterable." Here is how I can use it:

```
def main() -> None:
    args = get_args()
    seq = list(islice(fib(args.litter), args.generations + 1)) ❶
    print(seq[-1]) ❷
```

❶ The first argument to islice() is the function that will be called, and the second argument is the number of times to call it. The function is lazy, so I use list() to coerce the values.

❷ Print the last value.

Since I only use the seq variable one time, I could eschew that assignment. If benchmarking proved the following to be the best-performing version, I might be willing to write a one-liner:

```
def main() -> None:
    args = get_args()
    print(list(islice(fib(args.litter), args.generations + 1))[-1])
```

Clever code is fun but can become unreadable.[1] You have been warned.

Generators are cool but more complex than generating a list. They are the appropriate way to generate a very large or potentially infinite sequence of values because they are lazy, only computing the next value when your code requires it.

Solution 3: Using Recursion and Memoization

While there are many more fun ways to write an algorithm to produce an infinite series of numbers, I'll show just one more using *recursion*, which is when a function calls itself:

```
def main() -> None:
    args = get_args()

    def fib(n: int) -> int: ❶
        return 1 if n in (1, 2) \ ❷
            else fib(n - 2) * args.litter + fib(n - 1) ❸

    print(fib(args.generations)) ❹
```

❶ Define a function called fib() that takes the number of the generation wanted as an int and returns an int.

---

1 As the legendary David St. Hubbins and Nigel Tufnel observed, "It's such a fine line between stupid and clever."

❷ If the generation is 1 or 2, return 1. This is the all-important base case that does not make a recursive call.

❸ For all other cases, call the fib() function twice, once for two generations back and another for the previous generation. Factor in the litter size as before.

❹ Print the results of the fib() function for the given generations.

 Here's another instance where I define a fib() function as a closure *inside* the main() function so as to use the args.litter value inside the fib() function. This is to close around args.litter, effectively binding that value to the function. If I had defined the function outside the main() function, I would have had to pass the args.litter argument on the recursive calls.

This is a really elegant solution that gets taught in pretty much every introductory computer science class. It's fun to study, but it turns out to be wicked slow because I end up calling the function so many times. That is, fib(5) needs to call fib(4) and fib(3) to add those values. In turn, fib(4) needs to call fib(3) and fib(2), and so on. Figure 4-4 shows that fib(5) results in 14 function calls to produce 5 distinct values. For instance, fib(2) is calculated three times, but we only need to calculate it once.

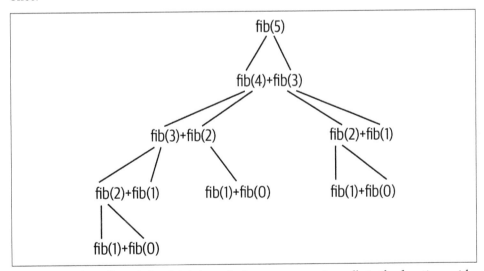

Figure 4-4. The call stack for fib(5) results in many recursive calls to the function, with their number increasing approximately exponentially as the input value increases

To illustrate the problem, I'll take a sampling of how long this program takes to finish up to the maximum n of 40. Again, I'll use a for loop in bash to show you how I would commonly benchmark such a program on the command line:

```
$ for n in 10 20 30 40;
> do echo "==> $n <==" && time ./solution3_recursion.py $n 1
> done
==> 10 <==
55

real    0m0.045s
user    0m0.032s
sys     0m0.011s
==> 20 <==
6765

real    0m0.041s
user    0m0.031s
sys     0m0.009s
==> 30 <==
832040

real    0m0.292s
user    0m0.281s
sys     0m0.009s
==> 40 <==
102334155

real    0m31.629s
user    0m31.505s
sys     0m0.043s
```

The jump from 0.29s for n=30 to 31s for n=40 is huge. Imagine going to 50 and beyond. I need to either find a way to speed this up or abandon all hope for recursion. The solution is to cache previously calculated results. This is called *memoization*, and there are many ways to implement this. The following is one method. Note you will need to import typing.Callable:

```
def memoize(f: Callable) -> Callable: ❶
    """ Memoize a function """

    cache = {} ❷

    def memo(x): ❸
        if x not in cache: ❹
            cache[x] = f(x) ❺
        return cache[x] ❻

    return memo ❼
```

➊ Define a function that takes a function (something that is *callable*) and returns a function.

➋ Use a dictionary to store cached values.

➌ Define `memo()` as a closure around the cache. The function will take some parameter x when called.

➍ See if the argument value is in the cache.

➎ If not, call the function with the argument and set the cache for that argument value to the result.

➏ Return the cached value for the argument.

➐ Return the new function.

Note that the `memoize()` function returns a new function. In Python, functions are considered *first-class objects*, meaning they can be used like other kinds of variables—you can pass them as arguments and overwrite their definitions. The `memoize()` function is an example of a *higher-order function* (HOF) because it takes other functions as arguments. I'll be using other HOFs, like `filter()` and `map()`, throughout the book.

To use the `memoize()` function, I will define `fib()` and then *redefine it* with the memoized version. If you run this, you will see an almost instantaneous result no matter how high n goes:

```
def main() -> None:
    args = get_args()

    def fib(n: int) -> int:
        return 1 if n in (1, 2) else fib(n - 2) * args.litter + fib(n - 1)

    fib = memoize(fib)  ➊

    print(fib(args.generations))
```

➊ Overwrite the existing `fib()` definition with the memoized function.

A preferred method to accomplish this uses a *decorator*, which is a function that modifies another function:

```
def main() -> None:
    args = get_args()

    @memoize  ➊
    def fib(n: int) -> int:
```

```
        return 1 if n in (1, 2) else fib(n - 2) * args.litter + fib(n - 1)

    print(fib(args.generations))
```

❶ Decorate the `fib()` function with the `memoize()` function.

As fun as writing memoization functions is, it again turns out that this is such a common need that others have already solved it for us. I can remove the `memoize()` function and instead import the `functools.lru_cache` (least-recently-used cache) function:

```
from functools import lru_cache
```

Decorate the `fib()` function with the `lru_cache()` function to get memoization with minimal distraction:

```
def main() -> None:
    args = get_args()

    @lru_cache()  ❶
    def fib(n: int) -> int:
        return 1 if n in (1, 2) else fib(n - 2) * args.litter + fib(n - 1)

    print(fib(args.generations))
```

❶ Memoize the `fib()` function via decoration with the `lru_cache()` function. Note that Python 3.6 requires the parentheses, but 3.8 and later versions do not.

Benchmarking the Solutions

Which is the fastest solution? I've shown you how to use a `for` loop in `bash` with the `time` command to compare the runtimes of commands:

```
$ for py in ./solution1_list.py ./solution2_generator_islice.py \
./solution3_recursion_lru_cache.py; do echo $py && time $py 40 5; done
./solution1_list.py
148277527396903091

real    0m0.070s
user    0m0.043s
sys     0m0.016s
./solution2_generator_islice.py
148277527396903091

real    0m0.049s
user    0m0.033s
sys     0m0.013s
./solution3_recursion_lru_cache.py
148277527396903091

real    0m0.041s
```

```
user    0m0.030s
sys     0m0.010s
```

It would appear that the recursive solution using LRU caching is the fastest, but again I have very little data—just one run per program. Also, I have to eyeball this data and figure out which is the fastest.

There's a better way. I have installed a tool called hyperfine (*https://oreil.ly/shqOS*) to run each command many times and compare the results:

```
$ hyperfine -L prg ./solution1_list.py,./solution2_generator_islice.py,\
./solution3_recursion_lru_cache.py '{prg} 40 5' --prepare 'rm -rf __pycache__'
Benchmark #1: ./solution1_list.py 40 5
  Time (mean ± σ):      38.1 ms ±   1.1 ms    [User: 28.3 ms, System: 8.2 ms]
  Range (min … max):    36.6 ms …  42.8 ms    60 runs

Benchmark #2: ./solution2_generator_islice.py 40 5
  Time (mean ± σ):      38.0 ms ±   0.6 ms    [User: 28.2 ms, System: 8.1 ms]
  Range (min … max):    36.7 ms …  39.2 ms    66 runs

Benchmark #3: ./solution3_recursion_lru_cache.py 40 5
  Time (mean ± σ):      37.9 ms ±   0.6 ms    [User: 28.1 ms, System: 8.1 ms]
  Range (min … max):    36.6 ms …  39.4 ms    65 runs

Summary
  './solution3_recursion_lru_cache.py 40 5' ran
    1.00 ± 0.02 times faster than './solution2_generator_islice.py 40 5'
    1.01 ± 0.03 times faster than './solution1_list.py 40 5'
```

It appears that hyperfine ran each command 60-66 times, averaged the results, and found that the solution3_recursion_lru_cache.py program is perhaps slightly faster. Another benchmarking tool you might find useful is bench (*https://oreil.ly/FKnmd*), but you can search for other benchmarking tools on the internet that might suit your tastes more. Whatever tool you use, benchmarking along with testing is vital to challenging assumptions about your code.

 I used the --prepare option to tell hyperfine to remove the *pycache* directory before running the commands. This is a directory created by Python to cache *bytecode* of the program. If a program's source code hasn't changed since the last time it was run, then Python can skip compilation and use the bytecode version that exists in the *pycache* directory. I needed to remove this as hyperfine detected statistical outliers when running the commands, probably due to caching effects.

Testing the Good, the Bad, and the Ugly

For every challenge, I hope you spend part of your time reading through the tests. Learning how to design and write tests is as important as anything else I'm showing you. As I mentioned before, my first tests check that the expected program exists and will produce usage statements on request. After that, I usually give invalid inputs to ensure the program fails. I would like to highlight the tests for bad n and k parameters. They are essentially the same, so I'll just show the first one as it demonstrates how to randomly select an invalid integer value—one that is possibly negative or too high:

```
def test_bad_n():
    """ Dies when n is bad """

    n = random.choice(list(range(-10, 0)) + list(range(41, 50)))  ❶
    k = random.randint(1, 5)  ❷
    rv, out = getstatusoutput(f'{RUN} {n} {k}')  ❸
    assert rv != 0  ❹
    assert out.lower().startswith('usage:')  ❺
    assert re.search(f'n "{n}" must be between 1 and 40', out)  ❻
```

❶ Join the two lists of invalid number ranges and randomly select one value.

❷ Select a random integer in the range from 1 to 5 (both bounds inclusive).

❸ Run the program with the arguments and capture the output.

❹ Make sure the program reported a failure (nonzero exit value).

❺ Check the output begins with a usage statement.

❻ Look for an error message describing the problem with the n argument.

I often like to use randomly selected invalid values when testing. This partially comes from writing tests for students so that they won't write programs that fail on a single bad input, but I also find it helps me to not accidentally code for a specific input value. I haven't yet covered the random module, but it gives you a way to make pseudorandom choices. First, you need to import the module:

```
>>> import random
```

For instance, you can use random.randint() to select a single integer from a given range:

```
>>> random.randint(1, 5)
2
>>> random.randint(1, 5)
5
```

Or use the `random.choice()` function to randomly select a single value from some sequence. Here I wanted to construct a discontiguous range of negative numbers separated from a range of positive numbers:

```
>>> random.choice(list(range(-10, 0)) + list(range(41, 50)))
46
>>> random.choice(list(range(-10, 0)) + list(range(41, 50)))
-1
```

The tests that follow all provide good inputs to the program. For example:

```
def test_2():
    """ Runs on good input """

    rv, out = getstatusoutput(f'{RUN} 30 4') ❶
    assert rv == 0 ❷
    assert out == '436390025825' ❸
```

❶ These are values I was given while attempting to solve the Rosalind challenge.

❷ The program should not fail on this input.

❸ This is the correct answer per Rosalind.

Testing, like documentation, is a love letter to your future self. As tedious as testing may seem, you'll appreciate failing tests when you try to add a feature and end up accidentally breaking something that previously worked. Assiduously writing and running tests can prevent you from deploying broken programs.

Running the Test Suite on All the Solutions

You've seen that in each chapter I write multiple solutions to explore various ways to solve the problems. I completely rely on my tests to ensure my programs are correct. You might be curious to see how I've automated the process of testing every single solution. Look at the *Makefile* and find the `all` target:

```
$ cat Makefile
.PHONY: test

test:
        python3 -m pytest -xv --flake8 --pylint --mypy fib.py tests/fib_test.py

all:
    ../bin/all_test.py fib.py
```

The all_test.py program will overwrite the fib.py program with each of the solutions before running the test suite. This could overwrite your solution. Be sure you commit your version to Git or at least make a copy before you run make all or you could lose your work.

The following is the all_test.py program that is run by the all target. I'll break it into two parts, starting with the first part up to get_args(). Most of this should be familiar by now:

```python
#!/usr/bin/env python3
""" Run the test suite on all solution*.py """

import argparse
import os
import re
import shutil
import sys
from subprocess import getstatusoutput
from functools import partial
from typing import NamedTuple


class Args(NamedTuple):
    """ Command-line arguments """
    program: str ❶
    quiet: bool ❷


# --------------------------------------------------
def get_args() -> Args:
    """ Get command-line arguments """

    parser = argparse.ArgumentParser(
        description='Run the test suite on all solution*.py',
        formatter_class=argparse.ArgumentDefaultsHelpFormatter)

    parser.add_argument('program', metavar='prg', help='Program to test') ❸

    parser.add_argument('-q', '--quiet', action='store_true', help='Be quiet') ❹

    args = parser.parse_args()

    return Args(args.program, args.quiet)
```

❶ The name of the program to test, which in this case is fib.py.

❷ A Boolean value of True or False to create more or less output.

❸ The default type is str.

❹ The action='store_true' makes this a Boolean flag. If the flag is present the value will be True, and it will be False otherwise.

The main() function is where the testing happens:

```
def main() -> None:
    args = get_args()
    cwd = os.getcwd()  ❶
    solutions = list(  ❷
        filter(partial(re.match, r'solution.*\.py'), os.listdir(cwd)))  ❸

    for solution in sorted(solutions):  ❹
        print(f'==> {solution} <==')
        shutil.copyfile(solution, os.path.join(cwd, args.program))  ❺
        subprocess.run(['chmod', '+x', args.program], check=True)  ❻
        rv, out = getstatusoutput('make test')  ❼
        if rv != 0:  ❽
            sys.exit(out)  ❾

        if not args.quiet:  ❿
            print(out)

    print('Done.')  ⓫
```

❶ Get the current working directory, which will be the *04_fib* directory if you are in that directory when running the command.

❷ Find all the solution*.py files in the current directory.

❸ Both filter() and partial() are HOFs; I'll explain them next.

❹ The filenames will be in random order, so iterate through the sorted files.

❺ Copy the solution*.py file to the testing filename.

❻ Make the program executable.

❼ Run the make test command, and capture the return value and output.

❽ See if the return value is not 0.

❾ Exit this program while printing the output from the testing and returning a nonzero value.

❿ Unless the program is supposed to be quiet, print the testing output.

⓫ Let the user know the program finishes normally.

In the preceding code, I'm using `sys.exit()` to immediately halt the program, print an error message, and return a nonzero exit value. If you consult the documentation, you'll find you can invoke `sys.exit()` with no arguments, an integer value, or a object like a string, which is what I'm using:

```
exit(status=None, /)
    Exit the interpreter by raising SystemExit(status).

    If the status is omitted or None, it defaults to zero (i.e., success).
    If the status is an integer, it will be used as the system exit status.
    If it is another kind of object, it will be printed and the system
    exit status will be one (i.e., failure).
```

The preceding program also uses the functions `filter()` or `partial()`, which I haven't covered yet. Both of these are HOFs. I'll explain how and why I'm using them. To start, the `os.listdir()` function will return the entire contents of a directory, including files and directories:

```
>>> import os
>>> files = os.listdir()
```

There's a lot there, so I'll import the `pprint()` function from the `pprint` module to *pretty-print* this:

```
>>> from pprint import pprint
>>> pprint(files)
['solution3_recursion_memoize_decorator.py',
 'solution2_generator_for_loop.py',
 '.pytest_cache',
 'Makefile',
 'solution2_generator_islice.py',
 'tests',
 '__pycache__',
 'fib.py',
 'README.md',
 'solution3_recursion_memoize.py',
 'bench.html',
 'solution2_generator.py',
 '.mypy_cache',
 '.gitignore',
 'solution1_list.py',
 'solution3_recursion_lru_cache.py',
 'solution3_recursion.py']
```

I want to filter those to names that start with *solution* and end with *.py*. On the command line, I would match this pattern using a *file glob* like `solution*.py`, where the `*` means *zero or more of any character* and the `.` is a literal dot. A regular expression version of this pattern is the slightly more complicated `solution.*\.py`, where `.` (dot) is a regex metacharacter representing *any character*, and `*` (star or asterisk) means *zero or more* (see Figure 4-5). To indicate a literal dot, I need to escape it with a backslash (\.). Note that it's prudent to use the r-string (*raw* string) to enclose this pattern.

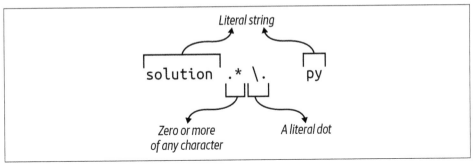

Figure 4-5. A regular expression to find files matching the file glob `solution*.py`

When the match is successful, a `re.Match` object is returned:

```
>>> import re
>>> re.match(r'solution.*\.py', 'solution1.py')
<re.Match object; span=(0, 12), match='solution1.py'>
```

When a match fails, the `None` value is returned. I have to use `type()` here because the `None` value is not displayed in the REPL:

```
>>> type(re.match(r'solution.*\.py', 'fib.py'))
<class 'NoneType'>
```

I want to apply this match to all the files returned by `os.listdir()`. I can use `filter()` and the `lambda` keyword to create an *anonymous* function. Each filename in `files` is passed as the `name` argument used in the match. The `filter()` function will only return elements that return a truthy value from the given function, so those filenames that return `None` when they fail to match are excluded:

```
>>> pprint(list(filter(lambda name: re.match(r'solution.*\.py', name), files)))
['solution3_recursion_memoize_decorator.py',
 'solution2_generator_for_loop.py',
 'solution2_generator_islice.py',
 'solution3_recursion_memoize.py',
 'solution2_generator.py',
 'solution1_list.py',
 'solution3_recursion_lru_cache.py',
 'solution3_recursion.py']
```

You see that the `re.match()` function takes two arguments—a pattern and a string to match. The `partial()` function allows me to *partially apply* the function, and the result is a new function. For example, the `operator.add()` function expects two values and returns their sum:

```
>>> import operator
>>> operator.add(1, 2)
3
```

I can create a function that adds 1 to any value, like so:

```
>>> from functools import partial
>>> succ = partial(op.add, 1)
```

The succ() function requires one argument, and will return the successor:

```
>>> succ(3)
4
>>> succ(succ(3))
5
```

Likewise, I can create a function f() that partially applies the re.match() function with its first argument, a regular expression pattern:

```
>>> f = partial(re.match, r'solution.*\.py')
```

The f() function is waiting for a string to apply the match:

```
>>> type(f('solution1.py'))
<class 're.Match'>
>>> type(f('fib.py'))
<class 'NoneType'>
```

If you call it without an argument, you will get an exception:

```
>>> f()
Traceback (most recent call last):
  File "<stdin>", line 1, in <module>
TypeError: match() missing 1 required positional argument: 'string'
```

I can replace the lambda with the partially applied function as the first argument to filter():

```
>>> pprint(list(filter(f, files)))
['solution3_recursion_memoize_decorator.py',
 'solution2_generator_for_loop.py',
 'solution2_generator_islice.py',
 'solution3_recursion_memoize.py',
 'solution2_generator.py',
 'solution1_list.py',
 'solution3_recursion_lru_cache.py',
 'solution3_recursion.py']
```

My programming style leans heavily on purely functional programming ideas. I find this style to be like playing with LEGO bricks—small, well-defined, and tested functions can be composed into larger programs that work well.

Going Further

There are many different styles of programming, such as procedural, functional, object-oriented, and so forth. Even within an object-oriented language like Python, I can use very different approaches to writing code. The first solution could be considered a *dynamic programming* approach because you try to solve the larger problem by first solving smaller problems. If you find recursive functions interesting, the Tower of Hanoi problem is another classic exercise. Purely functional languages like Haskell mostly avoid constructs like for loops and rely heavily on recursion and higher-order functions. Both spoken and programming languages shape the way we think, and I encourage you to try solving this problem using other languages you know to see how you might write different solutions.

Review

Key points from this chapter:

- Inside the get_args() function, you can perform manual validation of arguments and use the parser.error() function to manually generate argparse errors.

- You can use a list as a stack by pushing and popping elements.

- Using yield in a function turns it into a generator. When the function yields a value, the value is returned and the state of the function is preserved until the next value is requested. Generators can be used to create a potentially infinite stream of values.

- A recursive function calls itself, and the recursion can cause serious performance issues. One solution is to use memoization to cache values and avoid recomputation.

- Higher-order functions are functions that take other functions as arguments.

- Python's function decorators apply HOFs to other functions.

- Benchmarking is an important technique for determining the best-performing algorithm. The hyperfine and bench tools allow you to compare runtimes of commands over many iterations.

- The random module offers many functions for the pseudorandom selection of values.

## CHAPTER 5

Computing GC Content: Parsing FASTA and Analyzing Sequences

In Chapter 1, you counted all the bases in a string of DNA. In this exercise, you need to count the *G*s and *C*s in a sequence and divide by the length of the sequence to determine the GC content as described on the Rosalind GC page (*https://oreil.ly/ gv8V7*). GC content is informative in several ways. A higher GC content level indicates a relatively higher melting temperature in molecular biology, and DNA sequences that encode proteins tend to be found in GC-rich regions. There are many ways to solve this problem, and they all start with using Biopython to parse a FASTA file, a key file format in bioinformatics. I'll show you how to use the Bio.SeqIO module to iterate over the sequences in the file to identify the sequence with the highest GC content.

You will learn:

- How to parse FASTA format using Bio.SeqIO
- How to read STDIN (pronounced *standard in*)
- Several ways to express the notion of a for loop using list comprehensions, filter(), and map()
- How to address runtime challenges such as memory allocation when parsing large files
- More about the sorted() function
- How to include formatting instructions in format strings
- How to use the sum() function to add a list of numbers
- How to use regular expressions to count the occurrences of a pattern in a string

Getting Started

All the code and tests for this program are in the *05_gc* directory. While I'd like to name this program `gc.py`, it turns out that this conflicts with a very important Python module called `gc.py` (*https://oreil.ly/7eNBw*) which is used for garbage collection, such as freeing memory. Instead, I'll use `cgc.py` for *calculate GC*.

 If I called my program `gc.py`, my code would *shadow* the built-in `gc` module, making it unavailable. Likewise, I can create variables and functions with names like `len` or `dict` which would shadow those built-in functions. This will cause many bad things to happen, so it's best to avoid these names. Programs like `pylint` and `flake8` can find problems like this.

Start by copying the first solution and asking for the usage:

```
$ cp solution1_list.py cgc.py
$ ./cgc.py -h
usage: cgc.py [-h] [FILE] ❶

Compute GC content

positional arguments:
  FILE        Input sequence file (default: <_io.TextIOWrapper ❷
              name='<stdin>' mode='r' encoding='utf-8'>)

optional arguments:
  -h, --help  show this help message and exit
```

❶ Note that the positional [FILE] is in square brackets to indicate that it is optional.

❷ This is a rather ugly message that is trying to explain that the default input is STDIN.

As in Chapter 2, this program expects a file as input and will reject invalid or unreadable files. To illustrate this second point, create an empty file using `touch` and then use `chmod` (change mode) to set the permissions to 000 (all read/write/execute bits off):

```
$ touch cant-touch-this
$ chmod 000 cant-touch-this
```

Notice that the error message specifically tells me that I lack permission to read the file:

```
$ ./cgc.py cant-touch-this
usage: cgc.py [-h] [FILE]
```

```
cgc.py: error: argument FILE: can't open 'cant-touch-this': [Errno 13]
Permission denied: 'cant-touch-this'
```

Now run the program with valid input and observe that the program prints the ID of the record having the highest percentage of GC:

```
$ ./cgc.py tests/inputs/1.fa
Rosalind_0808 60.919540
```

This program can also read from STDIN. Simply because I think it's fun, I'll show you how, in the bash shell, I can use the pipe operator (|) to route the STDOUT from one program to the STDIN of another program. For instance, the cat program will print the contents of a file to STDOUT:

```
$ cat tests/inputs/1.fa
>Rosalind_6404
CCTGCGGAAGATCGGCACTAGAATAGCCAGAACCGTTTCTCTGAGGCTTCCGGCCTTCCC
TCCCACTAATAATTCTGAGG
>Rosalind_5959
CCATCGGTAGCGCATCCTTAGTCCAATTAAGTCCCTATCCAGGCGCTCCGCCGAAGGTCT
ATATCCATTTGTCAGCAGACACGC
>Rosalind_0808
CCACCCTCGTGGTATGGCTAGGCATTCAGGAACCGGAGAACGCTTCAGACCAGCCCGGAC
TGGGAACCTGCGGGCAGTAGGTGGAAT
```

Using the pipe, I can feed this to my program:

```
$ cat tests/inputs/1.fa | ./cgc.py
Rosalind_0808 60.919540
```

I can also use the < operator to redirect input from a file:

```
$ ./cgc.py < tests/inputs/1.fa
Rosalind_0808 60.919540
```

To get started, remove this program and start over:

```
$ new.py -fp 'Compute GC content' cgc.py
Done, see new script "cgc.py".
```

The following shows how to modify the first part of the program to accept a single positional argument that is a valid, readable file:

```
import argparse
import sys
from typing import NamedTuple, TextIO, List, Tuple
from Bio import SeqIO


class Args(NamedTuple):
    """ Command-line arguments """
    file: TextIO  ❶
```

```
def get_args() -> Args:
    """ Get command-line arguments """

    parser = argparse.ArgumentParser(
        description='Compute GC content',
        formatter_class=argparse.ArgumentDefaultsHelpFormatter)

    parser.add_argument('file',
                        metavar='FILE',
                        type=argparse.FileType('rt'),  ❷
                        nargs='?',
                        default=sys.stdin,
                        help='Input sequence file')

    args = parser.parse_args()

    return Args(args.file)
```

❶ The only attribute of the Args class is a filehandle.

❷ Create a positional file argument that, if provided, must be a readable text file.

It's rare to make a positional argument optional, but in this case, I want to either handle a single file input or read from STDIN. To do this, I use nargs='?' to indicate that the parameter should accept zero or one argument (see Table 2-2 in "Opening the Output Files" on page 54) and set default=sys.stdin. In Chapter 2, I mentioned that sys.stdout is a filehandle that is always open for writing. Similarly, sys.stdin is an open filehandle from which you can always read STDIN. This is all the code that is required to make your program read either from a file or from STDIN, and I think that's rather neat and tidy.

Modify your main() to print the name of the file:

```
def main() -> None:
    args = get_args()
    print(args.file.name)
```

Verify that it works:

```
$ ./cgc.py tests/inputs/1.fa
tests/inputs/1.fa
```

Run **pytest** to see how you're faring. You should pass the first three tests and fail on the fourth:

```
$ pytest -xv
============================= test session starts =============================
....

tests/cgc_test.py::test_exists PASSED                                   [ 20%]
tests/cgc_test.py::test_usage PASSED                                    [ 40%]
```

```
tests/cgc_test.py::test_bad_input PASSED                              [ 60%]
tests/cgc_test.py::test_good_input1 FAILED                            [ 80%]


=============================== FAILURES ===============================
_____ test_good_input1 _____

    def test_good_input1():
        """ Works on good input """

        rv, out = getstatusoutput(f'{RUN} {SAMPLE1}')  ❶
        assert rv == 0
>       assert out == 'Rosalind_0808 60.919540'  ❷
E       AssertionError: assert './tests/inputs/1.fa' == 'Rosalind_0808 60.919540'
E         - Rosalind_0808 60.919540  ❸
E         + ./tests/inputs/1.fa  ❹

tests/cgc_test.py:48: AssertionError
========================== short test summary info ==========================
FAILED tests/cgc_test.py::test_good_input1 - AssertionError: assert './tes...
!!!!!!!!!!!!!!!!!!!!!!!!!! stopping after 1 failures !!!!!!!!!!!!!!!!!!!!!!!!!!
========================== 1 failed, 3 passed in 0.34s ==========================
```

❶ The test is running the program using the first input file.

❷ The output is expected to be the given string.

❸ This is the expected string.

❹ This is the string that was printed.

So far you have created a syntactically correct, well-structured, and documented program that validates a file input, all by doing relatively little work. Next, you need to figure out how to find the sequence with the highest GC content.

Get Parsing FASTA Using Biopython

The data from the incoming file or STDIN should be sequence data in FASTA format, which is a common way to represent biological sequences. Let's look at the first file to understand the format:

```
$ cat tests/inputs/1.fa
>Rosalind_6404  ❶
CCTGCGGAAGATCGGCACTAGAATAGCCAGAACCGTTTCTCTGAGGCTTCCGGCCTTCCC  ❷
TCCCACTAATAATTCTGAGG
>Rosalind_5959
CCATCGGTAGCGCATCCTTAGTCCAATTAAGTCCCTATCCAGGCGCTCCGCCGAAGGTCT
ATATCCATTTGTCAGCAGACACGC
>Rosalind_0808
CCACCCTCGTGGTATGGCTAGGCATTCAGGAACCGGAGAACGCTTCAGACCAGCCCGGAC
TGGGAACCTGCGGGCAGTAGGTGGAAT
```

❶ A FASTA record starts with a > at the beginning of a line. The sequence ID is any following text up to the first space.

❷ A sequence can be any length and can span multiple lines or be placed on a single line.

 The header of a FASTA file can get very ugly, very quickly. I would encourage you to download real sequences from the National Center for Biotechnology Information (NCBI) or look at the files in the *17_synth/tests/inputs* directory for more examples.

While it would be fun (for certain values of fun) to teach you how to manually parse this file, I'll go straight to using Biopython's `Bio.SeqIO` module:

```
>>> from Bio import SeqIO
>>> recs = SeqIO.parse('tests/inputs/1.fa', 'fasta')  ❶
```

❶ The first argument is the name of the input file. As this function can parse many different record formats, the second argument is the format of the data.

I can check the type of `recs` using `type()`, as usual:

```
>>> type(recs)
<class 'Bio.SeqIO.FastaIO.FastaIterator'>
```

I've shown iterators a couple of times now, even creating one in Chapter 4. In that exercise, I used the `next()` function to retrieve the next value from the Fibonacci sequence generator. I'll do the same here to retrieve the first record and inspect its type:

```
>>> rec = next(recs)
>>> type(rec)
<class 'Bio.SeqRecord.SeqRecord'>
```

To learn more about a sequence record, I highly recommend you read the `SeqRecord` documentation (*https://biopython.org/wiki/SeqRecord*) in addition to the documentation in the REPL, which you can view using **help(rec)**. The data from the FASTA record must be *parsed*, which means discerning the meaning of the data from its syntax and structure. If you look at `rec` in the REPL, you'll see something that looks like a dictionary. This output is the same as that from `repr(seq)`, which is used to "return the canonical string representation of the object":

```
SeqRecord(
  seq=Seq('CCTGCGGAAGATCGGCACTAGAATAGCCAGAACCGTTTCTCTGAGGCTTCCGGC...AGG'),  ❶
  id='Rosalind_6404',  ❷
  name='Rosalind_6404',  ❸
  description='Rosalind_6404',
  dbxrefs=[])
```

❶ The multiple lines of the sequence are concatenated into a single sequence represented by a Seq object.

❷ The ID of a FASTA record is all the characters in the header starting *after* the >
and continuing up to the first space.

❸ The SeqRecord object is meant to also handle data with more fields, such as name,
description, and database cross-references (dbxrefs). Since those fields are not
present in FASTA records, the ID is duplicated for name and description, and
the dbxrefs value is the empty list.

If you print the sequence, this information will be *stringified* so it's a little easier to
read. This output is the same as that for str(rec), which is meant to provide a useful
string representation of an object:

```
>>> print(rec)
ID: Rosalind_6404
Name: Rosalind_6404
Description: Rosalind_6404
Number of features: 0
Seq('CCTGCGGAAGATCGGCACTAGAATAGCCAGAACCGTTTCTCTGAGGCTTCCGGC...AGG')
```

The most salient feature for this program is the record's sequence. You might expect
this would be a str, but it's actually another object:

```
>>> type(rec.seq)
<class 'Bio.Seq.Seq'>
```

Use **help(rec.seq)** to see what attributes and methods the Seq object offers. I only
want the DNA sequence itself, which I can get by coercing the sequence to a string
using the str() function:

```
>>> str(rec.seq)
'CCTGCGGAAGATCGGCACTAGAATAGCCAGAACCGTTTCTCTGAGGCTTCCGGCCTT...AGG'
```

Note this is the same class I used in the last solution of Chapter 3 to create a reverse
complement. I can use it here like so:

```
>>> rec.seq.reverse_complement()
Seq('CCTCAGAATTATTAGTGGGAGGGAAGGCCGGAAGCCTCAGAGAAACGGTTCTGG...AGG')
```

The Seq object has many other useful methods, and I encourage you to explore the
documentation as these can save you a lot of time.[1] At this point, you may feel you
have enough information to finish the challenge. You need to iterate through all the
sequences, determine what percentage of the bases are G or C, and return the ID and
GC content of the record with the maximum value. I would challenge you to write a

---

1 As the saying goes, "Weeks of coding can save you hours of planning."

solution on your own. If you need more help, I'll show you one approach, and then I'll cover several variations in the solutions.

Iterating the Sequences Using a for Loop

So far I've shown that `SeqIO.parse()` accepts a filename as the first argument, but the `args.file` argument will be an open filehandle. Luckily, the function will also accept this:

```
>>> from Bio import SeqIO
>>> recs = SeqIO.parse(open('./tests/inputs/1.fa'), 'fasta')
```

I can use a `for` loop to iterate through each record to print the ID and the first 10 bases of each sequence:

```
>>> for rec in recs:
...     print(rec.id, rec.seq[:10])
...
Rosalind_6404 CCTGCGGAAG
Rosalind_5959 CCATCGGTAG
Rosalind_0808 CCACCCTCGT
```

Take a moment to run those lines again and notice that nothing will be printed:

```
>>> for rec in recs:
...     print(rec.id, rec.seq[:10])
...
```

Earlier I showed that `recs` is a `Bio.SeqIO.FastaIO.FastaIterator`, and, like all iterators, it will produce values until exhausted. If you want to loop through the records again, you will need to recreate the `recs` object using the `SeqIO.parse()` function.

For the moment, assume the sequence is this:

```
>>> seq = 'CCACCCTCGTGGTATGGCT'
```

I need to find how many Cs and Gs occur in that string. I can use another `for` loop to iterate each base of the sequence and increment a counter whenever the base is a *G* or a *C*:

```
gc = 0 ❶
for base in seq: ❷
    if base in ('G', 'C'): ❸
        gc += 1 ❹
```

❶ Initialize a variable for the counts of the *G/C* bases.

❷ Iterate each base (character) in the sequence.

❸ See if the base is in the tuple containing G or C.

❹ Increment the GC counter.

To find the percentage of GC content, divide the GC count by the length of the sequence:

```
>>> gc
12
>>> len(seq)
19
>>> gc / len(seq)
0.631578947368421
```

The output from the program should be the ID of the sequence with the highest GC count, a single space, and the GC content truncated to six significant digits. The easiest way to format the number is to learn more about `str.format()`. The `help` doesn't have much in the way of documentation, so I recommend you read PEP 3101 (*https://oreil.ly/OIpEq*) on advanced string formatting.

In Chapter 1, I showed how I can use {} as placeholders for interpolating variables either using `str.format()` or f-strings. I can add formatting instructions after a colon (`:`) in the curly brackets. The syntax looks like that used with the `printf()` function in C-like languages, so {`:0.6f`} is a floating-point number to six places:

```
>>> '{:0.6f}'.format(gc * 100 / len(seq))
'63.157895'
```

Or, to execute the code directly inside an f-string:

```
>>> f'{gc * 100 / len(seq):0.06f}'
'63.157895'
```

To figure out the sequence with the maximum GC count, you have a couple of options, both of which I'll demonstrate in the solutions:

- Make a list of all the IDs and their GC content (a list of tuples would serve well). Sort the list by GC content and take the maximum value.

- Keep track of the ID and GC content of the maximum value. Overwrite this when a new maximum is found.

I think that should be enough for you to finish a solution. You can do this. Fear is the mind killer. Keep going until you pass *all* the tests, including those for linting and type checking. Your test output should look something like this:

```
$ make test
python3 -m pytest -xv --disable-pytest-warnings --flake8 --pylint
--pylint-rcfile=../pylintrc --mypy cgc.py tests/cgc_test.py
=========================== test session starts ===========================
...
collected 10 items
```

```
cgc.py::FLAKE8 SKIPPED                              [  9%]
cgc.py::mypy PASSED                                 [ 18%]
tests/cgc_test.py::FLAKE8 SKIPPED                   [ 27%]
tests/cgc_test.py::mypy PASSED                      [ 36%]
tests/cgc_test.py::test_exists PASSED               [ 45%]
tests/cgc_test.py::test_usage PASSED                [ 54%]
tests/cgc_test.py::test_bad_input PASSED            [ 63%]
tests/cgc_test.py::test_good_input1 PASSED          [ 72%]
tests/cgc_test.py::test_good_input2 PASSED          [ 81%]
tests/cgc_test.py::test_stdin PASSED                [ 90%]
::mypy PASSED                                       [100%]
================================ mypy ====================================

Success: no issues found in 2 source files
====================== 9 passed, 2 skipped in 1.67s ======================
```

Solutions

As before, all the solutions share the same get_args(), so only the differences will be shown.

Solution 1: Using a List

Let's look at my first solution. I always try to start with the most obvious and simple way, and you'll find this is often the most verbose. Once you understand the logic, I hope you'll be able to follow more powerful and terse ways to express the same ideas. For this first solution, be sure to also import List and Tuple from the typing module:

```
def main() -> None:
    args = get_args()
    seqs: List[Tuple[float, str]] = []  ❶

    for rec in SeqIO.parse(args.file, 'fasta'):  ❷
        gc = 0  ❸
        for base in rec.seq.upper():  ❹
            if base in ('C', 'G'):  ❺
                gc += 1  ❻
        pct = (gc * 100) / len(rec.seq)  ❼
        seqs.append((pct, rec.id))  ❽

    high = max(seqs)  ❾
    print(f'{high[1]} {high[0]:0.6f}')  ❿
```

❶ Initialize an empty list to hold the GC content and sequence IDs as tuples.

❷ Iterate through each record in the input file.

❸ Initialize a GC counter.

❹ Iterate through each sequence, uppercased to guard against possible mixed-case input.

❺ Check if the base is a *C* or *G*.

❻ Increment the GC counter.

❼ Calculate the GC content.

❽ Append a new tuple of the GC content and the sequence ID.

❾ Take the maximum value.

❿ Print the sequence ID and GC content of the highest value.

 The type annotation `List[Tuple[float, str]]` on the `seqs` variable provides not only a way to programmatically check the code using tools like mypy but also an added layer of documentation. The reader of this code doesn't have to jump ahead to see what kind of data will be added to the list because it has been explicitly described using types.

In this solution, I decided to make a list of all the IDs and GC percentages mostly so that I could show you how to create a list of tuples. Then I wanted to point out a few magical properties of Python's sorting. Let's start with the `sorted()` function, which works on strings as you might imagine:

```
>>> sorted(['McClintock', 'Curie', 'Doudna', 'Charpentier'])
['Charpentier', 'Curie', 'Doudna', 'McClintock']
```

When all the values are numbers, they will be sorted numerically, so I've got that going for me, which is nice:

```
>>> sorted([2, 10, 1])
[1, 2, 10]
```

Note that those same values *as strings* will sort in lexicographic order:

```
>>> sorted(['2', '10', '1'])
['1', '10', '2']
```

Python Lists Should Be Homogeneous

Comparing different types, such as strings and integers, will cause an exception:

```
>>> sorted([2, '10', 1])
Traceback (most recent call last):
  File "<stdin>", line 1, in <module>
TypeError: '<' not supported between instances of 'str' and 'int'
```

While it's an acceptable practice in Python to mix types in a list, it's likely to end in tears. This reminds me of a joke by Henny Youngman: A man goes to see his doctor. He says, "Doc, it hurts when I do this." The doctor says, "Then don't do that."

Essentially my advice is a bit like the doctor's: yeah, you can mix types in lists in Python, but doing so will lead to runtime exceptions if you try to sort them, so just don't do that. To avoid this, always use a type declaration to describe your data:

```
seqs: List[Tuple[float, str]] = []
```

Try adding this line:

```
seqs.append('foo')
```

Then check the program using mypy to see how clearly the error message shows how this violates the description of the data:

```
$ mypy solution1_list.py
solution1_list.py:38: error: Argument 1 to "append" of "list" has
incompatible type "str"; expected "Tuple[float, str]"
Found 1 error in 1 file (checked 1 source file)
```

Another option is to use the numpy module to create an *array*, which is like a Python list but where all the values are required (and coerced) to be of a common type. Arrays in numpy are both faster (due to memory management) and safer than Python lists. Note that mixing strings and numbers will result in a list of strings:

```
>>> import numpy as np
>>> nums = np.array([2, '10', 1])
>>> nums
array(['2', '10', '1'], dtype='<U21')
```

Although this is not correct—I want them to be integers—it is at least safe. I can use the int() function in a list comprehension to coerce all the values to integers:

```
>>> [int(n) for n in [2, '10', 1]]
[2, 10, 1]
```

Or the same thing expressed using the higher-order function map(), which takes a function like int() as the first argument and applies it to all the elements in the sequence to return a new list of elements transformed by that function:

```
>>> list(map(int, [2, '10', 1]))
[2, 10, 1]
```

Which makes for good sorting:

```
>>> sorted(map(int, [2, '10', 1]))
[1, 2, 10]
```

Now consider a list of tuples where the first element is a `float` and the second element is a `str`. How will `sorted()` handle this? By first sorting all the data by the first elements *numerically* and then the second elements *lexicographically*:

```
>>> sorted([(0.2, 'foo'), (.01, 'baz'), (.01, 'bar')])
[(0.01, 'bar'), (0.01, 'baz'), (0.2, 'foo')]
```

Structuring `seqs` as `List[Tuple[float, str]]` takes advantage of this built-in behavior of `sorted()`, allowing me to quickly sort the sequences by GC content and select the highest value:

```
>>> high = sorted(seqs)[-1]
```

This is the same as finding the highest value, which the `max()` function can do more easily:

```
>>> high = max(seqs)
```

`high` is a tuple where the first position is the sequence ID and the zeroth position is the GC content that needs to be formatted:

```
print(f'{high[1]} {high[0]:0.6f}')
```

Solution 2: Type Annotations and Unit Tests

Hidden inside the `for` loop is a kernel of code to compute GC content that needs to be extracted into a function with a test. Following the ideas of test-driven development (TDD), I will first define a `find_gc()` function:

```
def find_gc(seq: str) -> float:  ❶
    """ Calculate GC content """

    return 0.  ❷
```

❶ The function accepts a `str` and returns a `float`.

❷ For now, I return 0. Note the trailing . tells Python this is a `float`. This is shorthand for `0.0`.

Next, I'll define a function that will serve as a unit test. Since I'm using `pytest`, this function's name must start with `test_`. Because I'm testing the `find_gc()` function, I'll name the function `test_find_gc`. I will use a series of `assert` statements to test if the function returns the expected result for a given input. Note how this test function serves both as a formal test and as an additional piece of documentation, as the reader can see the inputs and outputs:

```
def test_find_gc():
    """ Test find_gc """

    assert find_gc('') == 0.      ❶
    assert find_gc('C') == 100.   ❷
    assert find_gc('G') == 100.   ❸
    assert find_gc('CGCCG') == 100.  ❹
    assert find_gc('ATTAA') == 0.
    assert find_gc('ACGT') == 50.
```

❶ If a function accepts a `str`, I always start by testing with the empty string to make sure it returns something useful.

❷ A single C should be 100% GC.

❸ Same for a single G.

❹ Various other tests mixing bases at various percentages.

It's rarely possible to exhaustively check every possible input to a function, so I often rely on spot-checking. Note that the `hypothesis` module (*https://hypothesis.readthe docs.io/en/latest*) can generate random values for testing. Presumably, the `find_gc()` function is simple enough that these tests are sufficient. My goal in writing functions is to make them as simple as possible, but no simpler. As Tony Hoare says, "There are two ways to write code: write code so simple there are obviously no bugs in it, or write code so complex that there are no obvious bugs in it."

The `find_gc()` and `test_find_gc()` functions are inside the `cgc.py` program, not in the *tests/cgc_test.py* module. To execute the unit test, I run `pytest` on the source code *expecting the test to fail*:

```
$ pytest -v cgc.py
=========================== test session starts ============================
...

cgc.py::test_find_gc FAILED                                      [100%] ❶

================================ FAILURES ==================================
_____ test_gc _____

    def test_find_gc():
        """ Test find_gc """

        assert find_gc('') == 0.      ❷
>       assert find_gc('C') == 100.   ❸
E       assert 0 == 100.0
```

```
E              +0
E              -100.0

cgc.py:74: AssertionError
=========================== short test summary info ===========================
FAILED cgc.py::test_gc - assert 0 == 100.0
============================== 1 failed in 0.32s ==============================
```

❶ The unit test fails as expected.

❷ The first test passes because it was expecting 0.

❸ This test fails because it should have returned 100.

Now I have established a baseline from which I can proceed. I know that my code fails to meet some expectation as formally defined using a test. To fix this, I move all the relevant code from main() into the function:

```
def find_gc(seq: str) -> float:
    """ Calculate GC content """

    if not seq: ❶
        return 0 ❷

    gc = 0 ❸
    for base in seq.upper():
        if base in ('C', 'G'):
            gc += 1

    return (gc * 100) / len(seq)
```

❶ This guards against trying to divide by 0 when the sequence is the empty string.

❷ If there is no sequence, the GC content is 0.

❸ This is the same code as before.

Then I run pytest again to check that the function works:

```
$ pytest -v cgc.py
=========================== test session starts ===========================
...

cgc.py::test_gc PASSED                                              [100%]

============================== 1 passed in 0.30s ==============================
```

This is TDD:

- Define a function to test.
- Write the test.
- Ensure the function fails the test.
- Make the function work.
- Ensure the function passes the test (and all your previous tests still pass).

If I later encounter sequences that trigger bugs in my code, I'll fix the code and add those as more tests. I shouldn't have to worry about weird cases like the find_gc() function receiving a None or a list of integers *because I used type annotations.* Testing is useful. Type annotations are useful. Combining tests and types leads to code that is easier to verify and comprehend.

I want to make one other addition to this solution: a custom type to document the tuple holding the GC content and sequence ID. I'll call it MySeq just to avoid any confusion with the Bio.Seq class. I add this below the Args definition:

```
class MySeq(NamedTuple):
    """ Sequence """
    gc: float ❶
    name: str ❷
```

❶   The GC content is a percentage.

❷   I would prefer to use the field name id, but that conflicts with the id() *identity* function which is built into Python.

Here is how it can be incorporated into the code:

```
def main() -> None:
    args = get_args()
    seqs: List[MySeq] = [] ❶

    for rec in SeqIO.parse(args.file, 'fasta'):
        seqs.append(MySeq(find_gc(rec.seq), rec.id)) ❷

    high = sorted(seqs)[-1] ❸
    print(f'{high.name} {high.gc:0.6f}') ❹
```

❶   Use MySeq as a type annotation.

❷   Create MySeq using the return value from the find_gc() function and the record ID.

❸ This still works because `MySeq` is a tuple.

❹ Use the field access rather than the index position of the tuple.

This version of the program is arguably easier to read. You can and should create as many custom types as you want to better document and test your code.

Solution 3: Keeping a Running Max Variable

The previous solution works well, but it's a bit verbose and needlessly keeps track of *all* the sequences when I only care about the maximum value. Given how small the test inputs are, this will never be a problem, but bioinformatics is always about scaling up. A solution that tries to store all the sequences will eventually choke. Consider processing 1 million sequences, or 1 billion, or 100 billion. Eventually, I'd run out of memory.

Here's a solution that would scale to any number of sequences, as it only ever allocates a single tuple to remember the highest value:

```
def main():
    args = get_args()
    high = MySeq(0., '') ❶

    for rec in SeqIO.parse(args.file, 'fasta'):
        pct = find_gc(rec.seq) ❷
        if pct > high.gc: ❸
            high = MySeq(pct, rec.id) ❹

    print(f'{high.name} {high.gc:0.6f}') ❺
```

❶ Initialize a variable to remember the highest value. Type annotation is superfluous as `mypy` will expect this variable to remain this type forever.

❷ Calculate the GC content.

❸ See if the percent GC is greater than the highest value.

❹ If so, overwrite the highest value using this percent GC and sequence ID.

❺ Print the highest value.

For this solution, I also took a slightly different approach to compute the GC content:

```
def find_gc(seq: str) -> float:
    """ Calculate GC content """

    return (seq.upper().count('C') + ❶
            seq.upper().count('G')) * 100 / len(seq) if seq else 0 ❷
```

❶ Use the `str.count()` method to find the Cs and Gs in the sequence.

❷ Since there are two conditions for the state of the sequence—the empty string or not—I prefer to write a single `return` using an `if` expression.

I'll benchmark the last solution against this one. First I need to generate an input file with a significant number of sequences, say 10K. In the *05_gc* directory, you'll find a `genseq.py` file similar to the one I used in the *02_rna* directory. This one generates a FASTA file:

```
$ ./genseq.py -h
usage: genseq.py [-h] [-l int] [-n int] [-s sigma] [-o FILE]

Generate long sequence

optional arguments:
  -h, --help           show this help message and exit
  -l int, --len int    Average sequence length (default: 500)
  -n int, --num int    Number of sequences (default: 1000)
  -s sigma, --sigma sigma
                       Sigma/STD (default: 0.1)
  -o FILE, --outfile FILE
                       Output file (default: seqs.fa)
```

Here's how I'll generate an input file:

```
$ ./genseq.py -n 10000 -o 10K.fa
Wrote 10,000 sequences of avg length 500 to "10K.fa".
```

I can use that with `hyperfine` to compare these two implementations:

```
$ hyperfine -L prg ./solution2_unit_test.py,./solution3_max_var.py '{prg} 10K.fa'
Benchmark #1: ./solution2_unit_test.py 10K.fa
  Time (mean ± σ):      1.546 s ±  0.035 s    [User: 2.117 s, System: 0.147 s]
  Range (min … max):    1.511 s …  1.625 s    10 runs

Benchmark #2: ./solution3_max_var.py 10K.fa
  Time (mean ± σ):     368.7 ms ±   3.0 ms    [User: 957.7 ms, System: 137.1 ms]
  Range (min … max):   364.9 ms … 374.7 ms    10 runs

Summary
  './solution3_max_var.py 10K.fa' ran
    4.19 ± 0.10 times faster than './solution2_unit_test.py 10K.fa'
```

It would appear that the third solution is about four times faster than the second running on 10K sequences. You can try generating more and longer sequences for your own benchmarking. I would recommend you create a file with at least one million sequences and compare your first solution with this version.

Solution 4: Using a List Comprehension with a Guard

Figure 5-1 shows that another way to find all the Cs and Gs in the sequence is to use a list comprehension and the `if` comparison from the first solution, which is called a *guard*.

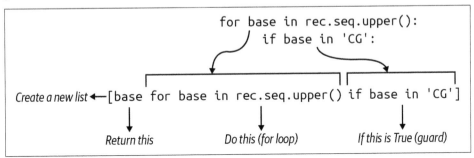

Figure 5-1. A list comprehension with a guard will select only those elements returning a truthy value for the if expression

The list comprehension only yields those elements passing the guard which checks that the `base` is in the string `'CG'`:

```
>>> gc = [base for base in 'CCACCCTCGTGGTATGGCT' if base in 'CG']
>>> gc
['C', 'C', 'C', 'C', 'C', 'C', 'G', 'G', 'G', 'G', 'G', 'C']
```

Since the result is a new list, I can use the `len()` function to find how many Cs and Gs are present:

```
>>> len(gc)
12
```

I can incorporate this idea into the `find_gc()` function:

```
def find_gc(seq: str) -> float:
    """ Calculate GC content """

    if not seq:
        return 0

    gc = len([base for base in seq.upper() if base in 'CG']) ❶
    return (gc * 100) / len(seq)
```

❶ Another way to count the Cs and Gs is to select them using a list comprehension with a guard.

Solution 5: Using the filter() Function

The idea of a list comprehension with a guard can be expressed with the higher-order function `filter()`. Earlier in the chapter I used the `map()` function to apply the `int()` function to all the elements of a list to produce a new list of integers. The `filter()` function works similarly, accepting a function as the first argument and an iterable as the second. It's different, though, as only those elements returning a truthy value when the function is applied will be returned. As this is a lazy function, I will need to coerce with `list()` in the REPL:

```
>>> list(filter(lambda base: base in 'CG', 'CCACCCTCGTGGTATGGCT'))
['C', 'C', 'C', 'C', 'C', 'C', 'G', 'G', 'G', 'G', 'G', 'C']
```

So here's another way to express the same idea from the last solution:

```
def find_gc(seq: str) -> float:
    """ Calculate GC content """

    if not seq:
        return 0

    gc = len(list(filter(lambda base: base in 'CG', seq.upper())))  ❶
    return (gc * 100) / len(seq)
```

❶ Use `filter()` to select only those bases matching *C* or *G*.

Solution 6: Using the map() Function and Summing Booleans

The `map()` function is a favorite of mine, so I want to show another way to use it. I could use `map()` to turn each base into a 1 if it's a *C* or *G*, and a 0 otherwise:

```
>>> seq = 'CCACCCTCGTGGTATGGCT'
>>> list(map(lambda base: 1 if base in 'CG' else 0, seq))
[1, 1, 0, 1, 1, 1, 0, 1, 1, 0, 1, 1, 0, 0, 0, 1, 1, 1, 0]
```

Counting the *C*s and *G*s would then be a matter of summing this list, which I can do using the `sum()` function:

```
>>> sum(map(lambda base: 1 if base in 'CG' else 0, seq))
12
```

Booleans Are Integers

Python can use Boolean algebra to combine values like `True`/`False` using and/or:

```
>>> True and False
False
>>> True or False
True
```

What do you think happens if you use + instead of and?

```
>>> True + True
2
>>> True + False
1
```

It turns out that Python's Boolean values lead a secret double life as integers, where True is 1 and False is 0.

I can shorten my map() to return the result of the comparison (which is a bool but also an int) and sum that instead:

```
>>> sum(map(lambda base: base in 'CG', seq))
12
```

Here is how I could incorporate this idea:

```
def find_gc(seq: str) -> float:
    """ Calculate GC content """

    if not seq:
        return 0

    gc = sum(map(lambda base: base in 'CG', seq.upper())) ❶
    return (gc * 100) / len(seq)
```

❶ Transform the sequence into Boolean values based on their comparison to the bases *C* or *G*, then sum the True values to get a count.

Solution 7: Using Regular Expressions to Find Patterns

So far I've been showing you multiple ways to manually iterate a sequence of characters in a string to pick out those matching C or G. This is pattern matching, and it's precisely what regular expressions do. The cost to you is learning another domain-specific language (DSL), but this is well worth the effort as regexes are widely used outside of Python. Begin by importing the re module:

```
>>> import re
```

You should read **help(re)**, as this is a fantastically useful module. I want to use the re.findall() function to find all occurrences of a pattern in a string. I can create a *character class* pattern for the regex engine by using square brackets to enclose any characters I want to include. The class [GC] means *match either G or C*:

```
>>> re.findall('[GC]', 'CCACCCTCGTGGTATGGCT')
['C', 'C', 'C', 'C', 'C', 'C', 'G', 'G', 'G', 'G', 'G', 'C']
```

As before, I can use the len() function to find how many Cs and Gs there are. The following code shows how I would incorporate this into my function. Note that I use the `if` expression to return 0 if the sequence is the empty string so I can avoid division when `len(seq)` is 0:

```
def find_gc(seq: str) -> float:
    """ Calculate GC content """

    return len(re.findall('[GC]', seq.upper()) * 100) / len(seq) if seq else 0
```

Note that it's important to change how this function is called from `main()` to explicitly coerce the `rec.seq` value (which is a Seq object) to a string by using `str()`:

```
def main() -> None:
    args = get_args()
    high = MySeq(0., '')

    for rec in SeqIO.parse(args.file, 'fasta'):
        pct = find_gc(str(rec.seq)) ❶
        if pct > high.gc:
            high = MySeq(pct, rec.id)

    print(f'{high.name} {high.gc:0.6f}')
```

❶ Coerce the sequence to a string value or the Seq object will be passed.

Solution 8: A More Complex find_gc() Function

In this final solution, I'll move almost all of the code from `main()` into the `find_gc()` function. I want the function to accept a `SeqRecord` object rather than a string of the sequence, and I want it to return the `MySeq` tuple.

First I'll change the tests:

```
def test_find_gc() -> None:
    """ Test find_gc """

    assert find_gc(SeqRecord(Seq(''), id='123')) == (0.0, '123')
    assert find_gc(SeqRecord(Seq('C'), id='ABC')) == (100.0, 'ABC')
    assert find_gc(SeqRecord(Seq('G'), id='XYZ')) == (100.0, 'XYZ')
    assert find_gc(SeqRecord(Seq('ACTG'), id='ABC')) == (50.0, 'ABC')
    assert find_gc(SeqRecord(Seq('GGCC'), id='XYZ')) == (100.0, 'XYZ')
```

These are essentially the same tests as before, but I'm now passing `SeqRecord` objects. To make this work in the REPL, you will need to import a couple of classes:

```
>>> from Bio.Seq import Seq
>>> from Bio.SeqRecord import SeqRecord
>>> seq = SeqRecord(Seq('ACTG'), id='ABC')
```

If you look at the object, it looks similar enough to the data I've been reading from input files because I only care about the `seq` field:

```
SeqRecord(seq=Seq('ACTG'),
  id='ABC',
  name='<unknown name>',
  description='<unknown description>',
  dbxrefs=[])
```

If you run **pytest**, your `test_find_gc()` function should fail because you haven't yet changed your `find_gc()` function. Here's how I wrote it:

```
def find_gc(rec: SeqRecord) -> MySeq: ❶
    """ Return the GC content, record ID for a sequence """

    pct = 0. ❷
    if seq := str(rec.seq): ❸
        gc = len(re.findall('[GC]', seq.upper())) ❹
        pct = (gc * 100) / len(seq)

    return MySeq(pct, rec.id) ❺
```

❶ The function accepts a `SeqRecord` and returns a `MySeq`.

❷ Initialize this to a floating-point 0..

❸ This syntax is new as of Python 3.8 and allows variable assignment (first) and testing (second) in one line using the *walrus* operator (`:=`).

❹ This is the same code as before.

❺ Return a `MySeq` object.

 The walrus operator `:=` was proposed in PEP 572 (*https://www.python.org/dev/peps/pep-0572*), which notes that the = operator allows you to name the result of an expression only in a statement form, "making it unavailable in list comprehensions and other expression contexts." This new operator combines two actions, that of *assigning* the value of an expression to a variable and then *evaluating* that variable. In the preceding code, `seq` is assigned the value of the stringified sequence. If that evaluates to something truthy, such as a nonempty string, then the following block of code will be executed.

This radically changes the `main()` function. The `for` loop can incorporate a `map()` function to turn each `SeqRecord` into a `MySeq`:

```
def main() -> None:
    args = get_args()
    high = MySeq(0., '') ❶
    for seq in map(find_gc, SeqIO.parse(args.file, 'fasta')): ❷
```

```
        if seq.gc > high.gc: ❸
            high = seq ❹

    print(f'{high.name} {high.gc:0.6f}') ❺
```

❶ Initialize the high variable.

❷ Use map() to turn each SeqRecord into a MySeq.

❸ Compare the current sequence's GC content against the running high.

❹ Overwrite the value.

❺ Print the results.

The point of the expanded find_gc() function was to hide more of the guts of the program so I can write a more expressive program. You may disagree, but I think this is the most readable version of the program.

Benchmarking

So which one is the winner? There is a bench.sh program that will run hyperfine on all the solution*.py with the *seqs.fa* file. Here is the result:

```
Summary
  './solution3_max_var.py seqs.fa' ran
    2.15 ± 0.03 times faster than './solution8_list_comp_map.py seqs.fa'
    3.88 ± 0.05 times faster than './solution7_re.py seqs.fa'
    5.38 ± 0.11 times faster than './solution2_unit_test.py seqs.fa'
    5.45 ± 0.18 times faster than './solution4_list_comp.py seqs.fa'
    5.46 ± 0.14 times faster than './solution1_list.py seqs.fa'
    6.22 ± 0.08 times faster than './solution6_map.py seqs.fa'
    6.29 ± 0.14 times faster than './solution5_filter.py seqs.fa'
```

Going Further

Try writing a FASTA parser. Create a new directory called *faparser*:

```
$ mkdir faparser
```

Change into that directory, and run **new.py** with the -t|--write_test option:

```
$ cd faparser/
$ new.py -t faparser.py
Done, see new script "faparser.py".
```

You should now have a structure that includes a *tests* directory along with a starting test file:

```
$ tree
.
├── Makefile
├── faparser.py
└── tests
    └── faparser_test.py

1 directory, 3 files
```

You can run **make test** or **pytest** to verify that everything at least runs. Copy the *tests/inputs* directory from *05_gc* to the new *tests* directory so you have some test input files. Now consider how you want your new program to work. I would imagine it would take one (or more) readable text files as inputs, so you could define your arguments accordingly. Then what will your program do with the data? Do you want it to print, for instance, the IDs and the length of each sequence? Now write the tests and code to manually parse the input FASTA files and print the output. Challenge yourself.

Review

Key points from this chapter:

- You can read STDIN from the open filehandle sys.stdin.

- The Bio.SeqIO.parse() function will parse FASTA-formatted sequence files into records, which provides access to the record's ID and sequence.

- You can use several constructs to visit all the elements of iterables, including for loops, list comprehensions, and the functions filter() and map().

- A list comprehension with a guard will only produce elements that return a truthy value for the guard. This can also be expressed using the filter() function.

- Avoid writing algorithms that attempt to store all the data from an input file, as you could exceed the available memory on your machine.

- The sorted() function will sort homogeneous lists of strings and numbers lexicographically and numerically, respectively. It can also sort homogeneous lists of tuples using each position of the tuples in order.

- Formatting templates for strings can include printf()-like instructions to control how output values are presented.

- The sum() function will add a list of numbers.

- Booleans in Python are actually integers.

- Regular expressions can find patterns of text.

## CHAPTER 6

Finding the Hamming Distance: Counting Point Mutations

The Hamming distance, named after the same Richard Hamming mentioned in the Preface, is the number of edits required to change one string into another. It's one metric for gauging sequence similarity. I have written a couple of other metrics for this, starting in Chapter 1 with tetranucleotide frequency and continuing in Chapter 5 with GC content. While the latter can be practically informative as coding regions tend to be GC-rich, tetranucleotide frequency falls pretty short of being useful. For example, the sequences *AAACCCGGGTTT* and *CGACGATATGTC* are wildly different yet produce the same base frequencies:

```
$ ./dna.py AAACCCGGGTTT
3 3 3 3
$ ./dna.py CGACGATATGTC
3 3 3 3
```

Taken alone, tetranucleotide frequency makes these sequences seem identical, but it's quite obvious that they would produce entirely different protein sequences and so would be functionally unlike. Figure 6-1 depicts an alignment of the 2 sequences indicating that only 3 of the 12 bases are shared, meaning they are only 25% similar.

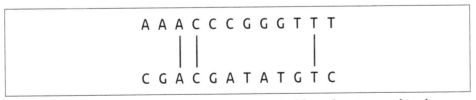

Figure 6-1. An alignment of two sequences with vertical bars showing matching bases

Another way to express this is to say that 9 of the 12 bases need to be changed to turn one of the sequences into the other. This is the Hamming distance, and it's somewhat equivalent in bioinformatics to single-nucleotide polymorphisms (SNPs, pronounced *snips*) or single-nucleotide variations (SNVs, pronounced *snivs*). This algorithm only accounts for the change of one base to another value and falls far short of something like sequence alignment that can identify insertions and deletions. For instance, Figure 6-2 shows that the sequences *AAACCCGGGTTT* and *AACCCGGGTTTA* are 92% similar when aligned (on the left), as they differ by a single base. The Hamming distance (on the right), though, shows only 8 bases are in common, which means they are only 66% similar.

Figure 6-2. The alignment of these sequences shows them to be nearly identical, while the Hamming distance finds they're only 66% similar

This program will always compare strings strictly from their beginnings, which limits the practical application to real-world bioinformatics. Still, it turns out that this naïve algorithm is a useful metric for sequence similarity, and writing the implementation presents many interesting solutions in Python.

In this chapter, you will learn:

- How to use the abs() and min() functions
- How to combine the elements from two lists of possibly unequal lengths
- How to write map() using lambda or existing functions
- How to use functions from the operator module
- How to use the itertools.starmap() function

Getting Started

You should work in the *06_hamm* directory of the repository. I suggest you start by getting a feel for how the solutions work, so copy one of them to the hamm.py program and request the help:

```
$ cp solution1_abs_iterate.py hamm.py
$ ./hamm.py -h
usage: hamm.py [-h] str str

Hamming distance
```

```
positional arguments:
  str         Sequence 1
  str         Sequence 2

optional arguments:
  -h, --help  show this help message and exit
```

The program requires two positional arguments, which are the two sequences to compare, and the program should print the Hamming distance. For example, I would need to make seven edits to change one of these sequences to the other:

```
$ ./hamm.py GAGCCTACTAACGGGAT CATCGTAATGACGGCCT
7
```

Run the tests (either with **pytest** or **make test**) to see a passing suite. Once you feel you understand what's expected, remove this file and start from scratch:

```
$ new.py -fp 'Hamming distance' hamm.py
Done, see new script "hamm.py".
```

Define the parameters so that the program requires two positional arguments which are the two sequences:

```
import argparse
from typing import NamedTuple


class Args(NamedTuple): ❶
    """ Command-line arguments """
    seq1: str
    seq2: str


# --------------------------------------------------
def get_args():
    """ Get command-line arguments """

    parser = argparse.ArgumentParser(
        description='Hamming distance',
        formatter_class=argparse.ArgumentDefaultsHelpFormatter)

    parser.add_argument('seq1', metavar='str', help='Sequence 1') ❷

    parser.add_argument('seq2', metavar='str', help='Sequence 2')

    args = parser.parse_args()

    return Args(args.seq1, args.seq2) ❸
```

❶ The program arguments will have two string values for the two sequences.

❷ The two sequences are required positional string values.

❸ Instantiate the Args object using the two sequences.

> The order in which you define positional parameters must match
> the order in which the arguments are provided on the command
> line. That is, the first positional parameter will hold the first posi-
> tional argument, the second positional parameter will match the
> second positional argument, etc. The order in which you define
> optional parameters does not matter, and optional parameters may
> be defined before or after positional parameters.

Change the main() function to print the two sequence:

```
def main():
    args = get_args()
    print(args.seq1, args.seq2)
```

By this point, you should have a program that prints the usage, validates that the user
supplies two sequences, and prints the sequences:

```
$ ./hamm.py GAGCCTACTAACGGGAT CATCGTAATGACGGCCT
GAGCCTACTAACGGGAT CATCGTAATGACGGCCT
```

If you run **pytest -xvv** (the two vs increase the verbosity of the output), you should
find that the program passes the first three tests. It should fail test_input1 with a
message like the following:

```
================================ FAILURES ======================================
_____ test_input1 _____

    def test_input1() -> None:
        """ Test with input1 """

>       run(INPUT1)

tests/hamm_test.py:47:
_ _ _ _ _ _ _ _ _ _ _ _ _ _ _ _ _ _ _ _ _ _ _ _ _ _ _ _ _ _ _ _ _ _ _ _ _ _ _ _

file = './tests/inputs/1.txt'  ❶

    def run(file: str) -> None:
        """ Run with input """

        assert os.path.isfile(file)
        seq1, seq2, expected = open(file).read().splitlines()  ❷

        rv, out = getstatusoutput(f'{RUN} {seq1} {seq2}')  ❸
        assert rv == 0
>       assert out.rstrip() == expected  ❹
```

```
E        AssertionError: assert 'GAGCCTACTAACGGGAT CATCGTAATGACGGCCT' == '7' ❺
E        - 7
E        + GAGCCTACTAACGGGAT CATCGTAATGACGGCCT

tests/hamm_test.py:40: AssertionError
=========================== short test summary info ============================
FAILED tests/hamm_test.py::test_input1 - AssertionError: assert 'GAGCCTACTAAC...
!!!!!!!!!!!!!!!!!!!!!!!!!! stopping after 1 failures !!!!!!!!!!!!!!!!!!!!!!!!!!
========================== 1 failed, 3 passed in 0.27s =========================
```

❶ The inputs for the test come from the file *./tests/inputs/1.txt*.

❷ The file is opened and read for the two sequences and the expected result.

❸ The program is run with the two sequences.

❹ The `assert` fails when it finds the output from the program does not match the expected answer.

❺ Specifically, the program printed the two sequences when it should have printed 7.

Iterating the Characters of Two Strings

Now to find the Hamming distance between the two sequences. To start, consider these two sequences:

```
>>> seq1, seq2 = 'AC', 'ACGT'
```

The distance is 2 because you would either need to add *GT* to the first sequence or remove *GT* from the second sequence to make them the same. I would suggest that the baseline distance is the difference in their lengths. Note that the Rosalind challenge assumes two strings of equal lengths, but I want to use this exercise to consider strings of different lengths.

Depending on the order in which you do the subtraction, you might end up with a negative number:

```
>>> len(seq1) - len(seq2)
-2
```

Use the `abs()` function to get the absolute value:

```
>>> distance = abs(len(seq1) - len(seq2))
>>> distance
2
```

Now I will consider how to iterate the characters they have in common. I can use the `min()` function to find the length of the shorter sequence:

```
>>> min(len(seq1), len(seq2))
2
```

And I can use this with the `range()` function to get the indexes of the common characters:

```
>>> for i in range(min(len(seq1), len(seq2))):
...     print(seq1[i], seq2[i])
...
...
A A
C C
```

When these two characters are *not* equal, the `distance` variable should be incremented because I would have to change one of the values to match the other. Remember that the Rosalind challenge always compares the two sequences from their beginnings. For instance, the sequences *ATTG* and *TTG* differ by one base, as I can either remove *A* from the first or add it to the second to make them match, but the rules of this particular challenge would say that the correct answer is 3:

```
$ ./hamm.py ATTG TTG
3
```

I believe this should be enough information for you to craft a solution that passes the test suite. Once you have a working solution, explore some other ways you might write your algorithm, and keep checking your work using the test suite. In addition to running the tests via **pytest**, be sure to use the **make test** option to verify that your code also passes the various linting and type-checking tests.

Solutions

This section works through eight variations on how to find the Hamming distance, starting with an entirely manual calculation that takes several lines of code and ending with a solution that combines several functions in a single line.

Solution 1: Iterating and Counting

The first solution follows from the suggestions in the previous section:

```
def main():
    args = get_args()
    seq1, seq2 = args.seq1, args.seq2   ❶

    l1, l2 = len(seq1), len(seq2)   ❷
    distance = abs(l1 - l2)   ❸

    for i in range(min(l1, l2)):   ❹
        if seq1[i] != seq2[i]:   ❺
```

```
            distance += 1  ❻

        print(distance)  ❼
```

❶ Copy the two sequences into variables.

❷ Since I'll use the lengths more than once, I store them in variables.

❸ The base distance is the difference between the two lengths.

❹ Use the shorter length to find the indexes in common.

❺ Check the letters at each position.

❻ Increment the distance by 1.

❼ Print the distance.

This solution is very explicit, laying out every individual step needed to compare all the characters of two strings. The following solutions will start to shorten many of the steps, so be sure you are comfortable with exactly what I've shown here.

Solution 2: Creating a Unit Test

The first solution leaves me feeling vaguely uncomfortable because the code to calculate the Hamming distance should be in a function with tests. I'll start by creating a function called hamming() after the main() function. As a matter of style, I like to put get_args() first so I can read it immediately when I open the program. My main() function always comes second, and all other functions and tests after that.

I'll start by imagining the inputs and output of my function:

```
def hamming(seq1: str, seq2: str) -> int:  ❶
    """ Calculate Hamming distance """

    return 0  ❷
```

❶ The function will accept two strings as positional arguments and will return an integer.

❷ To start, the function will always return 0.

I want to stress the fact that the function does not *print* the answer but rather *returns it as a result*. If you wrote this function to `print()` the distance, you would not be able to write a unit test. You would have to rely entirely on the integration test that looks to see if the program prints the correct answer. As much as possible, I would encourage you to write pure functions that act only on the arguments and have no side effects. Printing is a side effect, and, while the program does need to print the answer eventually, this function's job is solely to return an integer when given two strings.

I've already shown a few test cases I can encode. Feel free to add other tests of your own devising:

```python
def test_hamming() -> None:
    """ Test hamming """

    assert hamming('', '') == 0 ❶
    assert hamming('AC', 'ACGT') == 2 ❷
    assert hamming('GAGCCTACTAACGGGAT', 'CATCGTAATGACGGCCT') == 7 ❸
```

❶   I always think it's good practice to send empty strings for string inputs.

❷   The difference is due only to length.

❸   This is the example from the documentation.

I'm aware that this may seem a bit extreme, because this function is essentially the entire program. I'm almost duplicating the integration test, I know, but I'm using this to point out best practices for writing programs. The `hamming()` function is a good unit of code, and it belongs in a function with a test. In a much larger program, this would be one of perhaps dozens to hundreds of other functions, and each should be *encapsulated*, *documented*, and *tested*.

Following test-driven principles, run **pytest** on the program to ensure that the test fails:

```
$ pytest -v hamm.py
========================= test session starts =========================
...

hamm.py::test_hamming FAILED                                    [100%]

============================== FAILURES ===============================
_____ test_hamming _____

    def test_hamming() -> None:
        """ Test hamming """

        assert hamming('', '') == 0
```

```
>       assert hamming('AC', 'ACGT') == 2
E       assert 0 == 2
E         +0
E         -2
```

hamm.py:69: AssertionError
========================= short test summary info =========================
FAILED hamm.py::test_hamming - assert 0 == 2
========================= 1 failed in 0.13s =========================

Now copy the code from `main()` to fix the function:

```
def hamming(seq1: str, seq2: str) -> int:
    """ Calculate Hamming distance """

    l1, l2 = len(seq1), len(seq2)
    distance = abs(l1 - l2)

    for i in range(min(l1, l2)):
        if seq1[i] != seq2[i]:
            distance += 1

    return distance
```

Verify that your function is correct:

```
$ pytest -v hamm.py
========================= test session starts =========================
...

hamm.py::test_hamming PASSED                                [100%]

========================= 1 passed in 0.02s =========================
```

You can incorporate it into your `main()` function like so:

```
def main():
    args = get_args()
    print(hamming(args.seq1, args.seq2))  ❶
```

❶ Print the return value from the function for the two given sequences.

This hides the complexity of the program inside a named, documented, tested unit, shortening the main body of the program and improving the readability.

Solution 3: Using the zip() Function

The following solution uses the `zip()` function to combine the elements from two sequences. The result is a list of tuples containing the characters from each position (see Figure 6-3). Note that `zip()` is another lazy function, so I'll use `list()` to coerce the values in the REPL:

```
>>> list(zip('ABC', '123'))
[('A', '1'), ('B', '2'), ('C', '3')]
```

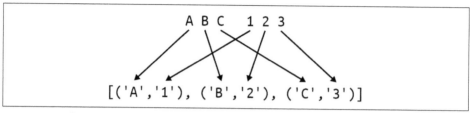

Figure 6-3. The tuples are composed of characters in common positions

If I use the *AC* and *ACGT* sequences, you'll notice that zip() stops with the shorter sequence, as shown in Figure 6-4:

```
>>> list(zip('AC', 'ACGT'))
[('A', 'A'), ('C', 'C')]
```

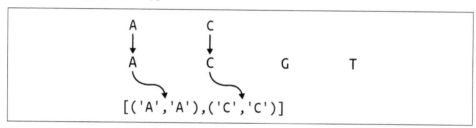

Figure 6-4. The zip() function will stop at the shortest sequence

I can use a for loop to iterate over each pair. So far in my for loops, I've used a single variable to represent each element in a list like this:

```
>>> for tup in zip('AC', 'ACGT'):
...     print(tup)
...
('A', 'A')
('C', 'C')
```

In Chapter 1, I showed how to *unpack* the values from a tuple into separate variables. The Python for loop allows me to unpack each tuple into the two characters, like so:

```
>>> for char1, char2 in zip('AC', 'ACGT'):
...     print(char1, char2)
...
A A
C C
```

The zip() function obviates a couple of lines from the first implementation:

```
def hamming(seq1: str, seq2: str) -> int:
    """ Calculate Hamming distance """

    distance = abs(len(seq1) - len(seq2))  ❶

    for char1, char2 in zip(seq1, seq2):  ❷
        if char1 != char2:  ❸
            distance += 1  ❹

    return distance
```

❶ Start with the absolute difference of the lengths.

❷ Use zip() to pair up the characters of the two strings.

❸ Check if the two characters are not equal.

❹ Increment the distance.

Solution 4: Using the zip_longest() Function

The next solution imports the zip_longest() function from the itertools module. As the name implies, it will zip the lists to the length of the longest list. Figure 6-5 shows that the function will insert None values when a shorter sequence has been exhausted:

```
>>> from itertools import zip_longest
>>> list(zip_longest('AC', 'ACGT'))
[('A', 'A'), ('C', 'C'), (None, 'G'), (None, 'T')]
```

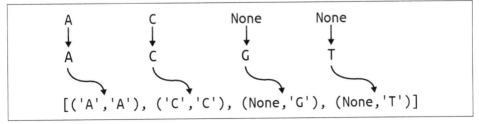

Figure 6-5. The zip_longest() function will stop at the longest sequence

I no longer need to start by subtracting the lengths of the sequences. Instead, I'll initialize a distance variable to 0 and then use zip_longest() to create tuples of bases to compare:

```
def hamming(seq1: str, seq2: str) -> int:
    """ Calculate Hamming distance """

    distance = 0  ❶
    for char1, char2 in zip_longest(seq1, seq2):  ❷
```

```
        if char1 != char2: ❸
            distance += 1 ❹

    return distance
```

❶ Initialize the distance to 0.

❷ Zip to the longest sequence.

❸ Compare the characters.

❹ Increment the counter.

Solution 5: Using a List Comprehension

All the solutions up to this point have used a `for` loop. I hope you're starting to antici-pate that I'm going to show you how to convert this into a list comprehension next. When the goal is to create a new list or reduce a list of values to some answer, it's often shorter and preferable to use a list comprehension.

The first version is going to use an `if` expression to return a 1 if the two characters are the same or a 0 if they are not:

```
>>> seq1, seq2, = 'GAGCCTACTAACGGGAT', 'CATCGTAATGACGGCCT'
>>> [1 if c1 != c2 else 0 for c1, c2 in zip_longest(seq1, seq2)]
[1, 0, 1, 0, 1, 0, 0, 1, 0, 1, 0, 0, 0, 0, 1, 1, 0]
```

The Hamming distance, then, is the sum of these:

```
>>> sum([1 if c1 != c2 else 0 for c1, c2 in zip_longest(seq1, seq2)])
7
```

Another way to express this idea is to only produce the 1s by using a *guard* clause, which is a conditional statement at the end of the list comprehension that decides whether or not a particular element is allowed:

```
>>> ones = [1 for c1, c2 in zip_longest(seq1, seq2) if c1 != c2] ❶
>>> ones
[1, 1, 1, 1, 1, 1, 1]
>>> sum(ones)
7
```

❶ The `if` statement is the guard that will produce the value 1 if the two characters are not equal.

You could also use the Boolean/integer coercion I showed in Chapter 5, where each `True` value will be treated as 1 and `False` is 0:

```
>>> bools = [c1 != c2 for c1, c2 in zip_longest(seq1, seq2)]
>>> bools
```

```
[True, False, True, False, True, False, False, True, False, True, False,
False, False, False, True, True, True, False]
>>> sum(bools)
7
```

Any of these ideas will reduce the function to a single line of code that passes the tests:

```
def hamming(seq1: str, seq2: str) -> int:
    """ Calculate Hamming distance """

    return sum([c1 != c2 for c1, c2 in zip_longest(seq1, seq2)])
```

Solution 6: Using the filter() Function

Chapters 4 and 5 show that a list comprehension with a guard can also be expressed using the filter() function. The syntax is a little ugly because Python doesn't allow the unpacking of the tuples from zip_longest() into separate variables. That is, I want to write a lambda that unpacks char1 and char2 into separate variables, but this is not possible:

```
>>> list(filter(lambda char1, char2: char1 != char2, zip_longest(seq1, seq2)))
Traceback (most recent call last):
  File "<stdin>", line 1, in <module>
TypeError: <lambda>() missing 1 required positional argument: 'char2'
```

Instead, I will usually call the lambda variable tup or t to remind me this is a tuple. I will use the positional tuple notation to compare the element in the zeroth position to the element in the first position. filter() will only produce those tuples where the elements are different:

```
>>> seq1, seq2 = 'AC', 'ACGT'
>>> list(filter(lambda t: t[0] != t[1], zip_longest(seq1, seq2)))
[(None, 'G'), (None, 'T')]
```

The Hamming distance then is the length of this list. Note that the len() function will not prompt filter() to produce values:

```
>>> len(filter(lambda t: t[0] != t[1], zip_longest(seq1, seq2)))
Traceback (most recent call last):
  File "<stdin>", line 1, in <module>
TypeError: object of type 'filter' has no len()
```

This is one of those instances where the code must use list() to force the lazy filter() function to generate the results. Here is how I can incorporate these ideas:

```
def hamming(seq1: str, seq2: str) -> int:
    """ Calculate Hamming distance """

    distance = filter(lambda t: t[0] != t[1], zip_longest(seq1, seq2)) ❶
    return len(list((distance))) ❷
```

❶ Use `filter()` to find tuple pairs of different characters.

❷ Return the length of the resulting list.

Solution 7: Using the map() Function with zip_longest()

This solution uses `map()` instead of `filter()` only to show you that the same inability to unpack the tuples also applies. I'd like to use `map()` to produce a list of Boolean values indicating whether the character pairs match or not:

```
>>> seq1, seq2 = 'AC', 'ACGT'
>>> list(map(lambda t: t[0] != t[1], zip_longest(seq1, seq2)))
[False, False, True, True]
```

The `lambda` is identical to the one to `filter()` that was used as the *predicate* to determine which elements are allowed to pass. Here the code *transforms* the elements into the result of applying the `lambda` function to the arguments, as shown in Figure 6-6. Remember that `map()` will always return the same number of elements it consumes, but `filter()` may return fewer or none at all.

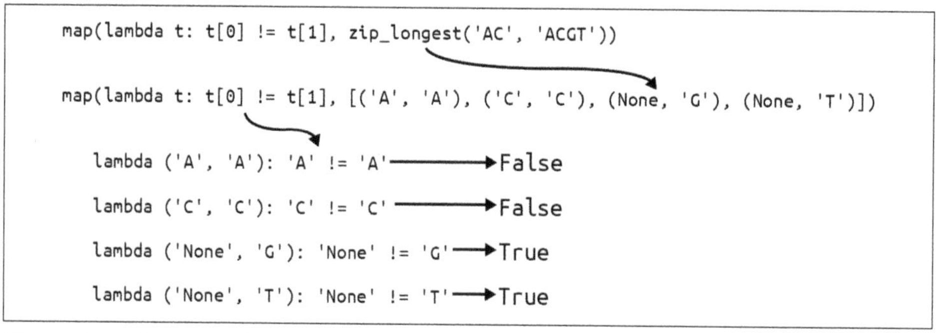

```
map(lambda t: t[0] != t[1], zip_longest('AC', 'ACGT'))

map(lambda t: t[0] != t[1], [('A', 'A'), ('C', 'C'), (None, 'G'), (None, 'T')])

    lambda ('A', 'A'): 'A' != 'A' ──────────▶ False

    lambda ('C', 'C'): 'C' != 'C' ──────────▶ False

    lambda ('None', 'G'): 'None' != 'G' ───▶ True

    lambda ('None', 'T'): 'None' != 'T' ───▶ True
```

Figure 6-6. The map() function transforms each tuple into a Boolean value representing the inequality of the two elements

I can sum these Booleans to get the number of mismatched pairs:

```
>>> seq1, seq2, = 'GAGCCTACTAACGGGAT', 'CATCGTAATGACGGCCT'
>>> sum(map(lambda t: t[0] != t[1], zip_longest(seq1, seq2)))
7
```

Here is the function with this idea:

```
def hamming(seq1: str, seq2: str) -> int:
    """ Calculate Hamming distance """

    return sum(map(lambda t: t[0] != t[1], zip_longest(seq1, seq2)))
```

Even though these functions have gone from 10 or more lines of code to a single line, it still makes sense for this to be a function with a descriptive name and tests. Eventually, you'll start creating modules of reusable code to share across your projects.

Solution 8: Using the starmap() and operator.ne() Functions

I confess that I showed the last few solutions solely to build up to this last solution. Let me start by showing how I can assign a lambda to a variable:

```
>>> not_same = lambda t: t[0] != t[1]
```

This is not recommended syntax, and pylint will definitely fail your code on this and recommend a def instead:

```
def not_same(t):
    return t[0] != t[1]
```

Both will create a function called not_same() that will accept a tuple and return whether the two elements are the same:

```
>>> not_same(('A', 'A'))
False
>>> not_same(('A', 'T'))
True
```

If, however, I wrote the function to accept two positional arguments, the same error I saw before would crop up:

```
>>> not_same = lambda a, b: a != b
>>> list(map(not_same, zip_longest(seq1, seq2)))
Traceback (most recent call last):
  File "<stdin>", line 1, in <module>
TypeError: <lambda>() missing 1 required positional argument: 'b'
```

What I need is a version of map() that can splat the incoming tuple (as I first showed in Chapter 1) by adding * (*star*, *asterisk*, or *splat*) to the tuple to expand it into its elements, which is exactly what the function itertools.starmap() does (see Figure 6-7):

```
>>> from itertools import zip_longest, starmap
>>> seq1, seq2 = 'AC', 'ACGT'
>>> list(starmap(not_same, zip_longest(seq1, seq2)))
[False, False, True, True]
```

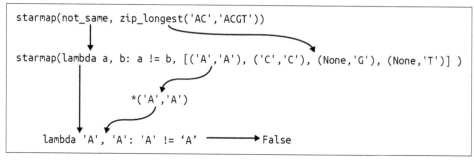

Figure 6-7. The `starmap()` function applies a splat to the incoming tuple to turn it into the two values that the `lambda` expects

But wait, there's more! I don't even need to write my own `not_same()` function because I already have `operator.ne()` (not equal), which I usually write using the `!=` operator:

```
>>> import operator
>>> operator.ne('A', 'A')
False
>>> operator.ne('A', 'T')
True
```

An *operator* is a special binary function (accepting two arguments) where the function name is usually some symbol like `+` that sits between the arguments. In the case of `+`, Python has to decide if this means `operator.add()`:

```
>>> 1 + 2
3
>>> operator.add(1, 2)
3
```

or `operator.concat()`:

```
>>> 'AC' + 'GT'
'ACGT'
>>> operator.concat('AC', 'GT')
'ACGT'
```

The point is that I already have an existing function that expects two arguments and returns whether they are equal, and I can use `starmap()` to properly expand the tuples into the needed arguments:

```
>>> seq1, seq2 = 'AC', 'ACGT'
>>> list(starmap(operator.ne, zip_longest(seq1, seq2)))
[False, False, True, True]
```

As before, the Hamming distance is the sum of the unmatched pairs:

```
>>> seq1, seq2, = 'GAGCCTACTAACGGGAT', 'CATCGTAATGACGGCCT'
>>> sum(starmap(operator.ne, zip_longest(seq1, seq2)))
7
```

To see it in action:

```
def hamming(seq1: str, seq2: str) -> int:
    """ Calculate Hamming distance """

    return sum(starmap(operator.ne, zip_longest(seq1, seq2))) ❶
```

❶ Zip the sequences, transform the tuples to Boolean comparisons, and sum.

This final solution relies entirely on fitting four functions that I didn't write. I believe the best code is code you don't write (or test or document). While I prefer this purely functional solution, you may feel this code is overly clever. You should use whatever version you'll be able to understand a year later.

Going Further

- Without looking at the source code, write a version of zip_longest(). Be sure to start with a test, then write the function that satisfies the test.

- Expand your program to handle more than two input sequences. Have your program print the Hamming distance between every pair of sequences. That means the program will print n choose k numbers which will be $n! / k!(n - k)!$. For three sequences, your program will print $3! / (2!(3 - 2)!) = 6 / 2 = 3$ distance pairs.

- Try writing a sequence alignment algorithm that will show there is, for instance, just one difference between the sequences *AAACCCGGGTTT* and *AACCCGGGTTTA*.

Review

This was a rather deep rabbit hole to go down just to find the Hamming distance, but it highlights lots of interesting bits about Python functions:

- The built-in zip() function will combine two or more lists into a list of tuples, grouping elements at common positions. It stops at the shortest sequence, so use the iterools.zip_longest() function if you want to go to the longest sequence.

- Both map() and filter() apply a function to some iterable of values. The map() function will return a new sequence transformed by the function, while filter() will only return those elements that return a truthy value when the function is applied.

- The function passed to map() and filter() can be an anonymous function created by lambda or an existing function.

- The operator module contains many functions like ne() (not equal) that can be used with map() and filter().

- The functools.starmap() function works just like map() but will splat the function's incoming values to expand them into a list of values.

## CHAPTER 7

Translating mRNA into Protein: More Functional Programming

According to the Central Dogma of molecular biology, *DNA makes mRNA, and mRNA makes protein*. In Chapter 2, I showed how to transcribe DNA to mRNA, so now it's time to translate mRNA into protein sequences. As described on the Rosalind PROT page (*https://oreil.ly/OgBcW*), I now need to write a program that accepts a string of mRNA and produces an amino acid sequence. I will show several solutions using lists, `for` loops, list comprehensions, dictionaries, and higher-order functions, but I confess I'll end with a Biopython function. Still, it will be tons of fun.

Mostly I'm going to focus on how to write, test, and compose small functions to create solutions. You'll learn:

- How to extract codons/k-mers from a sequence using string slices
- How to use a dictionary as a lookup table
- How to translate a `for` loop into a list comprehension and a `map()` expression
- How to use the `takewhile()` and `partial()` functions
- How to use the `Bio.Seq` module to translate mRNA into proteins

Getting Started

You will need to work in the *07_prot* directory for this exercise. I recommend you begin by copying the first solution to `prot.py` and asking for the usage:

```
$ cp solution1_for.py prot.py
$ ./prot.py -h
usage: prot.py [-h] RNA
```

```
Translate RNA to proteins

positional arguments:
  RNA           RNA sequence

optional arguments:
  -h, --help  show this help message and exit
```

The program requires an RNA sequence as a single positional argument. From here on, I'll use the term *RNA*, but know that I mean *mRNA*. Here's the result using the example string from the Rosalind page:

```
$ ./prot.py AUGGCCAUGGCGCCCAGAACUGAGAUCAAUAGUACCCGUAUUAACGGGUGA
MAMAPRTEINSTRING
```

Run **make test** to ensure the program works properly. When you feel you have a decent idea of how the program works, start from scratch:

```
$ new.py -fp 'Translate RNA to proteins' prot.py
Done, see new script "prot.py".
```

Here is how I define the parameters:

```
class Args(NamedTuple):
    """ Command-line arguments """
    rna: str ❶


def get_args() -> Args:
    """Get command-line arguments"""

    parser = argparse.ArgumentParser(
        description='Translate RNA to proteins',
        formatter_class=argparse.ArgumentDefaultsHelpFormatter)

    parser.add_argument('rna', type=str, metavar='RNA', help='RNA sequence') ❷

    args = parser.parse_args()

    return Args(args.rna)
```

❶ The only parameter is a string of mRNA.

❷ Define rna as a positional string.

Modify your arguments until the program will produce the correct usage, then modify your main() to print the incoming RNA string:

```
def main() -> None:
    args = get_args()
    print(args.rna)
```

Verify that it works:

```
$ ./prot.py AUGGCCAUGGCGCCCAGAACUGAGAUCAAUAGUACCCGUAUUAACGGGUGA
AUGGCCAUGGCGCCCAGAACUGAGAUCAAUAGUACCCGUAUUAACGGGUGA
```

Run **pytest** or **make test** to see how you fare. Your program should pass the first two tests and fail the third, where the output should be the protein translation. If you think you can figure this out, go ahead with your solution. It's perfectly fine to struggle. There's no hurry, so take a few days if you need to. Be sure to incorporate naps and walks (diffuse thinking time) in addition to your focused coding time. If you need some help, read on.

K-mers and Codons

So far you've seen many examples of how to iterate over the characters of a string, such as the bases of DNA. Here I need to group the bases of RNA into threes to read each *codon*, a sequence of three nucleotides that corresponds to an amino acid. There are 64 codons, as shown in Table 7-1.

Table 7-1. The RNA codon table describes how 3-mers/codons of RNA encode the 22 amino acids

AAA K	AAC N	AAG K	AAU N
ACA T	ACC T	ACG T	ACU T
AGA R	AGC S	AGG R	AGU S
AUA I	AUC I	AUG M	AUU I
CAA Q	CAC H	CAG Q	CAU H
CCA P	CCC P	CCG P	CCU P
CGA R	CGC R	CGG R	CGU R
CUA L	CUC L	CUG L	CUU L
GAA E	GAC D	GAG E	GAU D
GCA A	GCC A	GCG A	GCU A
GGA G	GGC G	GGG G	GGU G
GUA V	GUC V	GUG V	GUU V
UAC Y	UAU Y	UCA S	UCC S
UCG S	UCU S	UGC C	UGG W
UGU C	UUA L	UUC F	UUG L
UUU F	UAA Stop	UAG Stop	UGA Stop

Given some string of RNA:

```
>>> rna = 'AUGGCCAUGGCGCCCAGAACUGAGAUCAAUAGUACCCGUAUUAACGGGUGA'
```

I want to read the first three bases, *AUG*. As shown in Figure 7-1, I can use a string slice to manually grab the characters from indexes 0 to 3 (remembering that the upper bound is not inclusive):

```
>>> rna[0:3]
'AUG'
```

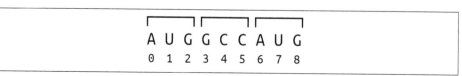

Figure 7-1. Extracting codons from RNA using string slices

The next codon can be found by adding 3 to the start and stop positions:

```
>>> rna[3:6]
'GCC'
```

Can you see a pattern emerging? For the first number, I need to start at 0 and add 3. For the second number, I need to add 3 to the first number (see Figure 7-2).

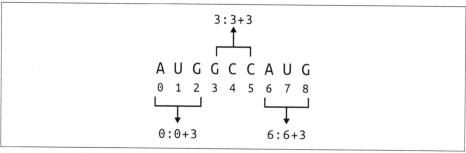

Figure 7-2. Each slice is a function of the start positions of the codons, which can be found using the range() function

I can handle the first part using the range() function, which can take one, two, or three arguments. Given just one argument, it will produce all the numbers from 0 up to but not including the given value. Note this is a lazy function which I'll coerce with list():

```
>>> list(range(10))
[0, 1, 2, 3, 4, 5, 6, 7, 8, 9]
```

Given two arguments, range() will assume the first is the start and the second is the stop:

```
>>> list(range(5, 10))
[5, 6, 7, 8, 9]
```

A third argument will be interpreted as the step size. In Chapter 3, I used `range()` with no start or stop positions and a step size of `-1` to reverse a string. In this case, I want to count from 0 up to the length of the RNA, stepping by 3. These are the starting positions for the codons:

```
>>> list(range(0, len(rna), 3))
[0, 3, 6, 9, 12, 15, 18, 21, 24, 27, 30, 33, 36, 39, 42, 45, 48]
```

I can use a list comprehension to generate the start and stop values as tuples. The stop positions are 3 more than the start positions. I'll show just the first five:

```
>>> [(n, n + 3) for n in range(0, len(rna), 3)][:5]
[(0, 3), (3, 6), (6, 9), (9, 12), (12, 15)]
```

I can use those values to take slices of the RNA:

```
>>> [rna[n:n + 3] for n in range(0, len(rna), 3)][:5]
['AUG', 'GCC', 'AUG', 'GCG', 'CCC']
```

The codons are subsequences of the RNA, and they are similar to *k-mers*. The *k* is the size, here 3, and *mer* is *a share* as in the word *polymer*. It's common to refer to k-mers by their size, so here I might call these *3-mers*. The k-mers overlap by one character, so the window shifts right by one base. Figure 7-3 shows the first seven 3-mers found in the first nine bases of the input RNA.

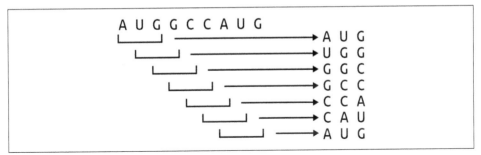

Figure 7-3. All the 3-mers in the first nine bases of the RNA sequence

The number *n* of k-mers in any sequence *s* is:

$$n = len(s) - k + 1$$

The length of this RNA sequence is 51, so it contains 49 3-mers:

```
>>> len(rna) - k + 1
49
```

Except when considering multiframe translation, which I'll show in Chapter 14, codons do not overlap and so shift 3 positions (see Figure 7-4), leaving 17 codons:

```
>>> len([rna[n:n + 3] for n in range(0, len(rna), 3)])
17
```

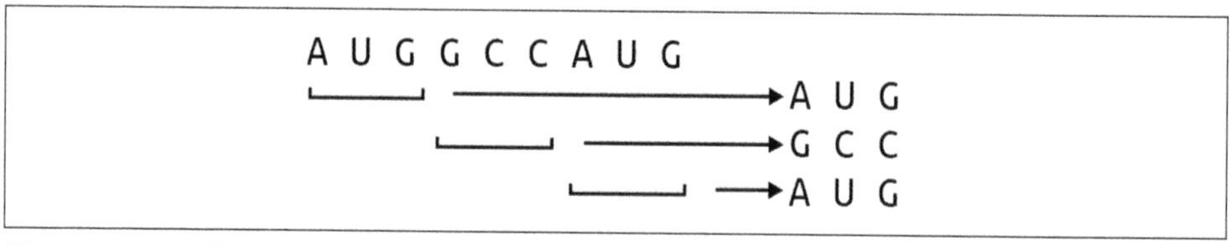

A U G G C C A U G
→ A U G
→ G C C
→ A U G

Figure 7-4. Codons are nonoverlapping 3-mers

Translating Codons

Now that you know how to extract the codons from the RNA, let's consider how to translate a codon into a protein. The Rosalind page provides the following translation table:

UUU F	CUU L	AUU I	GUU V
UUC F	CUC L	AUC I	GUC V
UUA L	CUA L	AUA I	GUA V
UUG L	CUG L	AUG M	GUG V
UCU S	CCU P	ACU T	GCU A
UCC S	CCC P	ACC T	GCC A
UCA S	CCA P	ACA T	GCA A
UCG S	CCG P	ACG T	GCG A
UAU Y	CAU H	AAU N	GAU D
UAC Y	CAC H	AAC N	GAC D
UAA Stop	CAA Q	AAA K	GAA E
UAG Stop	CAG Q	AAG K	GAG E
UGU C	CGU R	AGU S	GGU G
UGC C	CGC R	AGC S	GGC G
UGA Stop	CGA R	AGA R	GGA G
UGG W	CGG R	AGG R	GGG G

A dictionary would be a natural data structure to look up a string like AUG to find that it translates to the protein M, which happens to be the codon that indicates the beginning of a protein sequence. I will leave it to you to incorporate this data into your program. For what it's worth, I changed Stop to * in my dictionary for the stop codon, which indicates the end of the protein sequence. I called my dictionary codon_to_aa, and I can use it like so:

```
>>> rna = 'AUGGCCAUGGCGCCCAGAACUGAGAUCAAUAGUACCCGUAUUAACGGGUGA'
>>> aa = []
>>> for codon in [rna[n:n + 3] for n in range(0, len(rna), 3)]:
...     aa.append(codon_to_aa[codon])
...
>>> aa
['M', 'A', 'M', 'A', 'P', 'R', 'T', 'E', 'I', 'N', 'S', 'T', 'R', 'I',
 'N', 'G', '*']
```

The * codon indicates where the translation ends and is often shown so you know that a stop was found and the protein is complete. For the purposes of passing the Rosalind tests, the stop should not be included in the output. Note that the stop codon may occur before the end of the RNA string. This should be enough hints for you to create a solution that passes the tests. Be sure to run **pytest** and **make test** to ensure your program is logically and stylistically correct.

Solutions

In this section, I will show five variations on how to translate RNA into protein, moving from wholly manual solutions where I encode the RNA codon table with a dictionary to a single line of code that uses a function from Biopython. All solutions use the same get_args() as shown previously.

Solution 1: Using a for Loop

Here is the whole of my first solution, which uses a for loop to iterate the codons to translate them via a dictionary:

```
def main() -> None:
    args = get_args()
    rna = args.rna.upper() ❶
    codon_to_aa = { ❷
        'AAA': 'K', 'AAC': 'N', 'AAG': 'K', 'AAU': 'N', 'ACA': 'T',
        'ACC': 'T', 'ACG': 'T', 'ACU': 'T', 'AGA': 'R', 'AGC': 'S',
        'AGG': 'R', 'AGU': 'S', 'AUA': 'I', 'AUC': 'I', 'AUG': 'M',
        'AUU': 'I', 'CAA': 'Q', 'CAC': 'H', 'CAG': 'Q', 'CAU': 'H',
        'CCA': 'P', 'CCC': 'P', 'CCG': 'P', 'CCU': 'P', 'CGA': 'R',
        'CGC': 'R', 'CGG': 'R', 'CGU': 'R', 'CUA': 'L', 'CUC': 'L',
        'CUG': 'L', 'CUU': 'L', 'GAA': 'E', 'GAC': 'D', 'GAG': 'E',
        'GAU': 'D', 'GCA': 'A', 'GCC': 'A', 'GCG': 'A', 'GCU': 'A',
        'GGA': 'G', 'GGC': 'G', 'GGG': 'G', 'GGU': 'G', 'GUA': 'V',
        'GUC': 'V', 'GUG': 'V', 'GUU': 'V', 'UAC': 'Y', 'UAU': 'Y',
        'UCA': 'S', 'UCC': 'S', 'UCG': 'S', 'UCU': 'S', 'UGC': 'C',
        'UGG': 'W', 'UGU': 'C', 'UUA': 'L', 'UUC': 'F', 'UUG': 'L',
        'UUU': 'F', 'UAA': '*', 'UAG': '*', 'UGA': '*',
    }

    k = 3 ❸
    protein = '' ❹
    for codon in [rna[i:i + k] for i in range(0, len(rna), k)]: ❺
        aa = codon_to_aa.get(codon, '-') ❻
        if aa == '*': ❼
            break ❽
        protein += aa ❾

    print(protein) ❿
```

❶ Copy the incoming RNA, forcing to uppercase.

❷ Create a codon/AA lookup table using a dictionary.

❸ Establish the size of k for finding k-mers.

❹ Initialize the protein sequence to the empty string.

❺ Iterate through the codons of the RNA.

❻ Use `dict.get()` to look up the amino acid for this codon, and return a dash if it is not found.

❼ Check if this is the stop codon.

❽ Break out of the `for` loop.

❾ Append the amino acid to the protein sequence.

❿ Print the protein sequence.

Solution 2: Adding Unit Tests

The first solution works adequately well and, for such a short program, has a decent organization. The problem is that short programs usually become long programs. It's common to make functions longer and longer, so I'd like to show how I can break up the code in `main()` into a couple of smaller functions with tests. Generally speaking, I like to see a function fit into 50 lines or fewer on the high end. As for how short a function can be, I'm not opposed to a single line of code.

My first intuition is to extract the code that finds the codons and make that into a function with a unit test. I can start by defining a placeholder for the function with a type signature that helps me think about what the function accepts as arguments and will return as a result:

```
def codons(seq: str) -> List[str]: ❶
    """ Extract codons from a sequence """

    return [] ❷
```

❶ The function will accept a string and will return a list of strings.

❷ For now, just return an empty list.

Next, I define a `test_codons()` function to imagine how it might work. Whenever I have a string as a function parameter, I try passing the empty string. (Whenever I have an integer as a function parameter, I try passing 0.) Then I try other possible values and imagine what the function ought to do. As you can see, I'm making some judgment calls here by returning strings shorter than three bases. I only expect the function to break a string into substrings of *at least* three bases. There's no reason to let perfect be the enemy of good enough here:

```
def test_codons() -> None:
    """ Test codons """

    assert codons('') == []
    assert codons('A') == ['A']
    assert codons('ABC') == ['ABC']
    assert codons('ABCDE') == ['ABC', 'DE']
    assert codons('ABCDEF') == ['ABC', 'DEF']
```

Now to write the function that will satisfy these tests. If I move the relevant code from main() into the codons() function, I get this:

```
def codons(seq: str) -> List[str]:
    """ Extract codons from a sequence """

    k = 3
    ret = []
    for codon in [seq[i:i + k] for i in range(0, len(seq), k)]:
        ret.append(codon)

    return ret
```

If I try running pytest on this program, I see it passes. Huzzah! Since the for loop is being used to build up a return list, it would be stylistically better to use a list comprehension:

```
def codons(seq: str) -> List[str]:
    """ Extract codons from a sequence """

    k = 3
    return [seq[i:i + k] for i in range(0, len(seq), k)]
```

This is a nice little function that is documented and tested and which will make the rest of the code more readable:

```
def main() -> None:
    args = get_args()
    rna = args.rna.upper()
    codon_to_aa = {
        'AAA': 'K', 'AAC': 'N', 'AAG': 'K', 'AAU': 'N', 'ACA': 'T',
        'ACC': 'T', 'ACG': 'T', 'ACU': 'T', 'AGA': 'R', 'AGC': 'S',
        'AGG': 'R', 'AGU': 'S', 'AUA': 'I', 'AUC': 'I', 'AUG': 'M',
        'AUU': 'I', 'CAA': 'Q', 'CAC': 'H', 'CAG': 'Q', 'CAU': 'H',
        'CCA': 'P', 'CCC': 'P', 'CCG': 'P', 'CCU': 'P', 'CGA': 'R',
        'CGC': 'R', 'CGG': 'R', 'CGU': 'R', 'CUA': 'L', 'CUC': 'L',
        'CUG': 'L', 'CUU': 'L', 'GAA': 'E', 'GAC': 'D', 'GAG': 'E',
```

```
          'GAU': 'D', 'GCA': 'A', 'GCC': 'A', 'GCG': 'A', 'GCU': 'A',
          'GGA': 'G', 'GGC': 'G', 'GGG': 'G', 'GGU': 'G', 'GUA': 'V',
          'GUC': 'V', 'GUG': 'V', 'GUU': 'V', 'UAC': 'Y', 'UAU': 'Y',
          'UCA': 'S', 'UCC': 'S', 'UCG': 'S', 'UCU': 'S', 'UGC': 'C',
          'UGG': 'W', 'UGU': 'C', 'UUA': 'L', 'UUC': 'F', 'UUG': 'L',
          'UUU': 'F', 'UAA': '*', 'UAG': '*', 'UGA': '*',
      }

      protein = ''
      for codon in codons(rna): ❶
          aa = codon_to_aa.get(codon, '-')
          if aa == '*':
              break
          protein += aa

      print(protein)
```

❶ The complexity of finding the codons is hidden in a function.

Further, this function (and its test) will now be easier to incorporate into another program. The simplest case would be to copy and paste these lines, but a better solution would be to share the function. Let me demonstrate using the REPL. If you have the codons() function in your prot.py program, then import the function:

```
>>> from prot import codons
```

Now you can execute the codons() function:

```
>>> codons('AAACCCGGGTTT')
['AAA', 'CCC', 'GGG', 'TTT']
```

Or you can import the entire prot module and call the function like so:

```
>>> import prot
>>> prot.codons('AAACCCGGGTTT')
['AAA', 'CCC', 'GGG', 'TTT']
```

Python *programs* are also *modules* of reusable code. Sometimes you execute a source code file and it becomes a program, but there's not a big distinction in Python between a program and a module. This is the meaning of the couplet at the end of all the programs:

```
if __name__ == '__main__': ❶
    main() ❷
```

❶ When a Python program is being *executed* as a program, the value of __name__ is __main__.

❷ Call the main() function to start the program.

When a Python module is being *imported* by another piece of code, then the __name__ is the name of the module; for example, prot in the case of prot.py. If you simply called main() at the end of the program without checking the __name__, then it would be executed whenever your module is imported, which is not good.

As you write more and more Python, you'll likely find you're solving some of the same problems repeatedly. It would be far better to share common solutions by writing functions that you share across projects rather than copy-pasting bits of code. Python makes it pretty easy to put reusable functions into modules and import them into other programs.

Solution 3: Another Function and a List Comprehension

The codons() function is tidy and useful and makes the main() function easier to understand; however, all the code that's left in main() is concerned with translating the protein. I'd like to hide this away in a translate() function, and here is the test I'd like to use:

```python
def test_translate() -> None:
    """ Test translate """

    assert translate('') == '' ❶
    assert translate('AUG') == 'M' ❷
    assert translate('AUGCCGUAAUCU') == 'MP' ❸
    assert translate('AUGGCCAUGGCGCCCAGAACUGAGAU' ❹
                     'CAAUAGUACCCGUAUUAACGGGUGA') == 'MAMAPRTEINSTRING' ❺
```

❶ I usually test string parameters with the empty string.

❷ Test for a single amino acid.

❸ Test using a stop codon before the end of the sequence.

❹ Notice that adjacent string literals are joined into a single string. This is a useful way to break long lines in source code.

❺ Test using the example from Rosalind.

I move all the code from main() into this, changing the for loop to a list comprehension and using a list slice to truncate the protein at the stop codon:

```python
def translate(rna: str) -> str:
    """ Translate codon sequence """

    codon_to_aa = {
        'AAA': 'K', 'AAC': 'N', 'AAG': 'K', 'AAU': 'N', 'ACA': 'T',
```

```
         'ACC': 'T', 'ACG': 'T', 'ACU': 'T', 'AGA': 'R', 'AGC': 'S',
         'AGG': 'R', 'AGU': 'S', 'AUA': 'I', 'AUC': 'I', 'AUG': 'M',
         'AUU': 'I', 'CAA': 'Q', 'CAC': 'H', 'CAG': 'Q', 'CAU': 'H',
         'CCA': 'P', 'CCC': 'P', 'CCG': 'P', 'CCU': 'P', 'CGA': 'R',
         'CGC': 'R', 'CGG': 'R', 'CGU': 'R', 'CUA': 'L', 'CUC': 'L',
         'CUG': 'L', 'CUU': 'L', 'GAA': 'E', 'GAC': 'D', 'GAG': 'E',
         'GAU': 'D', 'GCA': 'A', 'GCC': 'A', 'GCG': 'A', 'GCU': 'A',
         'GGA': 'G', 'GGC': 'G', 'GGG': 'G', 'GGU': 'G', 'GUA': 'V',
         'GUC': 'V', 'GUG': 'V', 'GUU': 'V', 'UAC': 'Y', 'UAU': 'Y',
         'UCA': 'S', 'UCC': 'S', 'UCG': 'S', 'UCU': 'S', 'UGC': 'C',
         'UGG': 'W', 'UGU': 'C', 'UUA': 'L', 'UUC': 'F', 'UUG': 'L',
         'UUU': 'F', 'UAA': '*', 'UAG': '*', 'UGA': '*',
    }

    aa = [codon_to_aa.get(codon, '-') for codon in codons(rna)] ❶
    if '*' in aa: ❷
        aa = aa[:aa.index('*')] ❸

    return ''.join(aa) ❹
```

❶ Use a list comprehension to translate the list of codons to a list of amino acids.

❷ See if the stop (*) codon is present in the list of amino acids.

❸ Overwrite the amino acids using a list slice up to the index of the stop codon.

❹ Join the amino acids on the empty string and return the new protein sequence.

To understand this, consider the following RNA sequence:

```
>>> rna = 'AUGCCGUAAUCU'
```

I can use the codons() function to get the codons:

```
>>> from solution3_list_comp_slice import codons, translate
>>> codons(rna)
['AUG', 'CCG', 'UAA', 'UCU']
```

And use a list comprehension to turn those into amino acids:

```
>>> codon_to_aa = {
...     'AAA': 'K', 'AAC': 'N', 'AAG': 'K', 'AAU': 'N', 'ACA': 'T',
...     'ACC': 'T', 'ACG': 'T', 'ACU': 'T', 'AGA': 'R', 'AGC': 'S',
...     'AGG': 'R', 'AGU': 'S', 'AUA': 'I', 'AUC': 'I', 'AUG': 'M',
...     'AUU': 'I', 'CAA': 'Q', 'CAC': 'H', 'CAG': 'Q', 'CAU': 'H',
...     'CCA': 'P', 'CCC': 'P', 'CCG': 'P', 'CCU': 'P', 'CGA': 'R',
...     'CGC': 'R', 'CGG': 'R', 'CGU': 'R', 'CUA': 'L', 'CUC': 'L',
...     'CUG': 'L', 'CUU': 'L', 'GAA': 'E', 'GAC': 'D', 'GAG': 'E',
...     'GAU': 'D', 'GCA': 'A', 'GCC': 'A', 'GCG': 'A', 'GCU': 'A',
...     'GGA': 'G', 'GGC': 'G', 'GGG': 'G', 'GGU': 'G', 'GUA': 'V',
...     'GUC': 'V', 'GUG': 'V', 'GUU': 'V', 'UAC': 'Y', 'UAU': 'Y',
...     'UCA': 'S', 'UCC': 'S', 'UCG': 'S', 'UCU': 'S', 'UGC': 'C',
...     'UGG': 'W', 'UGU': 'C', 'UUA': 'L', 'UUC': 'F', 'UUG': 'L',
```

```
...        'UUU': 'F', 'UAA': '*', 'UAG': '*', 'UGA': '*',
... }
>>> aa = [codon_to_aa.get(c, '-') for c in codons(rna)]
>>> aa
['M', 'P', '*', 'S']
```

I can see the stop codon is present:

```
>>> '*' in aa
True
```

and so the sequence needs to be truncated at index 2:

```
>>> aa.index('*')
2
```

I can use a list slice to select up to the position of the stop codon. If no start position is supplied, then Python assumes the index 0:

```
>>> aa = aa[:aa.index('*')]
>>> aa
['M', 'P']
```

Finally, this list needs to be joined on the empty string:

```
>>> ''.join(aa)
'MP'
```

The main() function incorporates the new function and makes for a very readable program:

```
def main() -> None:
    args = get_args()
    print(translate(args.rna.upper()))
```

This is another instance where the unit test almost duplicates the integration test that verifies the program translates RNA into protein. The latter is still important as it ensures that the program works, produces documentation, handles the arguments, and so forth. As overengineered as this solution may seem, I want you to focus on how to break programs into smaller *functions* that you can understand, test, compose, and share.

Solution 4: Functional Programming with the map(), partial(), and takewhile() Functions

For this next solution, I want to show how to rewrite some of the logic using three HOFs, map(), partial(), and takewhile(). Figure 7-5 shows how the list comprehension can be rewritten as a map().

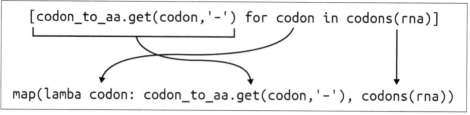

Figure 7-5. A list comprehension can be rewritten as a map()

I can use map() to get the amino acid sequence. You may or may not feel this is easier to read than a list comprehension; the point is to understand that they are functionally equivalent, both transforming one list into a new list:

```
>>> aa = list(map(lambda codon: codon_to_aa.get(codon, '-'), codons(rna)))
>>> aa
['M', 'P', '*', 'S']
```

The code to find the stop codon and slice the list can be rewritten using the itertools.takewhile() function:

```
>>> from itertools import takewhile
```

As the name implies, this function will *take* elements from a sequence *while* the predicate is met. Once the predicate fails, the function stops producing values. Here the condition is that the residue is not * (stop):

```
>>> list(takewhile(lambda residue: residue != '*', aa))
['M', 'P']
```

If you like using these kinds of HOFs, you can take this even further by using the functools.partial() function I showed in Chapter 4. Here I want to partially apply the operator.ne() (not equal) function:

```
>>> from functools import partial
>>> import operator
>>> not_stop = partial(operator.ne, '*')
```

The not_stop() function needs one more string value before it can return a value:

```
>>> not_stop('F')
True
>>> not_stop('*')
False
```

When I compose these functions, they almost read like an English sentence:

```
>>> list(takewhile(not_stop, aa))
['M', 'P']
```

Here is how I would write the translate() function with purely functional ideas:

```
def translate(rna: str) -> str:
    """ Translate codon sequence """

    codon_to_aa = {
        'AAA': 'K', 'AAC': 'N', 'AAG': 'K', 'AAU': 'N', 'ACA': 'T',
        'ACC': 'T', 'ACG': 'T', 'ACU': 'T', 'AGA': 'R', 'AGC': 'S',
        'AGG': 'R', 'AGU': 'S', 'AUA': 'I', 'AUC': 'I', 'AUG': 'M',
        'AUU': 'I', 'CAA': 'Q', 'CAC': 'H', 'CAG': 'Q', 'CAU': 'H',
        'CCA': 'P', 'CCC': 'P', 'CCG': 'P', 'CCU': 'P', 'CGA': 'R',
        'CGC': 'R', 'CGG': 'R', 'CGU': 'R', 'CUA': 'L', 'CUC': 'L',
        'CUG': 'L', 'CUU': 'L', 'GAA': 'E', 'GAC': 'D', 'GAG': 'E',
        'GAU': 'D', 'GCA': 'A', 'GCC': 'A', 'GCG': 'A', 'GCU': 'A',
        'GGA': 'G', 'GGC': 'G', 'GGG': 'G', 'GGU': 'G', 'GUA': 'V',
        'GUC': 'V', 'GUG': 'V', 'GUU': 'V', 'UAC': 'Y', 'UAU': 'Y',
        'UCA': 'S', 'UCC': 'S', 'UCG': 'S', 'UCU': 'S', 'UGC': 'C',
        'UGG': 'W', 'UGU': 'C', 'UUA': 'L', 'UUC': 'F', 'UUG': 'L',
        'UUU': 'F', 'UAA': '*', 'UAG': '*', 'UGA': '*',
    }

    aa = map(lambda codon: codon_to_aa.get(codon, '-'), codons(rna))
    return ''.join(takewhile(partial(operator.ne, '*'), aa))
```

Solution 5: Using Bio.Seq.translate()

As promised, the last solution uses Biopython. In Chapter 3, I used the
`Bio.Seq.reverse_complement()` function, and here I can use `Bio.Seq.translate()`.
First, import the `Bio.Seq` class:

```
>>> from Bio import Seq
```

Then call the `translate()` function. Note that the stop codon is represented by *:

```
>>> rna = 'AUGGCCAUGGCGCCCAGAACUGAGAUCAAUAGUACCCGUAUUAACGGGUGA'
>>> Seq.translate(rna)
'MAMAPRTEINSTRING*'
```

By default, this function does not stop translation at the stop codon:

```
>>> Seq.translate('AUGCCGUAAUCU')
'MP*S'
```

If you read `help(Seq.translate)` in the REPL, you'll find the `to_stop` option to
change this to the version expected by the Rosalind challenge:

```
>>> Seq.translate('AUGCCGUAAUCU', to_stop=True)
'MP'
```

Here is how I put it all together:

```
def main() -> None:
    args = get_args()
    print(Seq.translate(args.rna, to_stop=True))
```

This is the solution I would recommend because it relies on the widely used Biopython module. While it was fun and enlightening to explore how to manually code a solution, it's far better practice to use code that's already been written and tested by a dedicated team of developers.

Benchmarking

Which is the fastest solution? I can use the hyperfine benchmarking program I introduced in Chapter 4 to compare the runtimes of the programs. Because this is such a short program, I decided to run each program at least 1,000 times, as documented in the bench.sh program in the repository.

Although the second solution runs the fastest, perhaps as much at 1.5 times faster than the Biopython version, I'd still recommend using the latter because this is a thoroughly documented and tested module that is widely used in the community.

Going Further

Add a --frame-shift argument that defaults to 0 and allows values 0-2 (inclusive). Use the frameshift to start reading the RNA from an alternate position.

Review

The focus of this chapter was really on how to write, test, and compose functions to solve the problem at hand. I wrote functions to find codons in a sequence and translate RNA. Then I showed how to use higher-order functions to compose other functions, and finally, I used an off-the-shelf function from Biopython.

- K-mers are all the *k*-length subsequences of a sequence.
- Codons are 3-mers that do not overlap in a particular frame.
- Dictionaries are useful as lookup tables, such as for translating a codon to an amino acid.
- A for loop, a list comprehension, and map() are all methods for transforming one sequence into another.
- The takewhile() function is similar to the filter() function in accepting values from a sequence based on a predicate or test of the values.
- The partial() function allows one to partially apply the arguments to a function.
- The Bio.Seq.translate() function will translate an RNA sequence into a protein sequence.

## CHAPTER 8

Find a Motif in DNA:
Exploring Sequence Similarity

In the Rosalind SUBS challenge (*https://oreil.ly/hoUhB*), I'll be searching for any occurrences of one sequence inside another. A shared subsequence might represent a conserved element such as a marker, gene, or regulatory sequence. Conserved sequences between two organisms might suggest some inherited or convergent trait. I'll explore how to write a solution using the `str` (string) class in Python and will compare strings to lists. Then I'll explore how to express these ideas using higher-order functions and will continue the discussion of k-mers I started in Chapter 7. Finally, I'll show how regular expressions can find patterns and will point out problems with overlapping matches.

In this chapter, I'll demonstrate:

- How to use `str.find()`, `str.index()`, and string slices
- How to use sets to create unique collections of elements
- How to combine higher-order functions
- How to find subsequences using k-mers
- How to find possibly overlapping sequences using regular expressions

Getting Started

The code and tests for this chapter are in *08_subs*. I suggest you start by copying the first solution to the program `subs.py` and requesting help:

```
$ cd 08_subs/
$ cp solution1_str_find.py subs.py
```

```
$ ./subs.py -h
usage: subs.py [-h] seq subseq

Find subsequences

positional arguments:
  seq          Sequence
  subseq       subsequence

optional arguments:
  -h, --help   show this help message and exit
```

The program should report the starting locations where the subsequence can be found in the sequence. As shown in Figure 8-1, the subsequence *ATAT* can be found at positions 2, 4, and 10 in the sequence *GATATATGCATATACTT*:

```
$ ./subs.py GATATATGCATATACTT ATAT
2 4 10
```

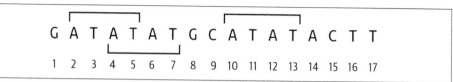

Figure 8-1. The subsequence ATAT can be found at positions 2, 4, and 10

Run the tests to see if you understand what will be expected, then start your program from scratch:

```
$ new.py -fp 'Find subsequences' subs.py
Done, see new script "subs.py".
```

Here is how I define the program's parameters:

```
class Args(NamedTuple): ❶
    """ Command-line arguments """
    seq: str
    subseq: str


def get_args() -> Args: ❷
    """ Get command-line arguments """

    parser = argparse.ArgumentParser(
        description='Find subsequences',
        formatter_class=argparse.ArgumentDefaultsHelpFormatter)

    parser.add_argument('seq', metavar='seq', help='Sequence')

    parser.add_argument('subseq', metavar='subseq', help='subsequence')

    args = parser.parse_args()
```

```
        return Args(args.seq, args.subseq)  ❸
```

❶ The Args class will have two string fields, seq and subseq.

❷ The function returns an Args object.

❸ Package and return the arguments using Args.

Have your main() print the sequence and subsequence:

```
def main() -> None:
    args = get_args()
    print(f'sequence = {args.seq}')
    print(f'subsequence = {args.subseq}')
```

Run the program with the expected inputs and verify that it prints the arguments correctly:

```
$ ./subs.py GATATATGCATATACTT ATAT
sequence = GATATATGCATATACTT
subsequence = ATAT
```

Now you have a program that should pass the first two tests. If you think you can finish this on your own, please proceed; otherwise, I'll show you one way to find the location of one string inside another.

Finding Subsequences

To demonstrate how to find the subsequence, I'll start by defining the following sequence and subsequence in the REPL:

```
>>> seq = 'GATATATGCATATACTT'
>>> subseq = 'ATAT'
```

I can use in to determine if one sequence is a subset of another. This also works for membership in lists, sets, or keys of a dictionary:

```
>>> subseq in seq
True
```

That's good information, but it doesn't tell me *where* the string can be found. Luckily there's the str.find() function, which says subseq can be found starting at index 1 (which is the second character):

```
>>> seq.find(subseq)
1
```

I know from the Rosalind description that the answer should be 2, 4, and 10. I just found 2, so how can I find the next? I can't just call the same function again because

I'll get the same answer. I need to look further into the sequence. Maybe help(str.find) could be of some use?

```
>>> help(str.find)
find(...)
    S.find(sub[, start[, end]]) -> int

    Return the lowest index in S where substring sub is found,
    such that sub is contained within S[start:end].  Optional
    arguments start and end are interpreted as in slice notation.

    Return -1 on failure.
```

It appears I can specify a *start* position. I'll use 1 greater than the position where the first subsequence was found, which was 1, so starting at 2:

```
>>> seq.find(subseq, 2)
3
```

Great. That's the next answer—well, 4 is the next answer, but you know what I mean. I'll try that again, this time starting at 4:

```
>>> seq.find(subseq, 4)
9
```

That was the last value I expected. What happens if I try using a start of 10? As the documentation shows, this will return -1 to indicate the subsequence cannot be found:

```
>>> seq.find(subseq, 10)
-1
```

Can you think of a way to iterate through the sequence, remembering the last position where the subsequence was found until it cannot be found?

Another option would be to use str.index(), but only if the subsequence is present:

```
>>> if subseq in seq:
...     seq.index(subseq)
...
1
```

To find the next occurrence, you could slice the sequence using the last known position. You'll have to add this position to the starting position, but you're essentially doing the same operation of moving further into the sequence to find if the subsequence is present and where:

```
>>> if subseq in seq[2:]:
...     seq.index(subseq[2:])
...
1
```

If you read **help(str.index)**, you'll see that, like str.find(), the function takes a second optional start position of the index at which to start looking:

```
>>> if subseq in seq[2:]:
...     seq.index(subseq, 2)
...
3
```

A third approach would be to use k-mers. If the subsequence is present, then it is by definition a k-mer, where *k* is the length of the subsequence. Use your code from Chapter 7 to extract all the k-mers and their positions from the sequence, and note the positions of the k-mers that match the subsequence.

Finally, since I'm looking for a pattern of text, I could use a regular expression. In Chapter 5, I used the re.findall() function to find all the Gs and Cs in DNA. I can similarly use this method to find all the subsequences in the sequence:

```
>>> import re
>>> re.findall(subseq, seq)
['ATAT', 'ATAT']
```

That seems to have a couple of problems. One is that it only returned two of the subsequences when I know there are three. The other problem is that this provides no information about *where* the matches are found. Fear not, the re.finditer() function solves this second problem:

```
>>> list(re.finditer(subseq, seq))
[<re.Match object; span=(1, 5), match='ATAT'>,
 <re.Match object; span=(9, 13), match='ATAT'>]
```

Now it's apparent that it finds the first and last subsequences. Why doesn't it find the second instance? It turns out regular expressions don't handle overlapping patterns very well, but some additions to the search pattern can fix this. I'll leave it to you and some internet searching to see if you can figure out a solution.

I've presented four different options for how to solve this problem. See if you can write solutions using each approach. The point is to explore the corners of Python, storing away tasty bits and tricks that might prove decisive in some future program you write. It's OK to spend hours or days working this out. Keep at it until you have solutions that pass both pytest and make test.

Solutions

All of the solutions share the same get_args() shown previously.

Solution 1: Using the str.find() Method

Here is my first solution using the str.find() method:

```
def main() -> None:
    args = get_args()
    last = 0 ❶
    found = [] ❷
    while True: ❸
        pos = args.seq.find(args.subseq, last) ❹
        if pos == -1: ❺
            break
        found.append(pos + 1) ❻
        last = pos + 1 ❼

    print(*found) ❽
```

❶ Initialize the last position to 0, the start of the sequence.

❷ Initialize a list to hold all the positions where the subsequence is found.

❸ Create an infinite loop using while.

❹ Use str.find() to look for the subsequence using the last known position.

❺ If the return is -1, the subsequence is not found, so exit the loop.

❻ Append one greater than the index to the list of found positions.

❼ Update the last known position with one greater than the found index.

❽ Print the found positions using * to expand the list into the elements. The function will use a space to separate multiple values.

This solution turns on keeping track of the *last* place the subsequence was found. I initialize this to 0:

```
>>> last = 0
```

I use this with str.find() to look for the subsequence starting at the last known position:

```
>>> seq = 'GATATATGCATATACTT'
>>> subseq = 'ATAT'
>>> pos = seq.find(subseq, last)
>>> pos
1
```

As long as `seq.find()` returns a value other than -1, I update the last position to one greater to search starting at the next character:

```
>>> last = pos + 1
>>> pos = seq.find(subseq, last)
>>> pos
3
```

Another call to the function finds the last instance:

```
>>> last = pos + 1
>>> pos = seq.find(subseq, last)
>>> pos
9
```

Finally, `seq.find()` returns -1 to indicate that the pattern can no longer be found:

```
>>> last = pos + 1
>>> pos = seq.find(subseq, last)
>>> pos
-1
```

This solution would be immediately understandable to someone with a background in the C programming language. It's a very *imperative* approach with lots of detailed logic for updating the state of the algorithm. *State* is how data in a program changes over time. For instance, properly updating and using the last known position is key to making this approach work. Later approaches use far less explicit coding.

Solution 2: Using the str.index() Method

This next solution is a variation that slices the sequence using the last known position:

```
def main() -> None:
    args = get_args()
    seq, subseq = args.seq, args.subseq    ❶
    found = []
    last = 0
    while subseq in seq[last:]:    ❷
        last = seq.index(subseq, last) + 1    ❸
        found.append(last)    ❹

    print(' '.join(map(str, found)))    ❺
```

❶ Unpack the sequence and subsequence.

❷ Ask if the subsequence appears in a slice of the sequence starting at the last found position. The `while` loop will execute as long as this condition is true.

❸ Use `str.index()` to get the starting position of the subsequence. The `last` variable gets updated by adding 1 to the subsequence index to create the next starting position.

❹ Append this position to the list of found positions.

❺ Use `map()` to coerce all the found integer positions to strings, then join them on spaces to print.

Here again, I rely on tracking the last place a subsequence was found. I start at position 0, or the beginning of the string:

```
>>> last = 0
>>> if subseq in seq[last:]:
...     last = seq.index(subseq, last) + 1
...
>>> last
2
```

The `while True` loop in the first solution is a common way to start an infinite loop. Here, the `while` loop will only execute as long as the subsequence is found in the slice of the sequence, meaning I don't have to manually decide when to `break` out of the loop:

```
>>> last = 0
>>> found = []
>>> while subseq in seq[last:]:
...     last = seq.index(subseq, last) + 1
...     found.append(last)
...
>>> found
[2, 4, 10]
```

The found positions, in this case, are a list of integer values. In the first solution, I used `*found` to splat the list and relied on `print()` to coerce the values to strings and join them on spaces. If instead I were to try to create a new string from `found` using `str.join()`, I would run into problems. The `str.join()` function joins many *strings* into a single string and so raises an exception when you give it nonstring values:

```
>>> ' '.join(found)
Traceback (most recent call last):
  File "<stdin>", line 1, in <module>
TypeError: sequence item 0: expected str instance, int found
```

I could use a list comprehension to turn each number n into a string using the `str()` function:

```
>>> ' '.join([str(n) for n in found])
'2 4 10'
```

This can also be written using a `map()`:

```
>>> ' '.join(map(lambda n: str(n), found))
'2 4 10'
```

I can leave out the `lambda` entirely because the `str()` function expects a single argument, and `map()` will naturally supply each value from `found` as the argument to `str()`. This is my preferred way to turn a list of integers into a list of strings:

```
>>> ' '.join(map(str, found))
'2 4 10'
```

Solution 3: A Purely Functional Approach

This next solution combines many of the preceding ideas using a purely functional approach. To start, consider the `while` loops in the first two solutions used to append nonnegative values to the `found` list. Does that sound like something a list comprehension could do? The range of values to iterate includes all the positions n from 0 to the end of the sequence minus the length of the subsequence:

```
>>> r = range(len(seq) - len(subseq))
>>> [n for n in r]
[0, 1, 2, 3, 4, 5, 6, 7, 8, 9, 10, 11, 12]
```

A list comprehension can use these values with `str.find()` to search for the subsequence in the sequence starting at each position n. Starting at positions 0 and 1, the subsequence can be found at index 1. Starting at positions 2 and 3, the subsequence can be found at index 3. This continues until -1 indicates the subsequence is not present for those positions n:

```
>>> [seq.find(subseq, n) for n in r]
[1, 1, 3, 3, 9, 9, 9, 9, 9, 9, -1, -1, -1]
```

I only want the nonnegative values, so I use `filter()` to remove them:

```
>>> list(filter(lambda n: n >= 0, [seq.find(subseq, n) for n in r]))
[1, 1, 3, 3, 9, 9, 9, 9, 9, 9]
```

Which could also be written by reversing the comparison in the `lambda`:

```
>>> list(filter(lambda n: 0 <= n, [seq.find(subseq, n) for n in r]))
[1, 1, 3, 3, 9, 9, 9, 9, 9, 9]
```

I show you this because I'd like to use `partial()` with the `operator.le()` (less than or equal) function because I don't like `lambda` expressions:

```
>>> from functools import partial
>>> import operator
>>> ok = partial(operator.le, 0)
>>> list(filter(ok, [seq.find(subseq, n) for n in r]))
[1, 1, 3, 3, 9, 9, 9, 9, 9, 9]
```

I'd like to change the list comprehension to a `map()`:

```
>>> list(filter(ok, map(lambda n: seq.find(subseq, n), r)))
[1, 1, 3, 3, 9, 9, 9, 9, 9, 9]
```

but again I want to get rid of the `lambda` by using `partial()`:

```
>>> find = partial(seq.find, subseq)
>>> list(filter(ok, map(find, r)))
[1, 1, 3, 3, 9, 9, 9, 9, 9, 9]
```

I can use `set()` to get a distinct list:

```
>>> set(filter(ok, map(find, r)))
{1, 3, 9}
```

These are almost the correct values, but they are the *index* positions, which are zero-based. I need the values one greater, so I can make a function to add 1 and apply this using a `map()`:

```
>>> add1 = partial(operator.add, 1)
>>> list(map(add1, set(filter(ok, map(find, r)))))
[2, 4, 10]
```

In these limited examples, the results are properly sorted; however, one can never rely on the order of values from a set. I must use the `sorted()` function to ensure they are properly sorted numerically:

```
>>> sorted(map(add1, set(filter(ok, map(find, r)))))
[2, 4, 10]
```

Finally, I need to print these values, which still exist as a list of integers:

```
>>> print(sorted(map(add1, set(filter(ok, map(find, r))))))
[2, 4, 10]
```

That's almost right. As in the first solution, I need to splat the results to get `print()` to see the individual elements:

```
>>> print(*sorted(map(add1, set(filter(ok, map(find, r))))))
2 4 10
```

That's a lot of closing parentheses. This code is starting to look a little like Lisp. If you combine all these ideas, you wind up with the same answer as the imperative solution but now by combining only functions:

```
def main() -> None:
    args = get_args()
    seq, subseq = args.seq, args.subseq ❶
    r = list(range(len(seq) - len(subseq))) ❷
    ok = partial(operator.le, 0) ❸
    find = partial(seq.find, subseq) ❹
    add1 = partial(operator.add, 1) ❺
    print(*sorted(map(add1, set(filter(ok, map(find, r)))))) ❻
```

❶ Unpack the sequence and subsequence.

❷ Generate a range of numbers up to the length of the sequence less the length of the subsequence.

❸ Create a partial ok() function that will return True if a given number is greater than or equal to 0.

❹ Create a partial find() function that will look for the subsequence in the sequence when provided with a start parameter.

❺ Create a partial add1() function that will return one greater than the argument.

❻ Apply all the numbers from the range to the find() function, filter out negative values, make the result unique by using the set() function, add one to the values, and sort the numbers before printing.

This solution uses only *pure* functions and would be fairly easy to understand for a person with a background in the Haskell programming language. If it seems like a jumble of confusion to you, I'd encourage you to spend some time working in the REPL with each piece until you understand how all these functions fit together perfectly.

Solution 4: Using K-mers

I mentioned that you might try finding the answer using k-mers, which I showed in Chapter 7. If the subsequence exists in the sequence, then it must be a k-mer, where *k* equals the length of the subsequence:

```
>>> seq = 'GATATATGCATATACTT'
>>> subseq = 'ATAT'
>>> k = len(subseq)
>>> k
4
```

Here are all the 4-mers in the sequence:

```
>>> kmers = [seq[i:i + k] for i in range(len(seq) - k + 1)]
>>> kmers
['GATA', 'ATAT', 'TATA', 'ATAT', 'TATG', 'ATGC', 'TGCA', 'GCAT', 'CATA',
 'ATAT', 'TATA', 'ATAC', 'TACT', 'ACTT']
```

Here are the 4-mers that are the same as the subsequence I'm looking for:

```
>>> list(filter(lambda s: s == subseq, kmers))
['ATAT', 'ATAT', 'ATAT']
```

I need to know the positions as well as the k-mers. The `enumerate()` function will return both the index and value of all the elements in a sequence. Here are the first four:

```
>>> kmers = list(enumerate([seq[i:i + k] for i in range(len(seq) - k + 1)]))
>>> kmers[:4]
[(0, 'GATA'), (1, 'ATAT'), (2, 'TATA'), (3, 'ATAT')]
```

I can use this with `filter()`, but now the `lambda` is receiving a tuple of the index and value so I will need to look at the second field (which is at index 1):

```
>>> list(filter(lambda t: t[1] == subseq, kmers))
[(1, 'ATAT'), (3, 'ATAT'), (9, 'ATAT')]
```

I really only care about getting the index for the matching k-mers. I could rewrite this using a `map()` with an `if` expression to return the index position when it's a match, and None otherwise:

```
>>> list(map(lambda t: t[0] if t[1] == subseq else None, kmers))
[None, 1, None, 3, None, None, None, None, None, 9, None, None, None, None]
```

I'm frustrated that the standard `map()` function can only pass a single value to the `lambda`. What I need is some way to splat the tuple, like `*t`, to turn it into two values. Luckily, I've studied the `itertools` module documentation and found the `starmap()` function, so named because it will add a *star* to the `lambda` argument to splat it. This allows me to unpack a tuple value like (0, 'GATA') into the variables `i` with the index value of 0 and `kmer` with the value 'GATA'. With these, I can compare the `kmer` to the subsequence and also add 1 to the index (`i`):

```
>>> from itertools import starmap
>>> list(starmap(lambda i, kmer: i + 1 if kmer == subseq else None, kmers))
[None, 2, None, 4, None, None, None, None, None, 10, None, None, None, None]
```

This probably seems like an odd choice until I show you that `filter()`, if passed None for the `lambda`, will use the truthiness of each value, so that None values will be excluded. Because this line of code is getting rather long, I'll write the function `f()` for `map()` on a separate line:

```
>>> f = lambda i, kmer: i + 1 if kmer == subseq else None
>>> list(filter(None, starmap(f, kmers)))
[2, 4, 10]
```

I can express a k-mer solution using imperative techniques:

```
def main() -> None:
    args = get_args()
    seq, subseq = args.seq, args.subseq
    k = len(subseq) ❶
    kmers = [seq[i:i + k] for i in range(len(seq) - k + 1)] ❷
    found = [i + 1 for i, kmer in enumerate(kmers) if kmer == subseq] ❸
    print(*found) ❹
```

❶ When looking for k-mers, k is the length of the subsequence.

❷ Use a list comprehension to generate all the k-mers from the sequence.

❸ Iterate through the index and value of all the k-mers, where the k-mer is equal to the subsequence. Return one greater than the index position.

❹ Print the found positions.

I can also express these ideas using purely functional techniques. Note that mypy insists on a type annotation for the found variable:

```
def main() -> None:
    args = get_args()
    seq, subseq = args.seq, args.subseq
    k = len(subseq)
    kmers = enumerate(seq[i:i + k] for i in range(len(seq) - k + 1)) ❶
    found: Iterator[int] = filter( ❷
        None, starmap(lambda i, kmer: i + 1 if kmer == subseq else None, kmers))
    print(*found) ❸
```

❶ Generate an enumerated list of the k-mers.

❷ Select the positions of those k-mers equal to the subsequence.

❸ Print the results.

I find the imperative version easier to read, but would recommend you use whichever you find most intuitive. Whichever solution you prefer, the interesting point is that k-mers can prove extremely useful in many situations, such as for partial sequence comparison.

Solution 5: Finding Overlapping Patterns Using Regular Expressions

To this point, I've been writing fairly complex solutions to find a pattern of characters inside a string. This is precisely the domain of regular expressions, and so it's a bit silly to write manual solutions. I showed earlier in this chapter that the re.finditer() function does not find overlapping matches and so returns just two hits when I know there are three:

```
>>> import re
>>> list(re.finditer(subseq, seq))
[<re.Match object; span=(1, 5), match='ATAT'>,
 <re.Match object; span=(9, 13), match='ATAT'>]
```

I'm going to show you that the solution is quite simple, but I want to stress that I did not know the solution until I searched the internet. The key to finding the answer was knowing what search terms to use—something like *regex overlapping patterns* turns up several useful results. The point of this aside is that no one knows all the answers, and you will constantly be searching for solutions to problems you never knew even existed. It's not what you know that's important, but what you can learn.

The problem turns out to be that the regex engine *consumes* strings as they match. That is, once the engine matches the first *ATAT*, it starts searching again at the end of the match. The solution is to wrap the search pattern in a *look-ahead assertion* using the syntax ?=(*<pattern>*) so that the engine won't consume the matched string. Note that this is a *positive* look-ahead; there are also *negative* look-ahead assertions as well as both positive and negative look-behind assertions.

So if the subsequence is *ATAT*, then I want the pattern to be ?=(ATAT). The problem now is that the regex engine won't *save* the match—I've just told it to look for this pattern but haven't told it to do anything with the text that is found. I need to further wrap the assertion in parentheses to create a *capture group*:

```
>>> list(re.finditer('(?=(ATAT))', 'GATATATGCATATACTT'))
[<re.Match object; span=(1, 1), match=''>,
 <re.Match object; span=(3, 3), match=''>,
 <re.Match object; span=(9, 9), match=''>]
```

I can use a list comprehension over this iterator to call the `match.start()` function on each of the `re.Match` objects, adding 1 to correct the position:

```
>>> [match.start() + 1 for match in re.finditer(f'(?=({subseq}))', seq)]
[2, 4, 10]
```

Here is the final solution that I would suggest as the best way to solve this problem:

```
def main() -> None:
    args = get_args()
    seq, subseq = args.seq, args.subseq
    print(*[m.start() + 1 for m in re.finditer(f'(?=({subseq}))', seq)])
```

Benchmarking

It's always interesting for me to see which solution runs the fastest. I'll use `hyperfine` again to run each version 1,000 times:

```
$ hyperfine -m 1000 -L prg ./solution1_str_find.py,./solution2_str_index.py,\
./solution3_functional.py,./solution4_kmers_functional.py,\
./solution4_kmers_imperative.py,./solution5_re.py \
'{prg} GATATATGCATATACTT ATAT' --prepare 'rm -rf __pycache__'
...
```

```
Summary
  './solution2_str_index.py GATATATGCATATACTT ATAT' ran
    1.01 ± 0.11 times faster than
        './solution4_kmers_imperative.py GATATATGCATATACTT ATAT'
    1.02 ± 0.14 times faster than
        './solution5_re.py GATATATGCATATACTT ATAT'
    1.02 ± 0.14 times faster than
        './solution3_functional.py GATATATGCATATACTT ATAT'
    1.03 ± 0.13 times faster than
        './solution4_kmers_functional.py GATATATGCATATACTT ATAT'
    1.09 ± 0.18 times faster than
        './solution1_str_find.py GATATATGCATATACTT ATAT'
```

The differences aren't significant enough, in my opinion, to sway me to choose based solely on performance. My preference would be to use regular expressions given that they are specifically designed to find patterns of text.

Going Further

Expand the program to look for a subsequence *pattern*. For example, you might search for simple sequence repeats (also known as *SSRs* or microsatellites) such as GA(26), which would mean "*GA* repeated 26 times." Or a repeat such as (GA)15GT(GA)2, which means "*GA* repeated 15 times, followed by *GT*, followed by *GA*, repeated 2 times." Also, consider how you might find subsequences expressed using the IUPAC codes mentioned in Chapter 1. For instance, *R* represents either *A* or *G*, so *ARC* can match the sequences *AAC* and *AGC*.

Review

Key points from this chapter:

- The str.find() and str.index() methods can determine if a subsequence is present in a given string.

- Sets can be used to create unique collections of elements.

- By definition, k-mers are subsequences and are relatively quick to extract and compare.

- Regular expressions can find overlapping sequences by using a look-ahead assertion combined with a capture group.

## CHAPTER 9

Overlap Graphs: Sequence Assembly Using Shared K-mers

A *graph* is a structure used to represent pairwise relationships between objects. As described in the Rosalind GRPH challenge (*https://oreil.ly/kDu52*), the goal of this exercise is to find pairs of sequences that can be joined using an overlap from the end of one sequence to the beginning of another. The practical application of this would be to join short DNA reads into longer contiguous sequences (*contigs*) or even whole genomes. To begin, I'll only be concerned about joining two sequences, but a second version of the program will use a graph structure that can join any number of sequences to approximate a complete assembly. In this implementation, the overlapping regions used to join sequences are required to be exact matches. Real-world assemblers must allow for variation in the size and composition of the overlapping sequences.

You will learn:

- How to use k-mers to create overlap graphs
- How to log runtime messages to a file
- How to use `collections.defaultdict()`
- How to use set intersection to find common elements between collections
- How to use `itertools.product()` to create Cartesian products of lists
- How to use the `iteration_utilities.starfilter()` function
- How to use Graphviz to model and visualize graph structures

Getting Started

The code and tests for this exercise are in the *09_grph* directory. Start by copying one of the solutions to the program grph.py and requesting the usage:

```
$ cd 09_grph/
$ cp solution1.py grph.py
$ ./grph.py -h
usage: grph.py [-h] [-k size] [-d] FILE

Overlap Graphs

positional arguments:
  FILE                  FASTA file ❶

optional arguments:
  -h, --help            show this help message and exit
  -k size, --overlap size
                        Size of overlap (default: 3) ❷
  -d, --debug           Debug (default: False) ❸
```

❶ The positional parameter is a required FASTA-formatted file of sequences.

❷ The -k option controls the length of the overlapping strings and defaults to 3.

❸ This is a *flag* or Boolean parameter. Its value will be True when the argument is present, and False otherwise.

The sample input shown on the Rosalind page is also the content of the first sample input file:

```
$ cat tests/inputs/1.fa
>Rosalind_0498
AAATAAA
>Rosalind_2391
AAATTTT
>Rosalind_2323
TTTTCCC
>Rosalind_0442
AAATCCC
>Rosalind_5013
GGGTGGG
```

The Rosalind problem always assumes an overlapping window of three bases. I see no reason for this parameter to be hardcoded, so my version includes a k parameter to indicate the size of the overlap window. When k is the default value of 3, for instance, three pairs of the sequences can be joined:

```
$ ./grph.py tests/inputs/1.fa
Rosalind_2391 Rosalind_2323
```

```
Rosalind_0498 Rosalind_2391
Rosalind_0498 Rosalind_0442
```

Figure 9-1 shows how these sequences overlap by three common bases.

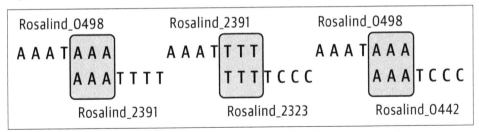

Figure 9-1. Three pairs of sequences form overlap graphs when joining on 3-mers

As shown in Figure 9-2, only one of these pairs can be joined when the overlap window increases to four bases:

```
$ ./grph.py -k 4 tests/inputs/1.fa
Rosalind_2391 Rosalind_2323
```

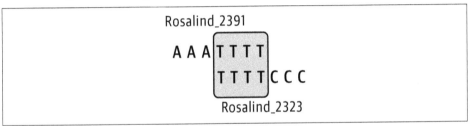

Figure 9-2. Only one pair of sequences forms an overlap graph when joining on 4-mers

Finally, the --debug option is a *flag*, which is a Boolean parameter that has a True value when the argument is present and False otherwise. When present, this option instructs the program to print runtime logging messages to a file called *.log* in the current working directory. This is not a requirement of the Rosalind challenge, but I think it's important for you to know how to log messages. To see it in action, run the program with the option:

```
$ ./grph.py tests/inputs/1.fa --debug
Rosalind_2391 Rosalind_2323
Rosalind_0498 Rosalind_2391
Rosalind_0498 Rosalind_0442
```

> The --debug flag can be placed before or after the positional argument, and argparse will correctly interpret its meaning. Other argument parsers require that all options and flags come before positional arguments. *Vive la différence.*

There should now be a *.log* file with the following contents, the meaning of which will become more apparent later:

```
$ cat .log
DEBUG:root:STARTS
defaultdict(<class 'list'>,
            {'AAA': ['Rosalind_0498', 'Rosalind_2391', 'Rosalind_0442'],
             'GGG': ['Rosalind_5013'],
             'TTT': ['Rosalind_2323']})
DEBUG:root:ENDS
defaultdict(<class 'list'>,
            {'AAA': ['Rosalind_0498'],
             'CCC': ['Rosalind_2323', 'Rosalind_0442'],
             'GGG': ['Rosalind_5013'],
             'TTT': ['Rosalind_2391']})
```

Once you understand how your program should work, start over with a new `grph.py` program:

```
$ new.py -fp 'Overlap Graphs' grph.py
Done, see new script "grph.py".
```

Here is how I define and validate the arguments:

```
from typing import List, NamedTuple, TextIO


class Args(NamedTuple):  ❶
    """ Command-line arguments """
    file: TextIO
    k: int
    debug: bool


# --------------------------------------------------
def get_args() -> Args:
    """ Get command-line arguments """

    parser = argparse.ArgumentParser(
        description='Overlap Graphs',
        formatter_class=argparse.ArgumentDefaultsHelpFormatter)

    parser.add_argument('file',  ❷
                        metavar='FILE',
                        type=argparse.FileType('rt'),
                        help='FASTA file')

    parser.add_argument('-k',  ❸
                        '--overlap',
                        help='Size of overlap',
                        metavar='size',
                        type=int,
                        default=3)
```

```
parser.add_argument('-d', '--debug', help='Debug', action='store_true') ❹

args = parser.parse_args()

if args.overlap < 1: ❺
    parser.error(f'-k "{args.overlap}" must be > 0') ❻

return Args(args.file, args.overlap, args.debug) ❼
```

❶ The `Args` class contains three fields: a `file` which is a filehandle; a `k` which should be a positive integer; and `debug`, which is a Boolean value.

❷ Use the `argparse.FileType` to ensure this is a readable text file.

❸ Define an integer argument that defaults to 3.

❹ Define a Boolean flag that will store a `True` value when present.

❺ Check if the `k` (overlap) value is negative.

❻ Use `parser.error()` to kill the program and generate a useful error message.

❼ Return the validated arguments.

I would like to stress how much is happening in these lines to ensure that the arguments to the program are correct. Argument values should be validated *as soon as possible* after the program starts. I've encountered too many programs that, for instance, never validate a file argument and then, deep in the bowels of the program, attempt to open a nonexistent file and wind up throwing a cryptic exception that no mere mortal could debug. If you want *reproducible* programs, the first order of business is documenting and validating all the arguments.

Modify your `main()` to the following:

```
def main() -> None:
    args = get_args()
    print(args.file.name)
```

Run your program with the first test input file and verify that you see this:

```
$ ./grph.py tests/inputs/1.fa
tests/inputs/1.fa
```

Try running your program with invalid values for `k` and the file input, then run **pytest** to verify that your program passes the first four tests. The failing test expects three pairs of sequence IDs that can be joined, but the program printed the name of the input file. Before I talk about how to create overlap graphs, I want to introduce *logging* as this can prove useful for debugging a program.

Managing Runtime Messages with STDOUT, STDERR, and Logging

I've shown how to print strings and data structures to the console. You just did it by printing the input filename to verify that the program is working. Printing such messages while writing and debugging a program might be facetiously called *log-driven development*. This is a simple and effective way to debug a program going back decades.[1]

By default, print() will emit messages to STDOUT (*standard out*), which Python represents using sys.stdout. I can use the print() function's file option to change this to STDERR (*standard error*) by indicating sys.stderr. Consider the following Python program:

```
$ cat log.py
#!/usr/bin/env python3

import sys

print('This is STDOUT.') ❶
print('This is also STDOUT.', file=sys.stdout) ❷
print('This is STDERR.', file=sys.stderr) ❸
```

❶ The default file value is STDOUT.

❷ I can specify standard out using the file option.

❸ This will print messages to standard error.

When I run this, it would appear that all output is printed to standard out:

```
$ ./log.py
This is STDOUT.
This is also STDOUT.
This is STDERR.
```

In the bash shell, I can separate and capture the two streams, however, using file redirection with >. Standard out can be captured using the filehandle 1 and standard error using 2. If you run the following command, you should see no output on the console:

---

1 Imagine debugging a program without even a console. In the 1950s, Claude Shannon was visiting Alan Turing's lab in England. During their conversation, a horn began sounding at regular intervals. Turing said this indicated his code was stuck in a loop. Without a console, this was how he monitored the progress of his programs.

```
$ ./log.py 1>out 2>err
```

There should now be two new files, one called *out* with the two lines that were printed to standard out:

```
$ cat out
This is STDOUT.
This is also STDOUT.
```

and another called *err* with the one line printed to standard error:

```
$ cat err
This is STDERR.
```

Just knowing how to print to and capture these two filehandles may prove sufficient for your debugging efforts. However, there may be times when you want more levels of printing than two, and you may want to control where these messages are written from your code rather than by using shell redirection. Enter *logging*, a way to control whether, when, how, and where runtime messages are printed. The Python `logging` module handles all of this, so start by importing this module:

```
import logging
```

For this program, I'll print debugging messages to a file called *.log* (in the current working directory) if the --debug flag is present. Modify your `main()` to this:

```
def main() -> None:
    args = get_args()

    logging.basicConfig( ❶
        filename='.log', ❷
        filemode='w', ❸
        level=logging.DEBUG if args.debug else logging.CRITICAL) ❹

    logging.debug('input file = "%s"', args.file.name) ❺
```

❶ This will *globally* affect all subsequent calls to the `logging` module's functions.

❷ All output will be written to the file *.log* in the current working directory. I chose a filename starting with a dot so that it will normally be hidden from view.

❸ The output file will be opened with the w (*write*) option, meaning it will be *overwritten* on each run. Use the a mode to *append*, but be warned that the file will grow for every run and will never be truncated or removed except by you.

❹ This sets the minimum logging level (see Table 9-1). Messages at any level below the set level will be ignored.

❺ Use the `logging.debug()` function to print a message to the log file when the logging level is set to DEBUG or higher.

In the previous example, I used the older `printf()` style of formatting for the call to `logging.debug()`. The placeholders are noted with symbols like `%s` for a string, and the values to substitute are passed as arguments. You can also use `str.format()` and f-strings for the log message, but `pylint` may suggest you use the `printf()` style.

Dotfiles

Files and directories with names starting with a dot are normally hidden when you use **ls**. You must use the `-a` option to `ls` to see *all* files. I'm naming the log file *.log* so I won't normally see it. You may also notice a *.gitignore* file in this directory. This file contains filenames and patterns of files and directories I do *not* want Git to add to my repo. Included is the *.log* file. Whenever you want to be sure data like configuration files, passwords, large sequence files, etc. will not be included by **git add**, put their names (or file globs that would match them) in this file.

A key concept to logging is the notion of logging levels. As shown in Table 9-1, the *critical* level is the highest, and the *debug* level is the lowest (the *notset* level has certain particularities). To learn more, I recommend you read **help(logging)** in the REPL or the module's online documentation (*https://oreil.ly/bWgOp*). For this program, I'll only use the lowest (debug) setting. When the `--debug` flag is present, the logging level is set to `logging.DEBUG` and all messages to `logging.debug()` are printed in the log file. When the flag is absent, the logging level is set to `logging.CRITICAL` and only messages logged with `logging.critical()` will pass through. You might think I should use the `logging.NOTSET` value, but note that this is lower than `logging.DEBUG` and so all debug messages would pass through.

Table 9-1. The logging levels available in Python's `logging` module

Level	Numeric value
CRITICAL	50
ERROR	40
WARNING	30
INFO	20
DEBUG	10
NOTSET	0

To see this in action, run your program as follows:

```
$ ./grph.py --debug tests/inputs/1.fa
```

It would appear the program did nothing, but there should now be a *.log* file with the following contents:

```
$ cat .log
DEBUG:root:input file = "tests/inputs/1.fa"
```

Run the program again without the --debug flag, and note that the *.log* file is empty as it was overwritten when opened but no content was ever logged. If you were to use the typical print-based debugging technique, then you'd have to find and remove (or comment out) all the print() statements in your program to turn off your debugging. If, instead, you use logging.debug(), then you can debug your program while logging at the debug level and then deploy your program to only log critical messages. Further, you can write the log messages to various locations depending on the environment, and all of this happens *programmatically* inside your code rather than relying on shell redirection to put log messages into the right place.

There are no tests to ensure your program creates log files. This is only to show you how to use logging. Note that calls to functions like logging.critical() and logging.debug() are controlled by the *global* scope of the logging module. I don't generally like programs to be controlled by global settings, but this is one exception I'll make, mostly because I don't have a choice. I encourage you to liberally sprinkle logging.debug() calls throughout your code to see the kinds of output you can generate. Consider how you could use logging while writing a program on your laptop versus deploying it to a remote computing cluster to run unattended.

Finding Overlaps

The next order of business is to read the input FASTA file. I first showed how to do this in Chapter 5. Again I'll use the Bio.SeqIO module for this by adding the following import:

```
from Bio import SeqIO
```

I can modify main() to the following (omitting any logging calls):

```
def main() -> None:
    args = get_args()

    for rec in SeqIO.parse(args.file, 'fasta'):
        print(rec.id, rec.seq)
```

And then run this on the first input file to ensure the program works properly:

```
$ ./grph.py tests/inputs/1.fa
Rosalind_0498 AAATAAA
Rosalind_2391 AAATTTT
Rosalind_2323 TTTTCCC
Rosalind_0442 AAATCCC
Rosalind_5013 GGGTGGG
```

In each exercise, I try to show how to write a program logically, step-by-step. I want you to learn to make very small changes to your program with some end goal in mind, then run your program to see the output. You should run the tests often to see what needs to be fixed, adding your own tests as you see fit. Also, consider making frequent commits of the program when it's working well so you can revert if you end up breaking it. Taking small steps and running your program often are key elements to learning to code.

Now think about how you might get the first and last *k* bases from each sequence. Could you use the code for extracting k-mers that I first showed in Chapter 7? For instance, try to get your program to print this:

```
$ ./grph.py tests/inputs/1.fa
Rosalind_0498 AAATAAA first AAA last AAA
Rosalind_2391 AAATTTT first AAA last TTT
Rosalind_2323 TTTTCCC first TTT last CCC
Rosalind_0442 AAATCCC first AAA last CCC
Rosalind_5013 GGGTGGG first GGG last GGG
```

Think about which *first* strings match which *end* strings. For instance, sequence 0498 ends with *AAA*, and sequence 0442 starts with *AAA*. These are sequences that can be joined into an overlap graph.

Change the value of k to 4:

```
$ ./grph.py tests/inputs/1.fa -k 4
Rosalind_0498 AAATAAA first AAAT last TAAA
Rosalind_2391 AAATTTT first AAAT last TTTT
Rosalind_2323 TTTTCCC first TTTT last TCCC
Rosalind_0442 AAATCCC first AAAT last TCCC
Rosalind_5013 GGGTGGG first GGGT last TGGG
```

Now you can see that only two sequences, 2391 and 2323, can be joined by their overlapping sequence *TTTT*. Vary k from 1 to 10 and examine the first and last regions. Do you have enough information to write a solution? If not, let's keep thinking about this.

Grouping Sequences by the Overlap

The for loop reads the sequences individually. While reading any one sequence to find the starting and ending overlap regions, I necessarily do not have enough information to say which other sequences can be joined. I'm going to have to create some data structure to hold the overlapping regions for *all* the sequences. Only then can I go back and figure out which ones can be joined. This gets at a key element of sequence assemblers—most need prodigious amounts of memory to gather all the information needed from all the input sequences, of which there may be millions to billions.

I chose to use two dictionaries, one for the *start* and one for the *end* regions. I decided the keys would be the *k*-length sequences, like *AAA* when k is 3, and the values would be a list of the sequence IDs sharing this region. I can use string slices with the value k to extract these leading and trailing sequences:

```
>>> k = 3
>>> seq = 'AAATTTT'
>>> seq[:k] ❶
'AAA'
>>> seq[-k:] ❷
'TTT'
```

❶ A slice of the first k bases.

❷ A slice of the last k bases, using negative indexing to start from the end of the sequence.

These are k-mers, which I used in the last chapter. They keep showing up, so it makes sense to write a find_kmers() function to extract k-mers from a sequence. I'll start by defining the function's signature:

```
def find_kmers(seq: str, k: int) -> List[str]: ❶
    """ Find k-mers in string """

    return [] ❷
```

❶ The function will accept a string (the sequence) and an integer value k and will return a list of strings.

❷ For now, return the empty list.

Now I write a test to imagine how I'd use this function:

```
def test_find_kmers() -> None:
    """Test find_kmers"""

    assert find_kmers('', 1) == [] ❶
    assert find_kmers('ACTG', 1) == ['A', 'C', 'T', 'G'] ❷
    assert find_kmers('ACTG', 2) == ['AC', 'CT', 'TG']
    assert find_kmers('ACTG', 3) == ['ACT', 'CTG']
    assert find_kmers('ACTG', 4) == ['ACTG']
    assert find_kmers('ACTG', 5) == [] ❸
```

❶ Pass the empty string as the sequence to ensure the function returns the empty list.

❷ Check all the values for k using a short sequence.

❸ There are no 5-mers for a string of length 4.

Try writing your version before you read ahead. Here is the function I wrote:

```
def find_kmers(seq: str, k: int) -> List[str]:
    """Find k-mers in string"""

    n = len(seq) - k + 1 ❶
    return [] if n < 1 else [seq[i:i + k] for i in range(n)] ❷
```

❶ Find the number n of *k*-length substrings in a string `seq`.

❷ If n is a negative number, return the empty list; otherwise, return the k-mers using a list comprehension.

Now I have a handy way to get the leading and trailing k-mers from a sequence:

```
>>> from grph import find_kmers
>>> kmers = find_kmers('AAATTTT', 3)
>>> kmers
['AAA', 'AAT', 'ATT', 'TTT', 'TTT']
>>> kmers[0] ❶
'AAA'
>>> kmers[-1] ❷
'TTT'
```

❶ The first element is the leading k-mer.

❷ The last element is the trailing k-mer.

The first and last k-mers give me the overlap sequences I need for the keys of my dictionary. I want the values of the dictionaries to be a list of the sequence IDs that share these k-mers. The `collections.defaultdict()` function I introduced in Chapter 1 is a good one to use for this because it allows me to easily instantiate each dictionary entry with an empty list. I need to import it and the `pprint.pformat()` function for logging purposes, so I add the following:

```
from collections import defaultdict
from pprint import pformat
```

Here is how I can use these ideas:

```
def main() -> None:
    args = get_args()

    logging.basicConfig(
        filename='.log',
        filemode='w',
        level=logging.DEBUG if args.debug else logging.CRITICAL)

    start, end = defaultdict(list), defaultdict(list) ❶
    for rec in SeqIO.parse(args.file, 'fasta'): ❷
        if kmers := find_kmers(str(rec.seq), args.k): ❸
```

```
            start[kmers[0]].append(rec.id)  ❹
            end[kmers[-1]].append(rec.id)  ❺

     logging.debug(f'STARTS\n{pformat(start)}')  ❻
     logging.debug(f'ENDS\n{pformat(end)}')
```

❶ Create dictionaries for the start and end regions that will have lists as the default values.

❷ Iterate the FASTA records.

❸ Coerce the Seq object to a string and find the k-mers. The := syntax assigns the return value to kmers, then the if evaluates if kmers is truthy. If the function returns no kmers, then the following block will not execute.

❹ Use the first k-mer as a key into the start dictionary and append this sequence ID to the list.

❺ Do likewise for the end dictionary using the last k-mer.

❻ Use the pprint.pformat() function to format the dictionaries for logging.

I've used the pprint.pprint() function in earlier chapters to print complex data structures in a prettier format than the default print() function. I can't use pprint() here because it would print to STDOUT (or STDERR). Instead, I need to format the data structure for the logging.debug() function to log.

Now run the program again with the first input and the --debug flag, then inspect the log file:

```
$ ./grph.py tests/inputs/1.fa -d
$ cat .log
DEBUG:root:STARTS  ❶
defaultdict(<class 'list'>,
            {'AAA': ['Rosalind_0498', 'Rosalind_2391', 'Rosalind_0442'],  ❷
             'GGG': ['Rosalind_5013'],
             'TTT': ['Rosalind_2323']})
DEBUG:root:ENDS  ❸
defaultdict(<class 'list'>,
            {'AAA': ['Rosalind_0498'],  ❹
             'CCC': ['Rosalind_2323', 'Rosalind_0442'],
             'GGG': ['Rosalind_5013'],
             'TTT': ['Rosalind_2391']})
```

❶ A dictionary of the various starting sequences and the IDs.

❷ Three sequences start with *AAA*: 0498, 2391, and 0442.

❸ A dictionary of the various ending sequences and the IDs.

❹ There is just one sequence ending with *AAA*, which is 0498.

The correct pairs for this input file and the overlapping 3-mers are as follows:

- Rosalind_0498, Rosalind_2391: *AAA*
- Rosalind_0498, Rosalind_0442: *AAA*
- Rosalind_2391, Rosalind_2323: *TTT*

When you combine, for instance, the sequence ending with *AAA* (0498) with those starting with this sequence (0498, 2391, 0442), you wind up with the following pairs:

- Rosalind_0498, Rosalind_0498
- Rosalind_0498, Rosalind_2391
- Rosalind_0498, Rosalind_0442

Since I can't join a sequence to itself, the first pair is disqualified. Find the next *end* and *start* sequence in common, then iterate all the sequence pairs. I'll leave you to finish this exercise by finding all the start and end keys that are in common and then combining all the sequence IDs to print the pairs that can be joined. The pairs can be in any order and still pass the tests. I just want to wish you good luck. We're all counting on you.

Solutions

I have two variations to share with you. The first solves the Rosalind problem to show how to combine any two sequences. The second extends the graphs to create a full assembly of all the sequences.

Solution 1: Using Set Intersections to Find Overlaps

In the following solution, I introduce how to use set intersections to find the k-mers shared between the start and end dictionaries:

```
def main() -> None:
    args = get_args()

    logging.basicConfig(
        filename='.log',
        filemode='w',
        level=logging.DEBUG if args.debug else logging.CRITICAL)

    start, end = defaultdict(list), defaultdict(list)
    for rec in SeqIO.parse(args.file, 'fasta'):
```

```
            if kmers := find_kmers(str(rec.seq), args.k):
                start[kmers[0]].append(rec.id)
                end[kmers[-1]].append(rec.id)

        logging.debug('STARTS\n{}'.format(pformat(start)))
        logging.debug('ENDS\n{}'.format(pformat(end)))

        for kmer in set(start).intersection(set(end)): ❶
            for pair in starfilter(op.ne, product(end[kmer], start[kmer])): ❷
                print(*pair) ❸
```

❶ Find the keys in common between the start and end dictionaries.

❷ Iterate through the pairs of the ending and starting sequences that are not equal to each other.

❸ Print the pair of sequences.

The last three lines took me a few attempts to write, so let me explain how I got there. Given these dictionaries:

```
>>> from pprint import pprint
>>> from Bio import SeqIO
>>> from collections import defaultdict
>>> from grph import find_kmers
>>> k = 3
>>> start, end = defaultdict(list), defaultdict(list)
>>> for rec in SeqIO.parse('tests/inputs/1.fa', 'fasta'):
...     if kmers := find_kmers(str(rec.seq), k):
...         start[kmers[0]].append(rec.id)
...         end[kmers[-1]].append(rec.id)
...
>>> pprint(start)
{'AAA': ['Rosalind_0498', 'Rosalind_2391', 'Rosalind_0442'],
 'GGG': ['Rosalind_5013'],
 'TTT': ['Rosalind_2323']}
>>> pprint(end)
{'AAA': ['Rosalind_0498'],
 'CCC': ['Rosalind_2323', 'Rosalind_0442'],
 'GGG': ['Rosalind_5013'],
```

I started with this idea:

```
>>> for kmer in end: ❶
...     if kmer in start: ❷
...         for seq_id in end[kmer]: ❸
...             for other in start[kmer]: ❹
...                 if seq_id != other: ❺
...                     print(seq_id, other) ❻
...
Rosalind_0498 Rosalind_2391
```

```
Rosalind_0498 Rosalind_0442
Rosalind_2391 Rosalind_2323
```

❶ Iterate over the k-mers (which are the *keys*) of the end dictionary.

❷ See if this k-mer is in the start dictionary.

❸ Iterate through each ending sequence ID for this k-mer.

❹ Iterate through each starting sequence ID for this k-mer.

❺ Make sure the sequences are not the same.

❻ Print the sequence IDs.

While that works just fine, I let this sit for a while and came back to it, asking myself exactly what I was trying to do. The first two lines are trying to find the keys that are in common between the two dictionaries. Set intersection is an easier way to achieve this. If I use the set() function on a dictionary, it creates a set using the keys of the dictionary:

```
>>> set(start)
{'TTT', 'GGG', 'AAA'}
>>> set(end)
{'TTT', 'CCC', 'AAA', 'GGG'}
```

I can then call the set.intersection() function to find the keys in common:

```
>>> set(start).intersection(set(end))
{'TTT', 'GGG', 'AAA'}
```

In the preceding code, the next lines find all the combinations of the ending and starting sequence IDs. This is more easily done using the itertools.product() function, which will create the Cartesian product of any number of lists. For example, consider the sequences that overlap on the k-mer *AAA*:

```
>>> from itertools import product
>>> kmer = 'AAA'
>>> pairs = list(product(end[kmer], start[kmer]))
>>> pprint(pairs)
[('Rosalind_0498', 'Rosalind_0498'),
 ('Rosalind_0498', 'Rosalind_2391'),
 ('Rosalind_0498', 'Rosalind_0442')]
```

I want to exclude any pairs where the two values are the same. I could write a filter() for this:

```
>>> list(filter(lambda p: p[0] != p[1], pairs)) ❶
[('Rosalind_0498', 'Rosalind_2391'), ('Rosalind_0498', 'Rosalind_0442')]
```

❶ The lambda receives the pair p and checks that the zeroth and first elements are not equal.

This works adequately, but I'm not satisfied with the code. I really hate that I can't unpack the tuple values in the lambda to filter(). Immediately I started thinking about how the itertools.starmap() function I used in Chapters 6 and 8 can do this, so I searched the internet for *Python starfilter* and found the function itera tion_utilities.starfilter() (*https://oreil.ly/c6KKV*). I installed this module and imported the function:

```
>>> from iteration_utilities import starfilter
>>> list(starfilter(lambda a, b: a != b, pairs))
[('Rosalind_0498', 'Rosalind_2391'), ('Rosalind_0498', 'Rosalind_0442')]
```

This is an improvement, but I can make it cleaner by using the operator.ne() (not equal) function, which will obviate the lambda:

```
>>> import operator as op
>>> list(starfilter(op.ne, pairs))
[('Rosalind_0498', 'Rosalind_2391'), ('Rosalind_0498', 'Rosalind_0442')]
```

Finally, I splat each of the pairs to make print() see the individual strings rather than the list container:

```
>>> for pair in starfilter(op.ne, pairs):
...     print(*pair)
...
Rosalind_0498 Rosalind_2391
Rosalind_0498 Rosalind_0442
```

I could have shortened this even more, but I fear this gets a little too dense:

```
>>> print('\n'.join(map(' '.join, starfilter(op.ne, pairs))))
Rosalind_0498 Rosalind_2391
Rosalind_0498 Rosalind_0442
```

In the end, there's a fair amount of code in the main() function that, in a larger program, I would probably move to a function with a unit test. In this case, the integration tests cover all the functionality, so it would be overkill.

Solution 2: Using a Graph to Find All Paths

This next solution approximates a full assembly of the sequences using a graph to link all the overlapping sequences. While not part of the original challenge, it is, nonetheless, interesting to contemplate while also proving surprisingly simple to implement and even visualize. Since *GRPH* is the challenge name, it makes sense to investigate how to represent a graph in Python code.

I can manually align all the sequences as shown in Figure 9-3. This reveals a graph structure where sequence Rosalind_0498 can join to either Rosalind_2391 or Rosalind_0442, and there follows a chain from Rosalind_0498 to Rosalind_2391 to Rosalind_2323.

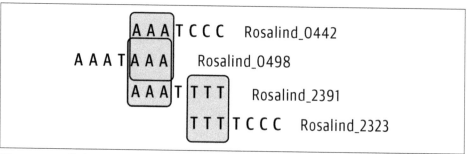

Figure 9-3. All the sequences in the first input file can be joined using 3-mers

To encode this, I use the Graphviz tool (*https://graphviz.org*) both to represent and to visualize a graph structure. Note that you will need to install Graphviz on your machine for this to work. For instance, on macOS you can use the Homebrew package manager (**brew install graphviz**), while on Ubuntu Linux you can use **apt install graphviz**.

The output from Graphviz will be a text file in the Dot language format (*https://graph viz.org/doc/info/lang.html*), which can be turned into a pictorial graph by the Graphviz dot tool. The second solution in the repository has options to control the output filename and whether the image should be opened:

```
$ ./solution2_graph.py -h
usage: solution2_graph.py [-h] [-k size] [-o FILE] [-v] [-d] FILE

Overlap Graphs

positional arguments:
  FILE                  FASTA file

optional arguments:
  -h, --help            show this help message and exit
  -k size, --overlap size
                        Size of overlap (default: 3)
  -o FILE, --outfile FILE
                        Output filename (default: graph.txt) ❶
  -v, --view            View outfile (default: False) ❷
  -d, --debug           Debug (default: False)
```

❶ The default output filename is *graph.txt*. A *.pdf* file will also be generated automatically, which is the visual rendition of the graph.

❷ This option controls whether the PDF should be opened automatically when the program finishes.

If you run this program on the first test input, you will see the same output as before so that it will pass the test suite:

```
$ ./solution2_graph.py tests/inputs/1.fa -o 1.txt
Rosalind_2391 Rosalind_2323
Rosalind_0498 Rosalind_2391
Rosalind_0498 Rosalind_0442
```

There should now also be a new output file called *1.txt* containing a graph structure encoded in the Dot language:

```
$ cat 1.txt
digraph {
        Rosalind_0498
        Rosalind_2391
        Rosalind_0498 -> Rosalind_2391
        Rosalind_0498
        Rosalind_0442
        Rosalind_0498 -> Rosalind_0442
        Rosalind_2391
        Rosalind_2323
        Rosalind_2391 -> Rosalind_2323
}
```

You can use the dot program to turn this into a visualization. Here is a command to save the graph to a PNG file:

```
$ dot -O -Tpng 1.txt
```

Figure 9-4 shows the resulting visualization of the graph joining all the sequences in the first FASTA file, recapitulating the manual alignment from Figure 9-3.

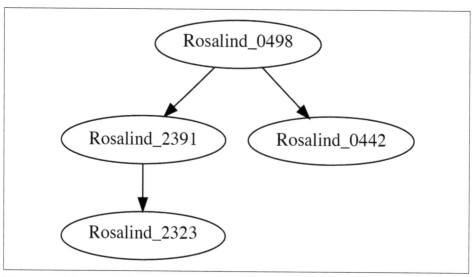

Figure 9-4. The output from the dot program showing an assembly of the sequences in the first input file when joined on 3-mers

If you run the program with the -v|--view flag, this image should be shown automatically. In graph terminology, each sequence is a *node*, and the relationship between two sequences is an *edge*.

Graphs may or may not have directionality. Figure 9-4 includes arrows implying that there is a relationship that flows from one node to another; therefore, this is a *directed graph*. The following code shows how I create and visualize this graph. Note that I import graphiz.Digraph to create the directed graph and that this code omits the logging code that is part of the actual solution:

```
def main() -> None:
    args = get_args()
    start, end = defaultdict(list), defaultdict(list)
    for rec in SeqIO.parse(args.file, 'fasta'):
        if kmers := find_kmers(str(rec.seq), args.k):
            start[kmers[0]].append(rec.id)
            end[kmers[-1]].append(rec.id)

    dot = Digraph()  ❶
    for kmer in set(start).intersection(set(end)):  ❷
        for s1, s2 in starfilter(op.ne, product(end[kmer], start[kmer])):  ❸
            print(s1, s2)  ❹
            dot.node(s1)  ❺
            dot.node(s2)
            dot.edge(s1, s2)  ❻

    args.outfile.close()  ❼
    dot.render(args.outfile.name, view=args.view)  ❽
```

❶ Create a directed graph.

❷ Iterate through the shared k-mers.

❸ Find sequence pairs sharing a k-mer, and unpack the two sequence IDs into s1 and s2.

❹ Print the output for the test.

❺ Add nodes for each sequence.

❻ Add an edge connecting the nodes.

❼ Close the output filehandle so that the graph can write to the filename.

❽ Write the graph structure to the output filename. Use the view option to open the image, depending on the args.view option.

These few lines of code have an outsized effect on the output of the program. For instance, Figure 9-5 shows that this program can essentially create a full assembly of the 100 sequences in the second input file.

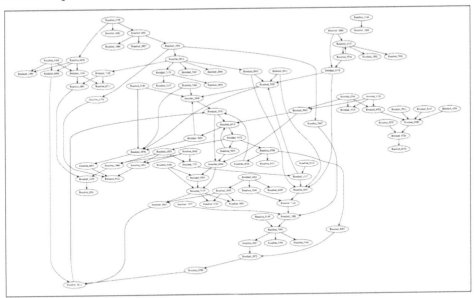

Figure 9-5. The graph of the second test input file

This image (though downsized to fit on the page) speaks volumes about the complexity and completeness of the data; for instance, the sequence pair in the upper-right corner—Rosalind_1144 and Rosalind_2208—cannot be joined to any other sequences. I would encourage you to try increasing k to 4 and inspecting the resulting graph to see a profoundly different result.

Graphs are truly powerful data structures. As noted in the introduction to this chapter, graphs encode pairwise relationships. It's amazing to see the assembly of the 100 sequences in Figure 9-5 emerge with so few lines of code. While it's possible to abuse Python lists and dictionaries to represent graphs, the Graphviz tool makes this much simpler.

I used a directed graph for this exercise, but that wasn't necessarily required. This could have been an undirected graph, too, but I like the arrows. I would note that you might encounter the term *directed acyclic graph* (DAG) to indicate a directed graph that has no cycles, which is when a node joins back to itself. Cycles might point to an incorrect assembly in the case of a linear genome but might be required for a circular genome, as in bacteria. If you find these ideas interesting, you should investigate De Bruijn graphs, which are often built from overlapping k-mers.

Going Further

Add a Hamming distance option that will allow the overlapping sequence to have the indicated edit distance. That is, a distance of 1 will allow for the overlap of sequences with a single base difference.

Review

Key points from this chapter:

- To find overlapping regions, I used k-mers to find the first and last *k* bases of each sequence.
- The `logging` module makes it easy to turn on and off the logging of runtime messages to a file.
- I used `defaultdict(list)` to create a dictionary that would auto-vivify any key not present with a default value of the empty list.
- Set intersection can find common elements between collections, such as the keys shared between two dictionaries.
- The `itertools.product()` function found all the possible pairs of sequences.
- The `iteration_utilities.starfilter()` function will splat the argument to the `lambda` for `filter()`, just as the `itertools.starmap()` function does for `map()`.

- The Graphviz tool can efficiently represent and visualize complex graph structures.

- Graphs can be textually represented using the Dot language, and the dot program can generate visualizations of graphs in various formats.

- Overlap graphs can be used to create a complete assembly of two or more sequences.

## CHAPTER 10

Finding the Longest Shared Subsequence: Finding K-mers, Writing Functions, and Using Binary Search

As described in the Rosalind LCSM challenge (*https://oreil.ly/SONgC*), the goal of this exercise is to find the longest substring that is shared by all sequences in a given FASTA file. In Chapter 8, I was searching for a given motif in some sequences. In this challenge, I don't know *a priori* that any shared motif is present—much less the size or composition of it—so I'll just look for any length of sequence that is present in every sequence. This is a challenging exercise that brings together many ideas I've shown in earlier chapters. I'll use the solutions to explore algorithm design, functions, tests, and code organization.

You will learn:

- How to use k-mers to find shared subsequences
- How to use `itertools.chain()` to concatenate lists of lists
- How and why to use a binary search
- One way to maximize a function
- How to use the key option with `min()` and `max()`

Getting Started

All the code and tests for this challenge are in the *10_lcsm* directory. Start by copying the first solution to the `lcsm.py` program and asking for help:

```
$ cp solution1_kmers_imperative.py lcsm.py
$ ./lcsm.py -h
usage: lcsm.py [-h] FILE

Longest Common Substring

positional arguments:
  FILE          Input FASTA

optional arguments:
  -h, --help  show this help message and exit
```

The only required argument is a single positional file of FASTA-formatted DNA sequences. As with other programs that accept files, the program will reject invalid or unreadable input. The following is the first input I'll use. The longest common subsequences in these sequences are *CA*, *TA*, and *AC*, with the last shown in bold in the output:

```
$ cat tests/inputs/1.fa
>Rosalind_1
GATTACA
>Rosalind_2
TAGACCA
>Rosalind_3
ATACA
```

Any of these answers are acceptable. Run the program with the first test input and see that it randomly selects one of the acceptable 2-mers:

```
$ ./lcsm.py tests/inputs/1.fa
CA
```

The second test input is much larger, and you'll notice that the program takes significantly longer to find the answer. On my laptop, it takes almost 40 seconds. In the solutions, I'll show you a way to significantly decrease the runtime using a binary search:

```
$ time ./lcsm.py tests/inputs/2.fa
GCCTTTTGATTTTAACGTTTATCGGGTGTAGTAAGATTGCGCGCTAATTCCAATAAACGTATGGAGGACATTCCCCGT

real    0m39.244s
user    0m33.708s
sys     0m6.202s
```

Although not a requirement of the challenge, I've included one input file that contains no shared subsequences for which the program should create a suitable response:

```
$ ./lcsm.py tests/inputs/none.fa
No common subsequence.
```

Start the lcsm.py program from scratch:

```
$ new.py -fp 'Longest Common Substring' lcsm.py
Done, see new script "lcsm.py".
```

Define the arguments like so:

```
class Args(NamedTuple): ❶
    """ Command-line arguments """
    file: TextIO


def get_args() -> Args:
    """ Get command-line arguments """

    parser = argparse.ArgumentParser(
        description='Longest Common Substring',
        formatter_class=argparse.ArgumentDefaultsHelpFormatter)

    parser.add_argument('file', ❷
                        help='Input FASTA',
                        metavar='FILE',
                        type=argparse.FileType('rt'))

    args = parser.parse_args()

    return Args(args.file) ❸
```

❶ The only input to this program is a FASTA-formatted file.

❷ Define a single file argument.

❸ Return the Args object containing the open filehandle.

Then update the main() function to print the incoming filename:

```
def main() -> None:
    args = get_args()
    print(args.file.name)
```

Verify that you see the correct usage and that the program correctly prints the filename:

```
$ ./lcsm.py tests/inputs/1.fa
tests/inputs/1.fa
```

At this point, your program should pass the first three tests. If you think you know how to complete the program, have at it. If you want a prod in the right direction, read on.

Finding the Shortest Sequence in a FASTA File

Reading a FASTA file should be familiar by now. I'll use Bio.SeqIO.parse() as before. My first idea on this problem was to find shared k-mers while maximizing for

k. The longest subsequence can be no longer than the shortest sequence in the file, so I decided to start with k equal to that. Finding the shortest sequence requires that I first scan through *all* the records. To review how to do this, the `Bio.SeqIO.parse()` function returns an iterator that gives me access to each FASTA record:

```
>>> from Bio import SeqIO
>>> fh = open('./tests/inputs/1.fa')
>>> recs = SeqIO.parse(fh, 'fasta')
>>> type(recs)
<class 'Bio.SeqIO.FastaIO.FastaIterator'>
```

I can use the `next()` function I first showed in Chapter 4 to force the iterator to produce the next value, the type of which is `SeqRecord`:

```
>>> rec = next(recs)
>>> type(rec)
<class 'Bio.SeqRecord.SeqRecord'>
```

In addition to the sequence itself, the FASTA record contains metadata such as the sequence ID, name, and such:

```
>>> rec
SeqRecord(seq=Seq('GATTACA'),
          id='Rosalind_1',
          name='Rosalind_1',
          description='Rosalind_1',
          dbxrefs=[])
```

The read information is wrapped in a `Seq` object, which has many interesting and useful methods you can explore in the REPL using **help(rec.seq)**. I'm only interested in the raw sequence, so I can use the `str()` function to coerce it to a string:

```
>>> str(rec.seq)
'GATTACA'
```

I need all the sequences in a list so that I can find the length of the shortest one. I can use a list comprehension to read the entire file into a list since I'll be using these many times:

```
>>> fh = open('./tests/inputs/1.fa') ❶
>>> seqs = [str(rec.seq) for rec in SeqIO.parse(fh, 'fasta')] ❷
>>> seqs
['GATTACA', 'TAGACCA', 'ATACA']
```

❶ Reopen the filehandle or the existing filehandle will continue from the second read.

❷ Create a list, coercing each record's sequence to a string.

Sequence files may hold millions of reads, and storing them in a list could easily exceed the available memory and crash your machine. (Ask me how I know.) The problem is that I need all the sequences in the next step to find the subsequence that is common to all of them. I have several *Makefile* targets that will use the genseq.py program in the *10_lcsm* directory to generate large FASTA inputs with a common motif for you to test. This program works adequately for the datasets provided by Rosalind.

The same idea can be expressed using the map() function:

```
>>> fh = open('./tests/inputs/1.fa')
>>> seqs = list(map(lambda rec: str(rec.seq), SeqIO.parse(fh, 'fasta')))
>>> seqs
['GATTACA', 'TAGACCA', 'ATACA']
```

To find the length of the shortest sequence, I need to find the lengths of all the sequences, which I can do using a list comprehension:

```
>>> [len(seq) for seq in seqs]
[7, 7, 5]
```

I prefer the shorter way to write this using a map():

```
>>> list(map(len, seqs))
[7, 7, 5]
```

Python has built-in min() and max() functions that will return the minimum or maximum value from a list:

```
>>> min(map(len, seqs))
5
>>> max(map(len, seqs))
7
```

So the shortest sequence is equal to the minimum of the lengths:

```
>>> shortest = min(map(len, seqs))
>>> shortest
5
```

Extracting K-mers from a Sequence

The longest shared subsequence can be no longer than the shortest sequence and must be shared by all the reads. Therefore, my next step is to find all the k-mers in all the sequences, starting with k equal to the length of the shortest sequence (5). In Chapter 9 I wrote a find_kmers() function and test, so I'll copy that code into this program. Remember to import typing.List for this:

```
def find_kmers(seq: str, k: int) -> List[str]:
    """ Find k-mers in string """
```

```
    n = len(seq) - k + 1
    return [] if n < 1 else [seq[i:i + k] for i in range(n)]


def test_find_kmers() -> None:
    """ Test find_kmers """

    assert find_kmers('', 1) == []
    assert find_kmers('ACTG', 1) == ['A', 'C', 'T', 'G']
    assert find_kmers('ACTG', 2) == ['AC', 'CT', 'TG']
    assert find_kmers('ACTG', 3) == ['ACT', 'CTG']
    assert find_kmers('ACTG', 4) == ['ACTG']
    assert find_kmers('ACTG', 5) == []
```

One logical approach is to start with the maximum possible value of k and count down, stopping when I find a k-mer shared by all the sequences. So far I've only used the range() function to count up. Can I reverse the start and stop values to count down? Apparently not. If the start value is greater than the stop value, then range() will produce an empty list:

```
>>> list(range(shortest, 0))
[]
```

When reading codons in Chapter 7, I mentioned that the range() function accepts up to three arguments, the last of which is the *step*, which I used there to jump three bases at a time. Here I need to use a step of -1 to count down. Remember that the stop value is not included:

```
>>> list(range(shortest, 0, -1))
[5, 4, 3, 2, 1]
```

Another way to count backward is to count up and reverse the results:

```
>>> list(reversed(range(1, shortest + 1)))
[5, 4, 3, 2, 1]
```

Either way, I want to iterate over decreasing values of k until I find a k-mer that is shared by all the sequences. A sequence might contain multiple copies of the same k-mer, so it's important to make the result unique by using the set() function:

```
>>> from lcsm import find_kmers
>>> from pprint import pprint
>>> for k in range(shortest, 0, -1):
...     print(f'==> {k} <==')
...     pprint([set(find_kmers(s, k)) for s in seqs])
...
==> 5 <==
[{'TTACA', 'GATTA', 'ATTAC'}, {'TAGAC', 'AGACC', 'GACCA'}, {'ATACA'}]
==> 4 <==
[{'ATTA', 'TTAC', 'TACA', 'GATT'},
 {'GACC', 'AGAC', 'TAGA', 'ACCA'},
```

```
  {'TACA', 'ATAC'}]
==> 3 <==
[{'ACA', 'TAC', 'GAT', 'ATT', 'TTA'},
 {'AGA', 'TAG', 'CCA', 'ACC', 'GAC'},
 {'ACA', 'ATA', 'TAC'}]
==> 2 <==
[{'AC', 'AT', 'CA', 'TA', 'TT', 'GA'},
 {'AC', 'CA', 'CC', 'TA', 'AG', 'GA'},
 {'AC', 'AT', 'CA', 'TA'}]
==> 1 <==
[{'G', 'C', 'T', 'A'}, {'G', 'C', 'T', 'A'}, {'C', 'T', 'A'}]
```

Can you see a way to use this idea to count all the k-mers for each value of k? Look for k-mers that have a frequency matching the number of sequences. If you find more than one, print any one of them.

Solutions

The two variations for this program use the same basic logic to find the longest shared subsequence. The first version proves to scale poorly as the input size increases because it uses a stepwise, linear approach to iterating over every possible k-length of sequence. The second version introduces a binary search to find a good starting value for k and then initiates a hill-climbing search to discover a maximum value for k.

Solution 1: Counting Frequencies of K-mers

In the previous section, I got as far as finding all the k-mers in the sequences for values of k, starting with the shortest sequence and moving down to 1. Here I'll start with k equal to 5, which was the length of the shortest sequence in the first FASTA file:

```
>>> fh = open('./tests/inputs/1.fa')
>>> seqs = [str(rec.seq) for rec in SeqIO.parse(fh, 'fasta')]
>>> shortest = min(map(len, seqs))
>>> kmers = [set(find_kmers(seq, shortest)) for seq in seqs]
>>> kmers
[{'TTACA', 'GATTA', 'ATTAC'}, {'TAGAC', 'AGACC', 'GACCA'}, {'ATACA'}]
```

I need a way to count how many times each k-mer appears across all the sequences. One approach is to use collections.Counter(), which I first showed in Chapter 1:

```
>>> from collections import Counter
>>> counts = Counter()
```

I can iterate over each set of k-mers from the sequences and use the Counter.update() method to add them:

```
>>> for group in kmers:
...     counts.update(group)
...
>>> pprint(counts)
Counter({'TTACA': 1,
         'GATTA': 1,
         'ATTAC': 1,
         'TAGAC': 1,
         'AGACC': 1,
         'GACCA': 1,
         'ATACA': 1})
```

Or I could concatenate the many lists of k-mers together into a single list using `itertools.chain()`:

```
>>> from itertools import chain
>>> list(chain.from_iterable(kmers))
['TTACA', 'GATTA', 'ATTAC', 'TAGAC', 'AGACC', 'GACCA', 'ATACA']
```

Using this as the input for the `Counter()` produces the same collection, showing that each 5-mer is unique, occurring once each:

```
>>> counts = Counter(chain.from_iterable(kmers))
>>> pprint(counts)
Counter({'TTACA': 1,
         'GATTA': 1,
         'ATTAC': 1,
         'TAGAC': 1,
         'AGACC': 1,
         'GACCA': 1,
         'ATACA': 1})
```

The `Counter()` is a regular dictionary underneath, which means I have access to all the dictionary methods. I want to iterate through the keys and values as pairs using the `dict.items()` method to find where the count of the k-mers is equal to the number of sequences:

```
>>> n = len(seqs)
>>> candidates = []
>>> for kmer, count in counts.items():
...     if count == n:
...         candidates.append(kmer)
...
>>> candidates
[]
```

When k is 5, there are no candidate sequences, so I need to try with a smaller value. Since I know the right answer is 2, I'll rerun this code with k=2 to produce this dictionary:

```
>>> k = 2
>>> kmers = [set(find_kmers(seq, k)) for seq in seqs]
>>> counts = Counter(chain.from_iterable(kmers))
```

```
>>> pprint(counts)
Counter({'CA': 3,
         'AC': 3,
         'TA': 3,
         'GA': 2,
         'AT': 2,
         'TT': 1,
         'AG': 1,
         'CC': 1})
```

From this, I find three candidate 2-mers have a frequency of 3, which equals the number of sequences:

```
>>> candidates = []
>>> for kmer, count in counts.items():
...     if count == n:
...         candidates.append(kmer)
...
>>> candidates
['CA', 'AC', 'TA']
```

It doesn't matter which of the candidates I choose, so I'll use the `random.choice()` function which returns one value from a list of choices:

```
>>> import random
>>> random.choice(candidates)
'AC'
```

I like where this is going, so I'd like to put it into a function so I can test it:

```
def common_kmers(seqs: List[str], k: int) -> List[str]:
    """ Find k-mers common to all sequences """

    kmers = [set(find_kmers(seq, k)) for seq in seqs]
    counts = Counter(chain.from_iterable(kmers))
    n = len(seqs) ❶
    return [kmer for kmer, freq in counts.items() if freq == n] ❷
```

❶ Find the number of sequences.

❷ Return the k-mers having a frequency equal to the number of sequences.

This makes for a pretty readable `main()`:

```
import random
import sys


def main() -> None:
    args = get_args()
    seqs = [str(rec.seq) for rec in SeqIO.parse(args.file, 'fasta')] ❶
    shortest = min(map(len, seqs)) ❷
```

```
    for k in range(shortest, 0, -1): ❸
        if kmers := common_kmers(seqs, k): ❹
            print(random.choice(kmers)) ❺
            sys.exit(0) ❻

    print('No common subsequence.') ❼
```

❶ Read all the sequences into a list.

❷ Find the length of the shortest sequence.

❸ Count down from the shortest sequence.

❹ Find all the common k-mers using this value of k.

❺ If any k-mers are found, print a random selection.

❻ Exit the program using an exit value of 0 (no errors).

❼ If I make it to this point, inform the user there is no shared sequence.

In the preceding code, I'm again using the walrus operator (:=) I introduced in Chapter 5 to first assign the result of calling common_kmers() to the variable kmers and then evaluate kmers for truthiness. Python will only enter the next block if kmers is truthy, meaning there were common k-mers found for this value of k. Before the addition of this language feature, I would have had to write the assignment and evaluation on two lines, like so:

```
    kmers = common_kmers(seqs, k)
    if kmers:
        print(random.choice(kmers))
```

Solution 2: Speeding Things Up with a Binary Search

As noted in the opening section of this chapter, this solution grows much slower as the size of the inputs increases. One way to track the progress of the program is to put a print(k) statement at the beginning of the for loop. Run this with the second input file, and you'll see that it starts counting down from 1,000 and doesn't reach the correct value for k until it hits 78.

Counting backward by 1 is taking too long. If your friend asked you to guess a number between 1 and 1,000, you wouldn't start at 1,000 and keep guessing 1 less each time your friend said, "Too high." It's much faster (and better for your friendship) to guess 500. If your friend chose 453, they'd say "Too high," so you'd be wise to choose 250. They'd reply, "Too low," and you'd keep splitting the differences between your last high and low guesses until you found the right answer. This is a *binary search*, and it's a great way to quickly find the location of a wanted value from a sorted list of values.

To understand this better, I've included a program in the *10_lcsm* directory called binsearch.py:

```
$ ./binsearch.py -h
usage: binsearch.py [-h] -n int -m int

Binary Search

optional arguments:
  -h, --help          show this help message and exit
  -n int, --num int   The number to guess (default: None)
  -m int, --max int   The maximum range (default: None)
```

The following is the relevant portion of the program. You can read the source code for the argument definitions if you like. The binary_search() function is recursive, like one solution to the Fibonacci sequence problem from Chapter 4. Note that the search values must be sorted for binary searches to work, which the range() function provides:

```
def main() -> None:
    args = get_args()
    nums = list(range(args.maximum + 1))
    pos = binary_search(args.num, nums, 0, args.maximum)
    print(f'Found {args.num}!' if pos > 0 else f'{args.num} not present.')


def binary_search(x: int, xs: List[int], low: int, high: int) -> int:
    print(f'{low:4} {high:4}', file=sys.stderr)

    if high >= low: ❶
        mid = (high + low) // 2 ❷

        if xs[mid] == x: ❸
            return mid

        if xs[mid] > x: ❹
            return binary_search(x, xs, low, mid - 1) ❺

        return binary_search(x, xs, mid + 1, high) ❻

    return -1 ❼
```

❶ The base case to exit the recursion is when this is false.

❷ The midpoint is halfway between high and low, using floor division.

❸ Return the midpoint if the element is in the middle.

❹ See if the value at the midpoint is greater than the desired value.

❺ Search the lower values.

❻ Search the higher values.

❼ The value was not found.

 The names x and xs in the binary_search() function are meant to be singular and plural. In my head, I pronounce them *ex* and *exes*. This kind of notation is common in purely functional programming because I'm not trying to describe what kind of value x is. It could be a string or a number or anything. The important point is that xs is some collection of comparable values all of the same type.

I included some print() statements so that, running with the previous numbers, you can see how low and high finally converge on the target number in 10 steps:

```
$ ./binsearch.py -n 453 -m 1000
    0 1000
    0  499
  250  499
  375  499
  438  499
  438  467
  453  467
  453  459
  453  455
  453  453
Found 453!
```

It takes just eight iterations to determine the number is not present:

```
$ ./binsearch.py -n 453 -m 100
    0  100
   51  100
   76  100
   89  100
   95  100
   98  100
  100  100
  101  100
453 not present.
```

The binary search can tell me if a value occurs in a list of values, but this is not quite my problem. While I'm reasonably sure there will be at least a 2- or 1-mer in common in most datasets, I have included one file that has none:

```
$ cat tests/inputs/none.fa
>Rosalind_1
GGGGGGG
>Rosalind_2
AAAAAAAA
>Rosalind_3
CCCC
>Rosalind_4
TTTTTTTT
```

If there is an acceptable value for k, then I need to find the *maximum* value. I decided to use the binary search to find a starting point for a hill-climbing search to find the maximum value. First I'll show `main()`, and then I'll break down the other functions:

```
def main() -> None:
    args = get_args()
    seqs = [str(rec.seq) for rec in SeqIO.parse(args.file, 'fasta')] ❶
    shortest = min(map(len, seqs)) ❷
    common = partial(common_kmers, seqs) ❸
    start = binary_search(common, 1, shortest) ❹

    if start >= 0: ❺
        candidates = [] ❻
        for k in range(start, shortest + 1): ❼
            if kmers := common(k): ❽
                candidates.append(random.choice(kmers)) ❾
            else:
                break ❿

        print(max(candidates, key=len)) ⓫
    else:
        print('No common subsequence.') ⓬
```

❶ Get a list of the sequences as strings.

❷ Find the length of the shortest sequence.

❸ Partially apply the `common_kmers()` function with the `seqs` input.

❹ Use the binary search to find a starting point for the given function, using 1 for the lowest value of k and the shortest sequence length for the maximum.

❺ Check that the binary search found something useful.

❻ Initialize a list of the candidate values.

❼ Start the hill climbing with the binary search result.

❽ Check if there are common k-mers.

❾ If so, randomly add one to the list of candidates.

❿ If there are no common k-mers, break out of the loop.

⓫ Choose the candidate sequence having the longest length.

⓬ Let the user know that there is no answer.

While there are many things to explain in the preceding code, I want to highlight the call to max(). I showed earlier that this function will return the maximum value from a list. Normally you might think to use this on a list of numbers:

```
>>> max([4, 2, 8, 1])
8
```

In the preceding code, I want to find the longest string in a list. I can map() the len() function to find their lengths:

```
>>> seqs = ['A', 'CC', 'GGGG', 'TTT']
>>> list(map(len, seqs))
[1, 2, 4, 3]
```

This shows that the third sequence, *GGGG*, is the longest. The max() function accepts an optional **key** argument, which is a function to apply to each element before comparing. If I use the len() function, then max() correctly identifies the longest sequence:

```
>>> max(seqs, key=len)
'GGGG'
```

Let's take a look at how I modified the binary_search() function to suit my needs:

```
def binary_search(f: Callable, low: int, high: int) -> int: ❶
    """ Binary search """

    hi, lo = f(high), f(low) ❷
    mid = (high + low) // 2 ❸

    if hi and lo: ❹
        return high

    if lo and not hi: ❺
        return binary_search(f, low, mid)

    if hi and not lo: ❻
        return binary_search(f, mid, high)

    return -1 ❼
```

❶ The function takes another function f() along with low and high values as arguments. In this instance, the function f() will return the common k-mers, but the function can perform any calculation you like.

❷ Call the function f() with the highest and lowest values for k.

❸ Find the midpoint value of k.

❹ If the function f() found common k-mers for both the high and low k values, return the highest k.

❺ If the high k found no k-mers but the low value did, recursively call the function searching in the lower values of k.

❻ If the low k found no k-mers but the high value did, recursively call the function searching in the higher values of k.

❼ Return -1 to indicate no k-mers were found using the high and low arguments to f().

Here is the test I wrote for this:

```
def test_binary_search() -> None:
    """ Test binary_search """

    seqs1 = ['GATTACA', 'TAGACCA', 'ATACA'] ❶
    f1 = partial(common_kmers, seqs1) ❷
    assert binary_search(f1, 1, 5) == 2 ❸

    seqs2 = ['GATTACTA', 'TAGACTCA', 'ATACTA'] ❹
    f2 = partial(common_kmers, seqs2)
    assert binary_search(f2, 1, 6) == 3 ❺
```

❶ These are the sequences I've been using that have three shared 2-mers.

❷ Define a function to find the k-mers in the first set of sequences.

❸ The search finds a k of 2 which is the right answer.

❹ The same sequences as before but now with a shared 3-mer.

❺ The search finds a k of 3.

Unlike the previous binary search, my version won't (necessarily) return the *exact* answer, just a decent starting point. If there are no shared sequences for any size k, then I let the user know:

```
$ ./solution2_binary_search.py tests/inputs/none.fa
No common subsequence.
```

If there is a shared subsequence, this version runs significantly faster—perhaps as much as 28 times faster:

```
$ hyperfine -L prg ./solution1_kmers_functional.py,./solution2_binary_search.py\
  '{prg} tests/inputs/2.fa'
Benchmark #1: ./solution1_kmers_functional.py tests/inputs/2.fa
  Time (mean ± σ):     40.686 s ±  0.443 s    [User: 35.208 s, System: 6.042 s]
  Range (min … max):   40.165 s … 41.349 s    10 runs

Benchmark #2: ./solution2_binary_search.py tests/inputs/2.fa
  Time (mean ± σ):      1.441 s ±  0.037 s    [User: 1.903 s, System: 0.255 s]
  Range (min … max):    1.378 s …  1.492 s    10 runs

Summary
  './solution2_binary_search.py tests/inputs/2.fa' ran
   28.24 ± 0.79 times faster than './solution1_kmers_functional.py
   tests/inputs/2.fa'
```

When I was searching from the maximum k value and iterating down, I was performing a *linear* search through all the possible values. This means the time to search grows in proportion (linearly) to the number n of values. A binary search, by contrast, grows at a rate of log n. It's common to talk about the runtime growth of algorithms using *Big O* notation, so you might see binary search described as O(log n), whereas linear searching is O(n)—which is much worse.

Going Further

As with the suggestion in Chapter 9, add a Hamming distance option that will allow for the indicated number of differences when deciding on a shared k-mer.

Review

Key points from this chapter:

- K-mers can be used to find conserved regions of sequences.
- Lists of lists can be combined into a single list using `itertools.chain()`.
- A binary search can be used on sorted values to find a value more quickly than searching through the list linearly.
- Hill climbing is one way to maximize the input to a function.
- The key option for `min()` and `max()` is a function that is applied to the values before comparing them.

## CHAPTER 11

Finding a Protein Motif: Fetching Data and Using Regular Expressions

We've spent quite a bit of time now looking for sequence motifs. As described in the Rosalind MPRT challenge (*https://oreil.ly/EAp3i*), shared or conserved sequences in proteins imply shared functions. In this exercise, I need to identify protein sequences that contain the N-glycosylation motif. The input to the program is a list of protein IDs that will be used to download the sequences from the UniProt website (*https://www.uniprot.org*). After demonstrating how to manually and programmatically download the data, I'll show how to find the motif using a regular expression and by writing a manual solution.

You will learn:

- How to programmatically fetch data from the internet
- How to write a regular expression to find the N-glycosylation motif
- How to manually find the N-glycosylation motif

Getting Started

All the code and tests for this program are located in the *11_mprt* directory. To begin, copy the first solution to the program mprt.py:

```
$ cd 11_mprt
$ cp solution1_regex.py mprt.py
```

Inspect the usage:

```
$ ./mprt.py -h
usage: mprt.py [-h] [-d DIR] FILE
```

```
Find locations of N-glycosylation motif

positional arguments:
  FILE                    Input text file of UniProt IDs ❶

optional arguments:
  -h, --help              show this help message and exit
  -d DIR, --download_dir DIR ❷
                          Directory for downloads (default: fasta)
```

❶ The required positional argument is a file of protein IDs.

❷ The optional download directory name defaults to *fasta*.

The input file will list protein IDs, one per line. The protein IDs provided in the Rosalind example comprise the first test input file:

```
$ cat tests/inputs/1.txt
A2Z669
B5ZC00
P07204_TRBM_HUMAN
P20840_SAG1_YEAST
```

Run the program using this as the argument. The output of the program lists each protein ID containing the N-glycosylation motif and the locations where it can be found:

```
$ ./mprt.py tests/inputs/1.txt
B5ZC00
85 118 142 306 395
P07204_TRBM_HUMAN
47 115 116 382 409
P20840_SAG1_YEAST
79 109 135 248 306 348 364 402 485 501 614
```

After running the preceding command, you should see that the default *fasta* directory has been created. Inside you should find four FASTA files. All subsequent runs using these protein IDs will be faster as the cached data will be used unless you remove the download directory, for instance by running **make clean**.

Take a look at the first two lines of each file using the command **head -2**. The headers for some of the FASTA records are quite long so I've broken them here so they won't wrap, but the actual headers must be on a single line:

```
$ head -2 fasta/*
==> fasta/A2Z669.fasta <==
>sp|A2Z669|CSPLT_ORYSI CASP-like protein 5A2 OS=Oryza sativa subsp.
 indica OX=39946 GN=OsI_33147 PE=3 SV=1
MRASRPVVHPVEAPPPAALAVAAAAVAVEAGVGAGGGAAAHGGENAQPRGVRMKDPPGAP

==> fasta/B5ZC00.fasta <==
```

```
>sp|B5ZC00|SYG_UREU1 Glycine--tRNA ligase OS=Ureaplasma urealyticum
 serovar 10 (strain ATCC 33699 / Western) OX=565575 GN=glyQS PE=3 SV=1
MKNKFKTQEELVNHLKTVGFVFANSEIYNGLANAWDYGPLGVLLKNNLKNLWWKEFVTKQ

==> fasta/P07204_TRBM_HUMAN.fasta <==
>sp|P07204|TRBM_HUMAN Thrombomodulin OS=Homo sapiens OX=9606 GN=THBD PE=1 SV=2
MLGVLVLGALALAGLGFPAPAEPQPGGSQCVEHDCFALYPGPATFLNASQICDGLRGHLM

==> fasta/P20840_SAG1_YEAST.fasta <==
>sp|P20840|SAG1_YEAST Alpha-agglutinin OS=Saccharomyces cerevisiae
 (strain ATCC 204508 / S288c) OX=559292 GN=SAG1 PE=1 SV=2
MFTFLKIILWLFSLALASAININDITFSNLEITPLTANKQPDQGWTATFDFSIADASSIR
```

Run **make test** to see the kinds of tests your program should pass. When you're ready, start the program from scratch:

```
$ new.py -fp 'Find locations of N-glycosylation motif' mprt.py
Done, see new script "mprt.py".
```

You should define a positional file argument and an optional download directory as the arguments to the program:

```
class Args(NamedTuple):
    """ Command-line arguments """
    file: TextIO ❶
    download_dir: str ❷


def get_args() -> Args:
    """Get command-line arguments"""

    parser = argparse.ArgumentParser(
        description='Find location of N-glycosylation motif',
        formatter_class=argparse.ArgumentDefaultsHelpFormatter)

    parser.add_argument('file',
                        help='Input text file of UniProt IDs',
                        metavar='FILE',
                        type=argparse.FileType('rt')) ❸

    parser.add_argument('-d',
                        '--download_dir',
                        help='Directory for downloads',
                        metavar='DIR',
                        type=str,
                        default='fasta') ❹

    args = parser.parse_args()

    return Args(args.file, args.download_dir)
```

❶ The file will be a filehandle.

❷   The download_dir will be a string.

❸   Ensure the file argument is a readable text file.

❹   The download_dir is an optional string with a reasonable default value.

Ensure your program can create the usage, then start by printing the protein IDs from the file. Each ID is terminated by a newline, so I'll use the str.rstrip() (*right strip*) method to remove any whitespace from the right side:

```
def main() -> None:
    args = get_args()
    for prot_id in map(str.rstrip, args.file):
        print(prot_id)
```

Run the program and make sure you see the protein IDs:

```
$ ./mprt.py tests/inputs/1.txt
A2Z669
B5ZC00
P07204_TRBM_HUMAN
P20840_SAG1_YEAST
```

If you run **pytest**, you should pass the first three tests and fail the fourth.

Downloading Sequences Files on the Command Line

The next order of business is fetching the protein sequences. The UniProt information for each protein is found by substituting the protein ID into the URL *http:// www.uniprot.org/uniprot/{uniprot_id}*. I'll change the program to print this string instead:

```
def main() -> None:
    args = get_args()
    for prot_id in map(str.rstrip, args.file):
        print(f'http://www.uniprot.org/uniprot/{prot_id}')
```

You should now see this output:

```
$ ./mprt.py tests/inputs/1.txt
http://www.uniprot.org/uniprot/A2Z669
http://www.uniprot.org/uniprot/B5ZC00
http://www.uniprot.org/uniprot/P07204_TRBM_HUMAN
http://www.uniprot.org/uniprot/P20840_SAG1_YEAST
```

Paste the first URL into your web browser and inspect the page. There is a wealth of data, all in a human-readable format. Scroll down to the sequence, and you should see 203 amino acids. It would be awful to have to parse this page to extract the sequence. Luckily, I can append *.fasta* to the URL and get a FASTA file of the sequence.

Before I show you how to download the sequences using Python, I think you should know how to do this using command-line tools. From the command line, you can use `curl` (which you may need to install) to download the sequence. By default, this will print the contents of the file to STDOUT:

```
$ curl https://www.uniprot.org/uniprot/A2Z669.fasta
>sp|A2Z669|CSPLT_ORYSI CASP-like protein 5A2 OS=Oryza sativa subsp.
 indica OX=39946 GN=OsI_33147 PE=3 SV=1
MRASRPVVHPVEAPPPAALAVAAAAVAVEAGVGAGGGAAAHGGENAQPRGVRMKDPPGAP
GTPGGLGLRLVQAFFAAAALAVMASTDDFPSVSAFCYLVAAAILQCLWSLSLAVVDIYAL
LVKRSLRNPQAVCIFTIGDGITGTLTLGAACASAGITVLIGNDLNICANNHCASFETATA
MAFISWFALAPSCVLNFWSMASR
```

You could either redirect this to a file:

```
$ curl https://www.uniprot.org/uniprot/A2Z669.fasta > A2Z669.fasta
```

or use the `-o|--output` option to name the output file:

```
$ curl -o A2Z669.fasta https://www.uniprot.org/uniprot/A2Z669.fasta
```

You can also use `wget` (*web get*, which may also need to be installed) to download the sequence file like so:

```
$ wget https://www.uniprot.org/uniprot/A2Z669.fasta
```

Whichever tool you use, you should now have a file called *A2Z669.fasta* with the sequence data:

```
$ cat A2Z669.fasta
>sp|A2Z669|CSPLT_ORYSI CASP-like protein 5A2 OS=Oryza sativa subsp.
 indica OX=39946 GN=OsI_33147 PE=3 SV=1
MRASRPVVHPVEAPPPAALAVAAAAVAVEAGVGAGGGAAAHGGENAQPRGVRMKDPPGAP
GTPGGLGLRLVQAFFAAAALAVMASTDDFPSVSAFCYLVAAAILQCLWSLSLAVVDIYAL
LVKRSLRNPQAVCIFTIGDGITGTLTLGAACASAGITVLIGNDLNICANNHCASFETATA
MAFISWFALAPSCVLNFWSMASR
```

I know this is a book on Python, but it's worth learning how to write a basic `bash` program. Just as some stories can be told in a haiku and others are sprawling novels, some tasks are easily expressed using a few shell commands and others require thousands of lines of code in a more complex language. Sometimes I can write 10 lines of `bash` to do what I need. When I hit about 30 lines of `bash`, I generally move to Python or Rust.

Here is how I could automate downloading the proteins with a `bash` script:

```
#!/usr/bin/env bash ❶

if [[ $# -ne 1 ]]; then ❷
    printf "usage: %s FILE\n" $(basename "$0") ❸
    exit 1 ❹
fi
```

```
OUT_DIR="fasta"  ❺
[[ ! -d "$OUT_DIR" ]] && mkdir -p "$OUT_DIR"  ❻

while read -r PROT_ID; do  ❼
    echo "$PROT_ID"  ❽
    URL="https://www.uniprot.org/uniprot/${PROT_ID}.fasta"  ❾
    OUT_FILE="$OUT_DIR/${PROT_ID}.fasta"  ❿
    wget -q -O "$OUT_FILE" "$URL"  ⓫
done < $1  ⓬

echo "Done, see output in \"$OUT_DIR\"."  ⓭
```

❶ The shebang (#!) should use the env (environment) to find bash.

❷ Check that the number of arguments ($#) is 1.

❸ Print a usage statement using the program basename ($0).

❹ Exit with a nonzero value.

❺ Define the output directory to be *fasta*. Note that in bash you can have no spaces around the = for variable assignment.

❻ Create the output directory if it does not exist.

❼ Read each line from the file into the PROT_ID variable.

❽ Print the current protein ID so the user knows something is happening.

❾ Construct the URL by using variable interpolation inside double quotes.

❿ Construct the output filename by combining the output directory and the protein ID.

⓫ Call wget with the -q (quiet) flag to fetch the URL into the output file.

⓬ This reads each line from the first positional argument ($1), which is the input filename.

⓭ Let the user know the program has finished and where to find the output.

I can run this like so:

```
$ ./fetch_fasta.sh tests/inputs/1.txt
A2Z669
B5ZC00
P07204_TRBM_HUMAN
```

```
P20840_SAG1_YEAST
Done, see output in "fasta".
```

Now there should be a *fasta* directory containing the four FASTA files. One way to write the `mprt.py` program would be to fetch all the input files first using something like this and then provide the FASTA files as arguments. This is a very common pattern in bioinformatics, and writing a shell script like this is a great way to document exactly how you retrieved the data for your analysis. Be sure you always commit programs like this to your source repository, and consider adding a *Makefile* target with a name like *fasta* that is flush-left followed by a colon and the command on the next line indented with a single tab character:

```
fasta:
        ./fetch_fasta.sh tests/inputs/1.txt
```

Now you should be able to run **make fasta** to automate the process of getting your data. By writing the program to accept the input file as an argument rather than hardcoding it, I can use this program and multiple *Makefile* targets to automate the process of downloading many different datasets. Reproducibility for the win.

Downloading Sequences Files with Python

I'll translate the `bash` utility to Python now. As you can see from the preceding program, there are several steps involved to fetch each sequence file. I don't want this to be a part of `main()` as it will clutter the program, so I'll write a function for this:

```
def fetch_fasta(fh: TextIO, fasta_dir: str) -> List[str]: ❶
    """ Fetch the FASTA files into the download directory """

    return [] ❷
```

❶ The function will accept a filehandle for the protein IDs and a download directory name, and will return a list of the files that were downloaded or were already present. Be sure to add `typing.List` to your imports.

❷ For now, return an empty list.

I want to call it like this:

```
def main() -> None:
    args = get_args()
    files = fetch_fasta(args.file, args.download_dir)
    print('\n'.join(files))
```

Run your program and ensure it compiles and prints nothing. Now add the following Python code to fetch the sequences. You'll need to import `os`, `sys`, and `requests` (*https://oreil.ly/nYSUM*), a library for making web requests:

```
def fetch_fasta(fh: TextIO, fasta_dir: str) -> List[str]:
    """ Fetch the FASTA files into the download directory """

    if not os.path.isdir(fasta_dir):      ❶
        os.makedirs(fasta_dir)            ❷

    files = []                            ❸
    for prot_id in map(str.rstrip, fh):   ❹
        fasta = os.path.join(fasta_dir, prot_id + '.fasta')  ❺
        if not os.path.isfile(fasta):     ❻
            url = f'http://www.uniprot.org/uniprot/{prot_id}.fasta'  ❼
            response = requests.get(url)  ❽
            if response.status_code == 200:  ❾
                print(response.text, file=open(fasta, 'wt'))  ❿
            else:
                print(f'Error fetching "{url}": "{response.status_code}"',
                      file=sys.stderr)    ⓫
                continue                  ⓬

        files.append(fasta)               ⓭

    return files                          ⓮
```

❶ Create the output directory if it does not exist.

❷ Create the directory and any needed parent directories.

❸ Initialize the return list of filenames.

❹ Read each protein ID from the file.

❺ Construct the output filename by combining the output directory plus the protein ID.

❻ Check if the file already exists.

❼ Construct the URL to the FASTA file.

❽ Make a *GET* request for the file.

❾ A response code of 200 indicates success.

❿ Write the text of the response to the output file.

⓫ Print a warning to STDERR that the file could not be fetched.

⓬ Skip to the next iteration.

⓮ Append the file to the return list.

⓭ Return the files that now exist locally.

 os.makedirs() is an example of a function that will throw an exception if it fails. This might happen due to the user having insufficient permissions to create a directory, or because of a disk error. What would be the point in my catching and handling such an error? If my program is unable to fix a problem, I feel it's better to let it crash loudly, producing an error code and a stacktrace of what went wrong. A human would have to fix the underlying problems before the program could work. Catching and mishandling the exception would be far worse than letting the program crash.

That logic almost exactly mirrors that of the bash program. If you run your program again, there should be a *fasta* directory with the four files, and the program should print the names of the downloaded files:

```
$ ./mprt.py tests/inputs/1.txt
fasta/A2Z669.fasta
fasta/B5ZC00.fasta
fasta/P07204_TRBM_HUMAN.fasta
fasta/P20840_SAG1_YEAST.fasta
```

Writing a Regular Expression to Find the Motif

The Rosalind page notes:

To allow for the presence of its varying forms, a protein motif is represented by a shorthand as follows: [XY] means *either X or Y* and {X} means *any amino acid except X*. For example, the N-glycosylation motif is written as N{P}[ST]{P}.

The Prosite website (*https://oreil.ly/aFwWe*) is a database of protein domains, families, and functional sites. The details for the N-glycosylation motif (*https://oreil.ly/VrQLl*) show a similar convention for the *consensus pattern* of N-{P}-[ST]-{P}. Both patterns are extremely close to the regular expression shown in Figure 11-1.

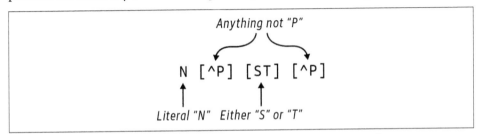

Figure 11-1. A regular expression for the N-glycosylation protein motif

In this regex, the N indicates the literal character *N*. The [ST] is a character class representing either the character *S* or *T*. It's the same as the regex [GC] I wrote in Chapter 5 to find either *G* or *C*. The [^P] is a *negated* character class, which means it will match any character that is *not P*.

Some people (OK, mostly me) like to represent regexes using the notation of finite state machines (FSMs), such as the one shown in Figure 11-2. Imagine the pattern entering on the left. It first needs to find the letter *N* to proceed to the next step. Next can be any character that is not the letter *P*. After that, the graph has two alternate paths through the letters *S* or *T*, which must be followed again by a not-*P* character. If the pattern makes it to the double circle, the match was successful.

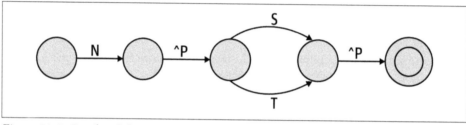

Figure 11-2. Graphical depiction of an FSM to identify the N-glycosylation motif

In Chapter 8, I pointed out a problem when using regular expressions to find overlapping text. There are no instances of this in the first test file, but another of the datasets I used to solve the problem did have two overlapping motifs. Let me demonstrate in the REPL:

```
>>> import re
>>> regex = re.compile('N[^P][ST][^P]')
```

I'm using the re.compile() function here to force the regex engine to parse the pattern and create the necessary internal code to do the matching. This is similar to how compiled languages like C use source code that humans can edit and read into machine code that computers can directly execute. This transformation happens once when you use re.compile(), whereas functions like re.search() must recompile the regex on each call.

Here is the relevant portion of the protein sequence for *P07204_TRBM_HUMAN* that has the pattern starting at both the first and second positions (see Figure 11-3). The re.findall() function shows that only the pattern starting at the first position is found:

```
>>> seq = 'NNTSYS'
>>> regex.findall(seq)
['NNTS']
```

Figure 11-3. This sequence contains two copies of the motif that overlap

As in Chapter 8, the solution is to wrap the regex in a look-ahead assertion using ?=(*<pattern>*), which itself will need to be wrapped in capturing parentheses:

```
>>> regex = re.compile('(?=(N[^P][ST][^P]))')
>>> regex.findall(seq)
['NNTS', 'NTSY']
```

I need to know the positions of the matches, which I can get from re.finditer(). This will return a list of re.Match objects, each of which has a match.start() function that will return the zero-offset index of the match's starting position. I need to add 1 to report the position using 1-based counting:

```
>>> [match.start() + 1 for match in regex.finditer(seq)]
[1, 2]
```

This should be enough for you to solve the rest of the problem. Keep hacking until you pass all the tests. Be sure to download a dataset from the Rosalind site and verify that your solution gives an answer that passes the test with that, too. See if you can also write a version that doesn't use regular expressions. Go back and study the FSM model and think about how you can implement those ideas in Python code.

Solutions

I will present two variations to solve this problem. Both use the same get_args() and fetch_fasta() functions shown previously. The first uses a regular expression to find the motif, and the second imagines how to solve the problem in a horrible, desolate intellectual wasteland where regular expressions don't exist.

Solution 1: Using a Regular Expression

The following is my final solution using a regular expression. Be sure to import re and Bio.SeqIO for this:

```
def main():
    args = get_args()
    files = fetch_fasta(args.file, args.download_dir) ❶
    regex = re.compile('(?=(N[^P][ST][^P]))') ❷

    for file in files: ❸
        prot_id, _ = os.path.splitext(os.path.basename(file)) ❹
        recs = SeqIO.parse(file, 'fasta') ❺
        if rec := next(recs): ❻
```

```
if matches := list(regex.finditer(str(rec.seq))):  ❼
    print(prot_id)  ❽
    print(*[match.start() + 1 for match in matches])  ❾
```

❶ Fetch the sequence files for the protein IDs in the given file. Put the files into the indicated download directory.

❷ Compile the regex for the N-glycosylation motif.

❸ Iterate through the files.

❹ Get the protein ID from the basename of the file minus the file extension.

❺ Create a lazy iterator to fetch the FASTA sequences from the file.

❻ Attempt to retrieve the first sequence record from the iterator.

❼ Coerce the sequence to a str, then try to find all the matches for the motif.

❽ Print the protein ID.

❾ Print all the matches, correcting to 1-based counting.

In this solution, I used the os.path.basename() and os.path.splitext() functions. I often use these, so I want to make sure you understand exactly what they do. I first introduced the os.path.basename() in Chapter 2. This function will return the filename from a path that might include directories:

```
>>> import os
>>> basename = os.path.basename('fasta/B5ZC00.fasta')
>>> basename
'B5ZC00.fasta'
```

The os.path.splitext() function will break a filename into the part before the file extension and the extension:

```
>>> os.path.splitext(basename)
('B5ZC00', '.fasta')
```

 File extensions can provide useful metadata about a file. For instance, your operating system may know to use Microsoft Excel to open files ending in *.xls* or *.xlsx*. There are many conventions for FASTA extensions, including *.fasta*, *.fa*, *.fna* (for nucleotides), and *.faa* (for amino acids). You can put whatever extension you like on a FASTA file or none at all, but remember that a FASTA file is always plain text and needs no special application to view it. Also, just because a file has a FASTA-like extension does not necessarily mean it's a FASTA file. *Caveat emptor.*

In the preceding code, I don't need the extension, so I assign it to the variable _ (underscore), which is a convention indicating that I don't intend to use the value. I could also use a list slice to grab the first element from the function:

```
>>> os.path.splitext(basename)[0]
'B5ZC00'
```

Solution 2: Writing a Manual Solution

If I were writing a program like this for production use, I would use a regular expression to find the motif. In this context, though, I wanted to challenge myself to find a manual solution. As usual, I want to write a function to encapsulate this idea, so I stub it out:

```
def find_motif(text: str) -> List[int]: ❶
    """ Find a pattern in some text """

    return [] ❷
```

❶ The function will take some text and return a list of integers where the motif can be found in the text.

❷ For now, return the empty list.

The biggest reason to have a function is to write a test where I encode examples I expect to match and fail:

```
def test_find_motif() -> None:
    """ Test find_pattern """

    assert find_motif('') == [] ❶
    assert find_motif('NPTX') == [] ❷
    assert find_motif('NXTP') == [] ❸
    assert find_motif('NXSX') == [0] ❹
    assert find_motif('ANXTX') == [1] ❺
    assert find_motif('NNTSYS') == [0, 1] ❻
    assert find_motif('XNNTSYS') == [1, 2] ❼
    assert find_motif('XNNTSYSXNNTSYS') == [1, 2, 8, 9] ❽
```

❶ Ensure the function does not do something silly like raise an exception when given the empty string.

❷ This should fail because it has a *P* in the second position.

❸ This should fail because it has a *P* in the fourth position.

❹ This should find the motif at the beginning of the string.

❺ This should find the motif not at the beginning of the string.

❻ This should find overlapping motifs at the beginning of the string.

❼ This should find overlapping motifs not at the beginning of the string.

❽ This is a slightly more complicated pattern containing four copies of the motif.

I can add these functions to my `mprt.py` program and I can run `pytest` on that source code to ensure that the tests *do* fail as expected. Now I need to write the `find_motif()` code that will pass these tests. I decided I would again use k-mers, so I will bring in the `find_kmers()` function (and test it, of course, but I'll omit that here) from Chapters 9 and 10:

```
def find_kmers(seq: str, k: int) -> List[str]:
    """ Find k-mers in string """

    n = len(seq) - k + 1
    return [] if n < 1 else [seq[i:i + k] for i in range(n)]
```

Since the motif is four characters long, I can use this to find all the 4-mers in a sequence:

```
>>> from solution2_manual import find_kmers
>>> seq = 'NNTSYS'
>>> find_kmers(seq, 4)
['NNTS', 'NTSY', 'TSYS']
```

I will also need their positions. The `enumerate()` function I introduced in Chapter 8 will provide both the indexes and values of the items in a sequence:

```
>>> list(enumerate(find_kmers(seq, 4)))
[(0, 'NNTS'), (1, 'NTSY'), (2, 'TSYS')]
```

I can unpack each position and k-mer while iterating like so:

```
>>> for i, kmer in enumerate(find_kmers(seq, 4)):
...     print(i, kmer)
...
...
0 NNTS
1 NTSY
2 TSYS
```

Take the first k-mer, *NNTS*. One way to test for this pattern is to manually check each index:

```
>>> kmer = 'NNTS'
>>> kmer[0] == 'N' and kmer[1] != 'P' and kmer[2] in 'ST' and kmer[3] != 'P'
True
```

I know the first two k-mers should match, and this is borne out:

```
>>> for i, kmer in enumerate(find_kmers(seq, 4)):
...     kmer[0] == 'N' and kmer[1] != 'P' and kmer[2] in 'ST' and kmer[3] != 'P'
...
```

```
True
True
False
```

While effective, this is tedious. I would like to hide this code in a function:

```
def is_match(seq: str) -> bool:
    """ Find the N-glycosylation """

    return len(seq) == 4 and (seq[0] == 'N' and seq[1] != 'P'
                              and seq[2] in 'ST' and seq[3] != 'P')
```

Here is a test I wrote for the function:

```
def test_is_match() -> None:
    """ Test is_match """

    assert not is_match('')       ❶
    assert is_match('NASA')       ❷
    assert is_match('NATA')
    assert not is_match('NATAN')  ❸
    assert not is_match('NPTA')   ❹
    assert not is_match('NASP')   ❺
```

❶   If a function accepts a string parameter, I always test with an empty string.

❷   The next two sequences should match.

❸   This sequence is too long and should be rejected.

❹   This sequence has a *P* in the second position and should be rejected.

❺   This sequence has a *P* in the fourth position and should be rejected.

That makes the code much more readable:

```
>>> for i, kmer in enumerate(find_kmers(seq, 4)):
...     print(i, kmer, is_match(kmer))
...
0 NNTS True
1 NTSY True
2 TSYS False
```

I only want the k-mers that match. I could write this using an `if` expression with a guard, which I showed in Chapters 5 and 6:

```
>>> kmers = list(enumerate(find_kmers(seq, 4)))
>>> [i for i, kmer in kmers if is_match(kmer)]
[0, 1]
```

Or using the `starfilter()` function I showed in Chapter 9:

```
>>> from iteration_utilities import starfilter
>>> list(starfilter(lambda i, s: is_match(s), kmers))
[(0, 'NNTS'), (1, 'NTSY')]
```

I only want the first elements from each of the tuples, so I could use a `map()` to select those:

```
>>> matches = starfilter(lambda i, s: is_match(s), kmers)
>>> list(map(lambda t: t[0], matches))
[0, 1]
```

For what it's worth, Haskell uses tuples extensively and includes two handy functions in the prelude: `fst()` to get the first element from a 2-tuple, and `snd()` to get the second. Be sure to import `typing.Tuple` for this code:

```
def fst(t: Tuple[Any, Any]) -> Any:
    return t[0]

def snd(t: Tuple[Any, Any]) -> Any:
    return t[1]
```

With these functions, I can eliminate the `starfilter()` like this:

```
>>> list(map(fst, filter(lambda t: is_match(snd(t)), kmers)))
[0, 1]
```

But notice a very subtle bug if I try to use the `filter()`/`starmap()` technique I've shown a couple of times:

```
>>> from itertools import starmap
>>> list(filter(None, starmap(lambda i, s: i if is_match(s) else None, kmers)))
[1]
```

It only returns the second match. Why is that? It's due to using `None` as the predicate to `filter()`. According to `help(filter)`, "If [the] function is `None`, return the items that are true." In Chapter 1, I introduced the ideas of truthy and falsey values. The Boolean values `True` and `False` are represented by the integer values 1 and 0, respectively; hence, the actual number zero (either `int` or `float`) is technically `False`, which means that any nonzero number is not-`False` or, if you will, truthy. Python will evaluate many data types in a Boolean context to decide if they are truthy or falsey.

In this case, using `None` as the predicate for `filter()` causes it to remove the number 0:

```
>>> list(filter(None, [1, 0, 2]))
[1, 2]
```

I came to Python from Perl and JavaScript, both of which also silently coerce values given different contexts, so I was not so surprised by this behavior. If you come from a language like Java, C, or Haskell that has stricter types, this is probably quite troubling. I often feel that Python is a very powerful language if you know exactly what you're doing at all times. This is a high bar, so it's extremely important when writing Python to use types and tests liberally.

In the end, I felt the list comprehension was the easiest to read. Here's how I wrote my function to manually identify the protein motif:

```python
def find_motif(text: str) -> List[int]:
    """ Find a pattern in some text """

    kmers = list(enumerate(find_kmers(text, 4)))  ❶
    return [i for i, kmer in kmers if is_match(kmer)]  ❷
```

❶ Get the positions and values of the 4-mers from the text.

❷ Select those positions for the k-mers matching the motif.

Using this function is almost identical to how I used the regular expression, which is the point of hiding complexities behind functions:

```python
def main() -> None:
    args = get_args()
    files = fetch_fasta(args.file, args.download_dir)

    for file in files:
        prot_id, _ = os.path.splitext(os.path.basename(file))
        recs = SeqIO.parse(file, 'fasta')
        if rec := next(recs):
            if matches := find_motif(str(rec.seq)):  ❶
                pos = map(lambda p: p + 1, matches)  ❷
                print('\n'.join([prot_id, ' '.join(map(str, pos))]))  ❸
```

❶ Try to find any matches to the motif.

❷ The matches are a list of 0-based indexes, so add 1 to each.

❸ Convert the integer values to strings and join them on spaces to print.

Although this works and was fun (your mileage may vary) to write, I would not want to use or maintain this code. I hope it gives you a sense of how much work the regular expression is doing for us. Regexes allow me to describe *what* I want, not *how* to get it.

Going Further

The Eukaryotic Linear Motifs database example (*http://elm.eu.org/elms*) provides regexes to find motifs that define functional sites in proteins. Write a program to search for any occurrence of any pattern in a given set of FASTA files.

Review

Key points from this chapter:

- You can use command-line utilities like `curl` and `wget` to fetch data from the internet. Sometimes it makes sense to write a shell script for such tasks, and sometimes it's better to encode this using a language like Python.

- A regular expression can find the N-glycosylation motif, but it's necessary to wrap it in a look-ahead assertion and capturing parentheses to find overlapping matches.

- It's possible to manually find the N-glycosylation motif, but it's not easy.

- The `os.path.splitext()` function is useful when you need to separate a filename from the extension.

- File extensions are conventions and may be unreliable.

## CHAPTER 12

Inferring mRNA from Protein: Products and Reductions of Lists

As described in the Rosalind mRNA challenge (*https://oreil.ly/ZYelo*), the goal of this program is to find the number of mRNA strings that could produce a given protein sequence. You'll see that this number can become exceedingly large, so the final answer will be the remainder after dividing by a given value. I hope to show that I can turn the tables on regular expressions by trying to generate all the strings that could be matched by a particular pattern. I'll also show how to create the products of numbers and lists as well as how to *reduce* any list of values to a single value, and along the way I'll talk about some memory issues that can cause problems.

You will learn:

- How to use the `functools.reduce()` function to create a mathematical `product()` function for multiplying numbers
- How to use Python's modulo (%) operator
- About buffer overflow problems
- What monoids are
- How to reverse a dictionary by flipping the keys and values

Getting Started

You should work in the *12_mrna* directory of the repository. Begin by copying the first solution to the program `mrna.py`:

```
$ cd 12_mrna/
$ cp solution1_dict.py mrna.py
```

As usual, inspect the usage first:

```
$ ./mrna.py -h
usage: mrna.py [-h] [-m int] protein

Inferring mRNA from Protein

positional arguments:
  protein              Input protein or file ❶

optional arguments:
  -h, --help           show this help message and exit
  -m int, --modulo int Modulo value (default: 1000000) ❷
```

❶ The required positional argument is a protein sequence or a file containing a protein sequence.

❷ The --modulo option defaults to 1,000,000.

Run the program with the Rosalind example of *MA* and verify that it prints 12, the number of possible mRNA sequences modulo 1,000,000 that could encode this protein sequence:

```
$ ./mrna.py MA
12
```

The program will also read an input file for the sequence. The first input file has a sequence that is 998 residues long, and the result should be 448832:

```
$ ./mrna.py tests/inputs/1.txt
448832
```

Run the program with other inputs and also execute the tests with **make test**. When you are satisfied you understand how the program should work, start over:

```
$ new.py -fp 'Infer mRNA from Protein' mrna.py
Done, see new script "mrna.py".
```

Define the parameters as described in the usage. The protein may be a string or a filename, but I chose to model the parameter as a string. If the user provides a file, I will read the contents and pass this to the program as I first demonstrated in Chapter 3:

```
class Args(NamedTuple):
    """ Command-line arguments """
    protein: str ❶
    modulo: int ❷


def get_args() -> Args:
    """ Get command-line arguments """

    parser = argparse.ArgumentParser(
```

```
            description='Infer mRNA from Protein',
            formatter_class=argparse.ArgumentDefaultsHelpFormatter)

    parser.add_argument('protein', ❶
                        metavar='protein',
                        type=str,
                        help='Input protein or file')

    parser.add_argument('-m', ❷
                        '--modulo',
                        metavar='int',
                        type=int,
                        default=1000000,
                        help='Modulo value')

    args = parser.parse_args()

    if os.path.isfile(args.protein): ❸
        args.protein = open(args.protein).read().rstrip()

    return Args(args.protein, args.modulo)
```

❶ The required protein argument should be a string which may be a filename.

❷ The modulo option is an integer that will default to 1000000.

❸ If the protein argument names an existing file, read the protein sequence from the file.

Change your main() to print the protein sequence:

```
def main() -> None:
    args = get_args()
    print(args.protein)
```

Verify that your program prints the protein both from the command line and a file:

```
$ ./mrna.py MA
MA
$ ./mrna.py tests/inputs/1.txt | wc -c ❶
  998
```

❶ The -c option to wc indicates I only want a count of the number of *characters* in the input.

Your program should pass the first two tests and fail the third.

Creating the Product of Lists

When the input is *MA*, the program should print the response 12, which is the number of possible mRNA strings that could have produced this protein sequence, as shown in Figure 12-1. Using the same RNA encoding table from Chapter 7, I see that the amino acid methionine (*M*) is encoded by the mRNA codon sequence *AUG*,[1] alanine (*A*) has four possible codons (*GCA, GCC, GCG, GCU*), and the stop codon has three (*UAA, UAG, UGA*). The product of these three groups is $1 \times 4 \times 3 = 12$.

M	AUG	AUG, GCA, UAA AUG, GCA, UAG AUG, GCA, UGA AUG, GCC, UAA
A	GCA GCC GCG GCU	AUG, GCC, UAG AUG, GCC, UGA AUG, GCG, UAA AUG, GCG, UAG
Stop	UAA UAG UGA	AUG, GCG, UGA AUG, GCU, UAA AUG, GCU, UAG AUG, GCU, UGA

Figure 12-1. The Cartesian product of all the codons that encode the protein sequence MA results in 12 mRNA sequences

In Chapter 9, I introduced the `itertools.product()` function that will generate the Cartesian product from lists of values. I can produce all possible combinations of the 12 codons in the REPL like so:

```
>>> from itertools import product
>>> from pprint import pprint
>>> combos = product(*codons)
```

If you try printing `combos` to see the contents, you'll see it's not a list of values but a *product object*. That is, this is another lazy object that will wait to produce the values until you need them:

```
>>> pprint(combos)
<itertools.product object at 0x7fbdd822dac0>
```

I can use the `list()` function to coerce the values:

```
>>> pprint(list(combos))
[('AUG', 'GCA', 'UAA'),
 ('AUG', 'GCA', 'UAG'),
```

---

1 While there are other possible start codons, this is the only one considered by the Rosalind problem.

```
('AUG', 'GCA', 'UGA'),
('AUG', 'GCC', 'UAA'),
('AUG', 'GCC', 'UAG'),
('AUG', 'GCC', 'UGA'),
('AUG', 'GCG', 'UAA'),
('AUG', 'GCG', 'UAG'),
('AUG', 'GCG', 'UGA'),
('AUG', 'GCU', 'UAA'),
('AUG', 'GCU', 'UAG'),
('AUG', 'GCU', 'UGA')]
```

I want to show you a sneaky little bug waiting for you. Try printing the combos again:

```
>>> pprint(list(combos))
[]
```

This product object, like a generator, will yield the values only once and then will be exhausted. All subsequent calls will produce an empty list. To save the results, I need to save the coerced list to a variable:

```
>>> combos = list(product(*codons))
```

The length of this product is 12, meaning there are 12 ways to combine those amino acids to produce the sequence *MA*:

```
>>> len(combos)
12
```

Avoiding Overflow with Modular Multiplication

As the length of the input protein sequence grows, the number of possible combinations will grow extremely large. For example, the second test uses a protein with a file of 998 residues, resulting in approximately 8.98×10^{29} putative mRNA sequences. The Rosalind challenge notes:

> Because of memory considerations, most data formats that are built into languages have upper bounds on how large an integer can be: in some versions of Python, an int variable may be required to be no larger than $2^{31}-1$, or 2,147,483,647. As a result, to deal with very large numbers in Rosalind, we need to devise a system that allows us to manipulate large numbers without actually having to store large numbers.

Very large numbers pose the risk of exceeding the memory limitations for the size of an integer, especially on older 32-bit systems. To avoid this, the final answer should be the number of combinations *modulo* 1,000,000. The modulo operation returns the remainder when one number is divided by another. For example, 5 modulo 2 = 1 because 5 divided by 2 is 2 with a remainder of 1. Python has the % operator to compute the modulo:

```
>>> 5 % 2
1
```

The answer for the 998-residue protein is 448,832, which is the remainder after dividing 8.98×10^{29} by 1,000,000:

```
$ ./mrna.py tests/inputs/1.txt
448832
```

In Chapter 5, I introduce the NumPy module for mathematical operations. As you might expect, there is a numpy.prod() function that will compute the product of a list of numbers. Unfortunately, it can quietly fail and return 0 when I try to compute something as large as the factorial of 1,000:

```
>>> import numpy as np
>>> np.prod(range(1, 1001))
0
```

The problem here is that NumPy is implemented in C, which is faster than Python, and the C code tries to store a larger number than will fit into the memory available for an integer. The unfortunate result is 0. It's common to call this type of error a *buffer overflow*, where here the buffer is an integer variable but it could be a string, floating-point number, list, or any other container. Generally speaking, Python programmers don't have to worry about memory allocations the way programmers do in other languages, but here I must be aware of the limitations of the underlying library. Because the maximum size of an int can differ depending on the machine, numpy.prod() is an unreliable solution and should be avoided.

Since Python 3.8, there exists a math.prod() function that can calculate incredibly large products such as the factorial of 1,000. This is because all the computation happens inside Python, and integers in Python are virtually *unbounded*, meaning they are limited only by the available memory on your machine. Try running this on your computer:

```
>>> import math
>>> math.prod(range(1, 1001))
```

Notice, however, the result is 0 when I apply the modulo operation:

```
>>> math.prod(range(1, 1001)) % 1000000
0
```

Once again, I've bumped up against an overflow that quietly fails, this one due to Python's use of a float in the division operation, which is a bounded type. For the provided tests, you should not encounter a problem if you use math.prod() and modulo the results. In the solutions, I will show a way to compute the product of an arbitrarily large set of numbers using the modulo operation to avoid integer overflow. This should be enough for you to solve the problem. Keep working until your program passes all the tests.

Solutions

I present three solutions that mostly differ in the structure of a dictionary used to represent the RNA translation information and in how to compute the mathematical product of a list of numbers.

Solution 1: Using a Dictionary for the RNA Codon Table

For my first solution, I used the RNA codon table from Chapter 7 to find the number of codons for each residue:

```
>>> c2aa = {
...     'AAA': 'K', 'AAC': 'N', 'AAG': 'K', 'AAU': 'N', 'ACA': 'T',
...     'ACC': 'T', 'ACG': 'T', 'ACU': 'T', 'AGA': 'R', 'AGC': 'S',
...     'AGG': 'R', 'AGU': 'S', 'AUA': 'I', 'AUC': 'I', 'AUG': 'M',
...     'AUU': 'I', 'CAA': 'Q', 'CAC': 'H', 'CAG': 'Q', 'CAU': 'H',
...     'CCA': 'P', 'CCC': 'P', 'CCG': 'P', 'CCU': 'P', 'CGA': 'R',
...     'CGC': 'R', 'CGG': 'R', 'CGU': 'R', 'CUA': 'L', 'CUC': 'L',
...     'CUG': 'L', 'CUU': 'L', 'GAA': 'E', 'GAC': 'D', 'GAG': 'E',
...     'GAU': 'D', 'GCA': 'A', 'GCC': 'A', 'GCG': 'A', 'GCU': 'A',
...     'GGA': 'G', 'GGC': 'G', 'GGG': 'G', 'GGU': 'G', 'GUA': 'V',
...     'GUC': 'V', 'GUG': 'V', 'GUU': 'V', 'UAC': 'Y', 'UAU': 'Y',
...     'UCA': 'S', 'UCC': 'S', 'UCG': 'S', 'UCU': 'S', 'UGC': 'C',
...     'UGG': 'W', 'UGU': 'C', 'UUA': 'L', 'UUC': 'F', 'UUG': 'L',
...     'UUU': 'F', 'UAA': '*', 'UAG': '*', 'UGA': '*',
... }
```

I want to iterate over each amino acid in the protein sequence *MA* plus the stop codon to find all the encoding codons. Note that sequences from Rosalind do not terminate with the stop codon, so I must append *. I can use a list comprehension with a guard to express this:

```
>>> protein = 'MA'
>>> for aa in protein + '*':
...     print(aa, [c for c, res in c2aa.items() if res == aa])
...
M ['AUG']
A ['GCA', 'GCC', 'GCG', 'GCU']
* ['UAA', 'UAG', 'UGA']
```

I don't need the actual list of codons that encode a given residue, only the number which I can find using the len() function:

```
>>> possible = [
...     len([c for c, res in c2aa.items() if res == aa])
...     for aa in protein + '*'
... ]
>>>
>>> possible
[1, 4, 3]
```

The answer lies in multiplying these values. In the previous section, I suggested you could use the `math.prod()` function:

```
>>> import math
>>> math.prod(possible)
12
```

Although this will work perfectly well, I'd like to take this opportunity to talk about *reducing* a sequence of values to a single value. In Chapter 5, I introduced the `sum()` function that will add the numbers 1, 4, and 3 to create the result 8:

```
>>> sum(possible)
8
```

It does this in pairs, first adding 1 + 4 to get 5, then adding 5 + 3 to get 8. If I change the + operator to *, then I get a product and the result is 12, as shown in Figure 12-2.

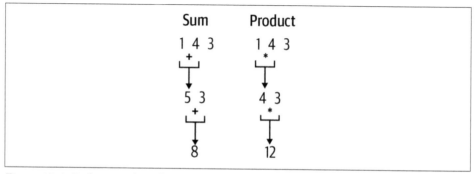

Figure 12-2. Reducing a list of numbers using addition and multiplication

This is the idea behind reducing a list of values, and it's precisely what the `functools.reduce()` function helps us to do. This is another higher-order function, like `filter()` and `map()` and others I've used throughout the book, but with an important difference: the `lambda` function will receive *two* arguments instead of only one. The documentation shows how to write `sum()`:

```
reduce(...)
    reduce(function, sequence[, initial]) -> value

    Apply a function of two arguments cumulatively to the items of a sequence,
    from left to right, so as to reduce the sequence to a single value.
    For example, reduce(lambda x, y: x+y, [1, 2, 3, 4, 5]) calculates
    ((((1+2)+3)+4)+5).  If initial is present, it is placed before the items
    of the sequence in the calculation, and serves as a default when the
    sequence is empty.
```

Here is how I could use this to write my own version of `sum()`:

```
>>> from functools import reduce
>>> reduce(lambda x, y: x + y, possible)
8
```

To create a product, I can change the addition to multiplication:

```
>>> reduce(lambda x, y: x * y, possible)
12
```

Monoids

For what it's worth, a homogeneous list of numbers or strings could be thought of as a *monoid*, which is an algebraic structure with a single associative binary operation and an identity element. For a list of numbers under addition, the binary operation is +, and the identity element is 0. The *identity element*, or *neutral element*, is a special value you could combine with another element under the binary operation that leaves the element unchanged. Since $0 + n = n$, then 0 is the identity element for addition and is the appropriate value for the *initial* argument to reduce():

```
>>> reduce(lambda x, y: x + y, [1, 2, 3, 4], 0)
10
```

It's also the appropriate return value when adding an empty list of numbers:

```
>>> reduce(lambda x, y: x + y, [], 0)
0
```

Under multiplication, the operator is * and the identity element is 1 because $1 * n = n$:

```
>>> reduce(lambda x, y: x * y, [1, 2, 3, 4], 1)
24
```

The product of an empty list is 1, so this is the correct initial value for reduce():

```
>>> reduce(lambda x, y: x * y, [], 1)
1
```

A list of strings under concatenation would also use +, and the result is a new string:

```
>>> reduce(lambda x, y: x + y, ['M', 'A'], '')
'MA'
```

The identity element for strings is the empty string:

```
>>> reduce(lambda x, y: x + y, [], '')
''
```

This is the same value returned when passing the empty list to str.join():

```
>>> ''.join([])
''
```

Even lists themselves are monoids. They can be combined using the + operator or operator.concat(), and the identity value is the empty list:

```
```

```
>>> reduce(operator.concat, [['A'], ['list', 'of'], ['values']], [])
['A', 'list', 'of', 'values']
```

These identity values are the defaults for `collections.defaultdict()`. That is, a `defaultdict(int)` initializes to 0, `defaultdict(str)` uses the empty string, and `defaultdict(list)` uses the empty list.

Monoids such as these can be reduced to a single value by progressively applying the *associative binary operation*, hence the name of the `reduce()` function. If you find this kind of theory interesting, you might be interested to learn more about the Haskell programming language and category theory.

Note that Python has an `id()` function that will "return the identity of an object," which is a unique numerical representation of a value in Python, akin to a memory address, and so is not at all like monoidal identity.

I can use `functools.reduce()` to write my own `product()` function:

```
def product(xs: List[int]) -> int: ❶
    """ Return the product """

    return reduce(lambda x, y: x * y, xs, 1) ❷
```

❶ Return the product of a list of integers.

❷ Use the `functools.reduce()` function to progressively multiply the values. Use 1 for the initial result to ensure that an empty list returns 1.

Why would I do this? Intellectual curiosity, for one, but I also want to show how I could use this to write a function that works without relying on Python's unbounded integers, which this version does. To avoid overflowing in any step of the reduction, I need to incorporate the modulo operation into the function itself rather than applying it to the end result. Given that I'm not a math wizard, I didn't know how to write such a function. I searched the internet and found some code which I modified into this:

```
def mulmod(a: int, b: int, mod: int) -> int: ❶
    """ Multiplication with modulo """

    def maybemod(x): ❷
        ret = (x % mod) if mod > 1 and x > mod else x
        return ret or x ❸

    res = 0 ❹
    a = maybemod(a) ❺
    while b > 0: ❻
        if b % 2 == 1: ❼
            res = maybemod(res + a) ❽
```

```
    a = maybemod(a * 2) ❾
    b //= 2 ❿

return res
```

❶ The `mulmod()` function accepts two integers a and b to multiply with an integer modulo value mod.

❷ This is a closure around the mod value to possibly return a value modulo mod.

❸ If the result is 0, return the original value; otherwise, return the computed value.

❹ Initialize the result.

❺ Possibly reduce the size of a.

❻ Loop while b is greater than 0.

❼ Check if b is an odd number.

❽ Add a to the result and possibly modulo the result.

❾ Double a and possibly modulo the value.

❿ Halve b using floor division, eventually resulting in 0 and terminating the loop.

Following is the test I wrote:

```
def test_mulmod() -> None:
    """ Text mulmod """

    assert mulmod(2, 4, 3) == 2
    assert mulmod(9223372036854775807, 9223372036854775807, 1000000) == 501249
```

I chose those large numbers because they are the `sys.maxsize` on my machine:

```
>>> import sys
>>> sys.maxsize
9223372036854775807
```

Note that this is the same answer that I can get from `math.prod()`, but my version does not rely on Python's dynamic integer sizing and is not tied (as much) to the available memory on my machine:

```
>>> import math
>>> math.prod([9223372036854775807, 9223372036854775807]) % 1000000
501249
```

To integrate this, I wrote a `modprod()` function and added a test as follows:

```
def modprod(xs: List[int], modulo: int) -> int:
    """ Return the product modulo a value """

    return reduce(lambda x, y: mulmod(x, y, modulo), xs, 1)


def test_modprod() -> None:
    """ Test modprod """

    assert modprod([], 3) == 1
    assert modprod([1, 4, 3], 1000000) == 12
    n = 9223372036854775807
    assert modprod([n, n], 1000000) == 501249
```

Note that it can handle the earlier example of the factorial of 1,000. The answer to this is still too large to print, but the point is that the answer is not 0:

```
>>> modprod(range(1, 1001), 1000000)
```

The final answer is the products of these numbers modulo the given argument. Here is how I put it all together:

```
def main() -> None:
    args = get_args()
    codon_to_aa = { ❶
        'AAA': 'K', 'AAC': 'N', 'AAG': 'K', 'AAU': 'N', 'ACA': 'T',
        'ACC': 'T', 'ACG': 'T', 'ACU': 'T', 'AGA': 'R', 'AGC': 'S',
        'AGG': 'R', 'AGU': 'S', 'AUA': 'I', 'AUC': 'I', 'AUG': 'M',
        'AUU': 'I', 'CAA': 'Q', 'CAC': 'H', 'CAG': 'Q', 'CAU': 'H',
        'CCA': 'P', 'CCC': 'P', 'CCG': 'P', 'CCU': 'P', 'CGA': 'R',
        'CGC': 'R', 'CGG': 'R', 'CGU': 'R', 'CUA': 'L', 'CUC': 'L',
        'CUG': 'L', 'CUU': 'L', 'GAA': 'E', 'GAC': 'D', 'GAG': 'E',
        'GAU': 'D', 'GCA': 'A', 'GCC': 'A', 'GCG': 'A', 'GCU': 'A',
        'GGA': 'G', 'GGC': 'G', 'GGG': 'G', 'GGU': 'G', 'GUA': 'V',
        'GUC': 'V', 'GUG': 'V', 'GUU': 'V', 'UAC': 'Y', 'UAU': 'Y',
        'UCA': 'S', 'UCC': 'S', 'UCG': 'S', 'UCU': 'S', 'UGC': 'C',
        'UGG': 'W', 'UGU': 'C', 'UUA': 'L', 'UUC': 'F', 'UUG': 'L',
        'UUU': 'F', 'UAA': '*', 'UAG': '*', 'UGA': '*',
    }

    possible = [ ❷
        len([c for c, res in codon_to_aa.items() if res == aa])
        for aa in args.protein + '*'
    ]
    print(modprod(possible, args.modulo)) ❸
```

❶ A dictionary encoding RNA codons to amino acids.

❷ Iterate through the residues of the protein plus the stop codon, then find the number of codons matching the given amino acid.

❸ Print the product of the possibilities modulo the given value.

Solution 2: Turn the Beat Around

For this next solution, I decided to reverse the keys and values of the RNA codons dictionary so that the unique amino acids form the keys and the values are the lists of codons. It's handy to know how to flip a dictionary like this, but it only works if the values are unique. For instance, I can create a lookup table to go from DNA bases like *A* or *T* to their names:

```
>>> base_to_name = dict(A='adenine', G='guanine', C='cytosine', T='thymine')
>>> base_to_name['A']
'adenine'
```

To turn that around so I could go from the name to the base, I can use dict.items() to get the key/value pairs:

```
>>> list(base_to_name.items())
[('A', 'adenine'), ('G', 'guanine'), ('C', 'cytosine'), ('T', 'thymine')]
```

I then map() those through reversed() to flip them, and finally pass the result to the dict() function to create a dictionary:

```
>>> dict(map(reversed, base_to_name.items()))
{'adenine': 'A', 'guanine': 'G', 'cytosine': 'C', 'thymine': 'T'}
```

If I try that on the RNA codons table from the first solution, however, I'll get this:

```
>>> pprint(dict(map(reversed, c2aa.items())))
{'*': 'UGA',
 'A': 'GCU',
 'C': 'UGU',
 'D': 'GAU',
 'E': 'GAG',
 'F': 'UUU',
 'G': 'GGU',
 'H': 'CAU',
 'I': 'AUU',
 'K': 'AAG',
 'L': 'UUG',
 'M': 'AUG',
 'N': 'AAU',
 'P': 'CCU',
 'Q': 'CAG',
 'R': 'CGU',
 'S': 'UCU',
 'T': 'ACU',
 'V': 'GUU',
 'W': 'UGG',
 'Y': 'UAU'}
```

You can see that I'm missing most of the codons. Only *M* and *W* have just one codon. What happened to the rest? When creating the dictionary, Python overwrote any existing values for a key with the newest value. In the original table, for instance,

UUG was the last value indicated for *L*, so that was the value that was left standing. Just remember this trick for reversing dictionary key/values and ensure that the values are unique. For what it's worth, if I needed to do this, I would use the `collections.defaultdict()` function:

```
>>> from collections import defaultdict
>>> aa2codon = defaultdict(list)
>>> for k, v in c2aa.items():
...     aa2codon[v].append(k)
...
>>> pprint(aa2codon)
defaultdict(<class 'list'>,
            {'*': ['UAA', 'UAG', 'UGA'],
             'A': ['GCA', 'GCC', 'GCG', 'GCU'],
             'C': ['UGC', 'UGU'],
             'D': ['GAC', 'GAU'],
             'E': ['GAA', 'GAG'],
             'F': ['UUC', 'UUU'],
             'G': ['GGA', 'GGC', 'GGG', 'GGU'],
             'H': ['CAC', 'CAU'],
             'I': ['AUA', 'AUC', 'AUU'],
             'K': ['AAA', 'AAG'],
             'L': ['CUA', 'CUC', 'CUG', 'CUU', 'UUA', 'UUG'],
             'M': ['AUG'],
             'N': ['AAC', 'AAU'],
             'P': ['CCA', 'CCC', 'CCG', 'CCU'],
             'Q': ['CAA', 'CAG'],
             'R': ['AGA', 'AGG', 'CGA', 'CGC', 'CGG', 'CGU'],
             'S': ['AGC', 'AGU', 'UCA', 'UCC', 'UCG', 'UCU'],
             'T': ['ACA', 'ACC', 'ACG', 'ACU'],
             'V': ['GUA', 'GUC', 'GUG', 'GUU'],
             'W': ['UGG'],
             'Y': ['UAC', 'UAU']})
```

This is the data structure I used in the following solution. I also show how to use the `math.prod()` function rather than rolling my own:

```
def main():
    args = get_args()
    aa_to_codon = { ❶
        'A': ['GCA', 'GCC', 'GCG', 'GCU'],
        'C': ['UGC', 'UGU'],
        'D': ['GAC', 'GAU'],
        'E': ['GAA', 'GAG'],
        'F': ['UUC', 'UUU']
        'G': ['GGA', 'GGC', 'GGG', 'GGU'],
        'H': ['CAC', 'CAU'],
        'I': ['AUA', 'AUC', 'AUU'],
        'K': ['AAA', 'AAG'],
        'L': ['CUA', 'CUC', 'CUG', 'CUU', 'UUA', 'UUG'],
        'M': ['AUG'],
        'N': ['AAC', 'AAU'],
```

```
        'P': ['CCA', 'CCC', 'CCG', 'CCU'],
        'Q': ['CAA', 'CAG'],
        'R': ['AGA', 'AGG', 'CGA', 'CGC', 'CGG', 'CGU'],
        'S': ['AGC', 'AGU', 'UCA', 'UCC', 'UCG', 'UCU'],
        'T': ['ACA', 'ACC', 'ACG', 'ACU'],
        'V': ['GUA', 'GUC', 'GUG', 'GUU'],
        'W': ['UGG'],
        'Y': ['UAC', 'UAU'],
        '*': ['UAA', 'UAG', 'UGA'],
    }

    possible = [len(aa_to_codon[aa]) for aa in args.protein + '*']  ❷
    print(math.prod(possible) % args.modulo)  ❸
```

❶ Represent the dictionary using the residues as the keys and the codons for the values.

❷ Find the number of codons encoding each amino acid in the protein sequence plus the stop codon.

❸ Use `math.prod()` to calculate the product, then apply the modulo operator.

This version is much shorter and assumes that the machine will have enough memory to compute the product. (Python will handle the memory requirements to represent astronomically large numbers.) For all the datasets given to me by Rosalind, this was true, but you may one day encounter the need to use something like the `mulmod()` function in your travels.

Solution 3: Encoding the Minimal Information

The previous solution encoded more information than was necessary to find the solution. Since I only need the *number* of codons that encode a given amino acid, not the actual list, I could instead create this lookup table:

```
>>> codons = {
...     'A': 4, 'C': 2, 'D': 2, 'E': 2, 'F': 2, 'G': 4, 'H': 2, 'I': 3,
...     'K': 2, 'L': 6, 'M': 1, 'N': 2, 'P': 4, 'Q': 2, 'R': 6, 'S': 6,
...     'T': 4, 'V': 4, 'W': 1, 'Y': 2, '*': 3,
... }
```

A list comprehension will return the numbers needed for the product. I will use 1 for the default argument to `dict.get()` here in case I find a residue not present in my dictionary:

```
>>> [codons.get(aa, 1) for aa in 'MA*']
[1, 4, 3]
```

Leading to this code:

```
def main():
    args = get_args()
    codons = {  ❶
        'A': 4, 'C': 2, 'D': 2, 'E': 2, 'F': 2, 'G': 4, 'H': 2, 'I': 3,
        'K': 2, 'L': 6, 'M': 1, 'N': 2, 'P': 4, 'Q': 2, 'R': 6, 'S': 6,
        'T': 4, 'V': 4, 'W': 1, 'Y': 2, '*': 3,
    }
    nums = [codons.get(aa, 1) for aa in args.protein + '*']  ❷
    print(math.prod(nums) % args.modulo)  ❸
```

❶ Encode the number of codons for each amino acid.

❷ Find the number of codons for each amino acid plus the stop.

❸ Print the product of the combinations modulo the given value.

Going Further

In a sense, I've reversed the idea of a regular expression match by creating all the possible strings for a match. That is, the 12 patterns that could produce the protein *MA* are as follows:

```
$ ./show_patterns.py MA
     1: AUGGCAUAA
     2: AUGGCAUAG
     3: AUGGCAUGA
     4: AUGGCCUAA
     5: AUGGCCUAG
     6: AUGGCCUGA
     7: AUGGCGUAA
     8: AUGGCGUAG
     9: AUGGCGUGA
    10: AUGGCUUAA
    11: AUGGCUUAG
    12: AUGGCUUGA
```

Essentially, I could try to use this information to create a single unified regular expression. That might not be easy or even possible, but it's an idea that might help me find a genomic source for a protein. For example, the first two sequences differ by their last base. The alternation between *A* and *G* can be expressed with the character class [AG]:

```
  AUGGCAUAA
+ AUGGCAUAG
  ---------
  AUGGCAUA[AG]
```

Could you write a tool that would combine many regular expression patterns into a single one?

Review

Key points from this chapter:

- The `itertools.product()` function will create the Cartesian product of a list of iterables.
- `functools.reduce()` is a higher-order function that provides a way to combine progressive pairs of elements from an iterable.
- Python's `%` (modulo) operator will return the remainder after division.
- Homogeneous lists of numbers and strings can be reduced under monoidal operations like addition, multiplication, and concatenation to a single value.
- A dictionary with unique values can be reversed by flipping the keys and values.
- The size of integer values in Python is limited only by the available memory.

## CHAPTER 13

Location Restriction Sites: Using, Testing, and Sharing Code

A *palindromic* sequence in DNA is one in which the 5' to 3' base pair sequence is identical on both strands. For example, Figure 13-1 shows that the reverse complement of the DNA sequence *GCATGC* is the sequence itself.

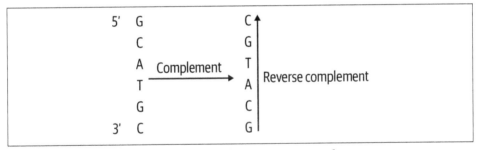

Figure 13-1. A reverse palindrome is equal to its reverse complement

I can verify this in code:

```
>>> from Bio import Seq
>>> seq = 'GCATGC'
>>> Seq.reverse_complement(seq) == seq
True
```

As described in the Rosalind REVP challenge (*https://oreil.ly/w3Tdm*), restriction enzymes recognize and cut within specific palindromic sequences of DNA known as restriction sites. They typically have a length of between 4 and 12 nucleotides. The goal of this exercise is to find the locations in a DNA sequence of every putative restriction enzyme. The code to solve this problem could be massively complicated, but a clear understanding of some functional programming techniques helps to create a short, elegant solution. I will explore `map()`, `zip()`, and `enumerate()` as well as many small, tested functions.

You will learn:

- How to find a reverse palindrome
- How to create modules to share common functions
- About the `PYTHONPATH` environment variable

Getting Started

The code and tests for this exercise are in the *13_revp* directory. Start by copying a solution to the program `revp.py`:

```
$ cd 13_revp
$ cp solution1_zip_enumerate.py revp.py
```

Inspect the usage:

```
$ ./revp.py -h
usage: revp.py [-h] FILE

Locating Restriction Sites

positional arguments:
   FILE        Input FASTA file ❶

optional arguments:
  -h, --help  show this help message and exit
```

❶ The only required argument is a single positional file of FASTA-formatted DNA sequences.

Have a look at the first test input file. The contents are identical to the example on the Rosalind page:

```
$ cat tests/inputs/1.fa
>Rosalind_24
TCAATGCATGCGGGTCTATATGCAT
```

Run the program with this input and verify that you see the position (using 1-based counting) and length of every reverse palindrome in the string having a length between 4 and 12, as illustrated in Figure 13-2. Note that the order of the results is unimportant:

```
$ ./revp.py tests/inputs/1.fa
5 4
7 4
17 4
```

```
18 4
21 4
4 6
6 6
20 6
```

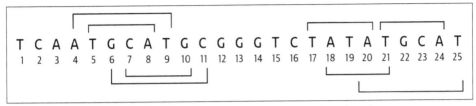

Figure 13-2. The locations of the eight reverse palindromes found in the sequence TCAATGCATGCGGGTCTATATGCAT.

Run the tests to verify that the program passes, then start over:

```
$ new.py -fp 'Locating Restriction Sites' revp.py
Done, see new script "revp.py".
```

Here is a way to define the program's parameter:

```
class Args(NamedTuple):
    """ Command-line arguments """
    file: TextIO ❶


def get_args() -> Args:
    """ Get command-line arguments """

    parser = argparse.ArgumentParser(
        description='Locating Restriction Sites',
        formatter_class=argparse.ArgumentDefaultsHelpFormatter)

    parser.add_argument('file', ❷
                        help='Input FASTA file',
                        metavar='FILE',
                        type=argparse.FileType('rt'))

    args = parser.parse_args()

    return Args(args.file)
```

❶ The only parameter is a file.

❷ Define a parameter that must be a readable text file.

Have the main() function print the input filename for now:

```
def main() -> None:
    args = get_args()
    print(args.file.name)
```

Manually verify that the program will produce the correct usage, will reject bogus files, and will print a valid input's name:

```
$ ./revp.py tests/inputs/1.fa
tests/inputs/1.fa
```

Run **make test** and you should find you pass some tests. Now you're ready to write the bones of the program.

Finding All Subsequences Using K-mers

The first step is to read the sequences from the FASTA input file. I can use SeqIO.parse() to create a lazy iterator and then use next() to get the first sequence:

```
>>> from Bio import SeqIO
>>> recs = SeqIO.parse(open('tests/inputs/1.fa'), 'fasta')
>>> rec = next(recs)
>>> seq = str(rec.seq)
>>> seq
'TCAATGCATGCGGGTCTATATGCAT'
```

 The preceding code is unsafe to use if the file is empty, such as *tests/inputs/empty.fa*. If you try to open this file in the same way and call next(), Python will raise a StopIteration exception. In your code, I recommend you use a for loop that detects the exhaustion of the iterator and gracefully exits.

```
>>> empty = SeqIO.parse(open('tests/inputs/empty.fa'), 'fasta')
>>> next(empty)
Traceback (most recent call last):
  File "<stdin>", line 1, in <module>
  File "/Library/Frameworks/Python.framework/Versions/3.9/lib/python3.9/
  site-packages/Bio/SeqIO/Interfaces.py", line 73, in __next__
    return next(self.records)
StopIteration
```

I need to find all the sequences between 4 and 12 bases long. This sounds like another job for k-mers, so I'll bring in the find_kmers() function from Chapter 9:

```
>>> def find_kmers(seq, k):
...     n = len(seq) - k + 1
...     return [] if n < 1 else [seq[i:i + k] for i in range(n)]
...
```

I can use `range()` to generate all the numbers between 4 and 12, remembering that the end position is not included so I have to go up to 13. As there are many k-mers for each k, I'll print the value of k and how many k-mers are found:

```
>>> for k in range(4, 13):
...     print(k, len(find_kmers(seq, k)))
...
4 22
5 21
6 20
7 19
8 18
9 17
10 16
11 15
12 14
```

Finding All Reverse Complements

I showed many ways to find the reverse complement in Chapter 3, with the conclusion that `Bio.Seq.reverse_complement()` is probably the easiest method. Start by finding all the 12-mers:

```
>>> kmers = find_kmers(seq, 12)
>>> kmers
['TCAATGCATGCG', 'CAATGCATGCGG', 'AATGCATGCGGG', 'ATGCATGCGGGT',
 'TGCATGCGGGTC', 'GCATGCGGGTCT', 'CATGCGGGTCTA', 'ATGCGGGTCTAT',
 'TGCGGGTCTATA', 'GCGGGTCTATAT', 'CGGGTCTATATG', 'GGGTCTATATGC',
 'GGTCTATATGCA', 'GTCTATATGCAT']
```

To create a list of the reverse complements, you could use a list comprehension:

```
>>> from Bio import Seq
>>> revc = [Seq.reverse_complement(kmer) for kmer in kmers]
```

Or use `map()`:

```
>>> revc = list(map(Seq.reverse_complement, kmers))
```

Either way, you should have 12 reverse complements:

```
>>> revc
['CGCATGCATTGA', 'CCGCATGCATTG', 'CCCGCATGCATT', 'ACCCGCATGCAT',
 'GACCCGCATGCA', 'AGACCCGCATGC', 'TAGACCCGCATG', 'ATAGACCCGCAT',
 'TATAGACCCGCA', 'ATATAGACCCGC', 'CATATAGACCCG', 'GCATATAGACCC',
 'TGCATATAGACC', 'ATGCATATAGAC']
```

Putting It All Together

You should have just about everything you need to complete this challenge. First, pair all the k-mers with their reverse complements, find those that are the same, and print their positions. You could iterate through them with a `for` loop, or you might consider using the zip() function that we first looked at in Chapter 6 to create the pairs. This is an interesting challenge, and I'm sure you can figure out a working solution before you read my versions.

Solutions

I'll show three variations to find the restriction sites, which increasingly rely on functions to hide the complexities of the program.

Solution 1: Using the zip() and enumerate() Functions

In my first solution, I first use zip() to pair the k-mers and reverse complements. Assume k=4:

```
>>> seq = 'TCAATGCATGCGGGTCTATATGCAT'
>>> kmers = find_kmers(seq, 4)
>>> revc = list(map(Seq.reverse_complement, kmers))
>>> pairs = list(zip(kmers, revc))
```

I also need to know the positions of the pairs, which I can get from enumerate(). If I inspect the pairs, I see that some of them (4, 6, 16, 17, and 20) are the same:

```
>>> pprint(list(enumerate(pairs)))
[(0, ("TCAA", "TTGA")),
 (1, ("CAAT", "ATTG")),
 (2, ("AATG", "CATT")),
 (3, ("ATGC", "GCAT")),
 (4, ("TGCA", "TGCA")),
 (5, ("GCAT", "ATGC")),
 (6, ("CATG", "CATG")),
 (7, ("ATGC", "GCAT")),
 (8, ("TGCG", "CGCA")),
 (9, ("GCGG", "CCGC")),
 (10, ("CGGG", "CCCG")),
 (11, ("GGGT", "ACCC")),
 (12, ("GGTC", "GACC")),
 (13, ("GTCT", "AGAC")),
 (14, ("TCTA", "TAGA")),
 (15, ("CTAT", "ATAG")),
 (16, ("TATA", "TATA")),
 (17, ("ATAT", "ATAT")),
 (18, ("TATG", "CATA")),
 (19, ("ATGC", "GCAT")),
 (20, ("TGCA", "TGCA")),
 (21, ("GCAT", "ATGC"))]
```

I can use a list comprehension with a guard to find all the positions where the pairs are the same. Note I add 1 to the index values to get 1-based positions:

```
>>> [pos + 1 for pos, pair in enumerate(pairs) if pair[0] == pair[1]]
[5, 7, 17, 18, 21]
```

In Chapter 11, I introduced the functions `fst()` and `snd()` for getting the first or second elements from a 2-tuple. I'd like to use those here so I don't have to use indexing with the tuples. I also keep using the `find_kmers()` function from previous chapters. It seems like it's time to put these functions into a separate module so I can import them as needed rather than copying them.

If you inspect the `common.py` module, you'll see these functions and their tests. I can run `pytest` to ensure they all pass:

```
$ pytest -v common.py
============================ test session starts =============================
...

common.py::test_fst PASSED                                             [ 33%]
common.py::test_snd PASSED                                             [ 66%]
common.py::test_find_kmers PASSED                                      [100%]

============================= 3 passed in 0.01s =============================
```

Because `common.py` is in the current directory, I can import any functions I like from it:

```
>>> from common import fst, snd
>>> [pos + 1 for pos, pair in enumerate(pairs) if fst(pair) == snd(pair)]
[5, 7, 17, 18, 21]
```

PYTHONPATH

You can also place modules of reusable code into a directory that is shared across all your projects. You can use the PYTHONPATH environment variable to indicate the location of additional directories where Python should look for modules. According to the PYPATH documentation (*https://oreil.ly/0MpPP*), it will:

> Augment the default search path for module files. The format is the same as the shell's PATH: one or more directory pathnames separated by `os.pathsep` (e.g., colons on Unix or semicolons on Windows). Nonexistent directories are silently ignored.

In Appendix B, I recommend that you install binaries and scripts to a location like `$HOME/.local/bin` and use something like `$HOME/.bashrc` to set your PATH to include this directory. (I prefer *.local* so that it is hidden from the normal directory listing.) I would likewise suggest you define a location for sharing common Python functions and modules and set your PYTHONPATH to include this location: perhaps something like `$HOME/.local/lib`.

Here is how I incorporated these ideas in the first solution:

```
def main() -> None:
    args = get_args()
```

```
for rec in SeqIO.parse(args.file, 'fasta'): ❶
    for k in range(4, 13): ❷
        kmers = find_kmers(str(rec.seq), k) ❸
        revc = list(map(Seq.reverse_complement, kmers)) ❹

        for pos, pair in enumerate(zip(kmers, revc)): ❺
            if fst(pair) == snd(pair): ❻
                print(pos + 1, k) ❼
```

❶ Iterate over the records in the FASTA file.

❷ Iterate through all the values of k.

❸ Find the k-mers for this k.

❹ Find the reverse complements of the k-mers.

❺ Iterate through the positions and pairs of k-mer/reverse complement.

❻ Check if the first element of the pair is the same as the second element.

❼ Print the position plus 1 (to correct for 0-based indexing) and the size of the sequence k.

Solution 2: Using the operator.eq() Function

Though I like the fst() and snd() functions and want to highlight how to share modules and functions, I'm duplicating the operator.eq() function. I first introduced this module in Chapter 6 to use the operator.ne() (not equal) function, and I've also used the operator.le() (less than or equal) and operator.add() functions elsewhere.

I can rewrite part of the preceding solution like so:

```
for pos, pair in enumerate(zip(kmers, revc)):
    if operator.eq(*pair): ❶
        print(pos + 1, k)
```

❶ Use the functional version of the == operator to compare the elements of the pair. Note the need to splat the pair to expand the tuple into its two values.

I prefer a list comprehension with a guard to condense this code:

```
def main() -> None:
    args = get_args()
    for rec in SeqIO.parse(args.file, 'fasta'):
        for k in range(4, 13):
            kmers = find_kmers(str(rec.seq), k)
```

```
revc = map(Seq.reverse_complement, kmers)
pairs = enumerate(zip(kmers, revc))

for pos in [pos + 1 for pos, pair in pairs if operator.eq(*pair)]: ❶
    print(pos, k)
```

❶ Use a guard for the equality comparison, and correct the position inside a list comprehension.

Solution 3: Writing a revp() Function

In this final solution, it behooves me to write a revp() function and create a test. This will make the program more readable and will also make it easier to move this function into something like the common.py module for sharing in other projects.

As usual, I imagine the signature of my function:

```
def revp(seq: str, k: int) -> List[int]: ❶
    """ Return positions of reverse palindromes """

    return [] ❷
```

❶ I want to pass in a sequence and a value for k to get back a list of locations where reverse palindromes of the given size are found.

❷ For now, return the empty list.

Here is the test I wrote. Note that I decided that the function should correct the indexes to 1-based counting:

```
def test_revp() -> None:
    """ Test revp """

    assert revp('CGCATGCATTGA', 4) == [3, 5]
    assert revp('CGCATGCATTGA', 5) == []
    assert revp('CGCATGCATTGA', 6) == [2, 4]
    assert revp('CGCATGCATTGA', 7) == []
    assert revp('CCCGCATGCATT', 4) == [5, 7]
    assert revp('CCCGCATGCATT', 5) == []
    assert revp('CCCGCATGCATT', 6) == [4, 6]
```

If I add these to my revp.py program and run pytest revp.py, I'll see that the test fails as it should. Now I can fill in the code:

```
def revp(seq: str, k: int) -> List[int]:
    """ Return positions of reverse palindromes """

    kmers = find_kmers(seq, k)
    revc = map(Seq.reverse_complement, kmers)
    pairs = enumerate(zip(kmers, revc))
    return [pos + 1 for pos, pair in pairs if operator.eq(*pair)]
```

If I run `pytest` again, I should get a passing test. The `main()` function is now more readable:

```
def main() -> None:
    args = get_args()
    for rec in SeqIO.parse(args.file, 'fasta'):
        for k in range(4, 13): ❶
            for pos in revp(str(rec.seq), k): ❷
                print(pos, k) ❸
```

❶ Iterate through each value of k.

❷ Iterate through each reverse palindrome of size k found in the sequence.

❸ Print the position and size of the reverse palindrome.

Note that it's possible to use more than one iterator inside a list comprehension. I can collapse the two `for` loops into a single one, like so:

```
for k, pos in [(k, pos) for k in range(4, 13) for pos in revp(seq, k)]: ❶
    print(pos, k)
```

❶ First iterate the k values, then use those to iterate the `revp()` values, returning both as a tuple.

I would probably not use this construct. It reminds me of my old coworker, Joe, who would joke: "If it was hard to write, it should be hard to read!"

Testing the Program

I'd like to take a moment to look at the integration test in *tests/revp_test.py*. The first two tests are always the same, checking for the existence of the expected program and that the program will produce some usage statement when requested. For a program that accepts files as inputs such as this one, I include a test that the program rejects an invalid file. I usually challenge other inputs too, like passing strings when integers are expected, to ensure the arguments are rejected.

After I've checked that the arguments to the program are all validated, I start passing good input values to see that the program works as expected. This requires that I use valid, known input and verify that the program produces the correct, expected output. In this case, I encode the inputs and outputs using files in the *tests/inputs* directory. For instance, the expected output for the input file *1.fa* is found in *1.fa.out*:

```
$ ls tests/inputs/
1.fa        2.fa        empty.fa
1.fa.out    2.fa.out    empty.fa.out
```

The following is the first input:

```
$ cat tests/inputs/1.fa
>Rosalind_24
TCAATGCATGCGGGTCTATATGCAT
```

and the expected output is:

```
$ cat tests/inputs/1.fa.out
5 4
7 4
17 4
18 4
21 4
4 6
6 6
20 6
```

The second input file is significantly larger than the first. This is common with the Rosalind problems, and so it would be ugly to try to include the input and output values as literal strings in the test program. The expected output for the second file is 70 lines long. The last test is for an empty file, and the expected output is the empty string. While that may seem obvious, the point is to check that the program does not throw an exception on an empty input file.

In *tests/revp_test.py*, I wrote a run() helper function that takes the name of the input file, reads the expected output filename, and runs the program with the input to check the output:

```
def run(file: str) -> None: ❶
    """ Run the test """

    expected_file = file + '.out' ❷
    assert os.path.isfile(expected_file) ❸

    rv, out = getstatusoutput(f'{PRG} {file}') ❹
    assert rv == 0 ❺

    expected = set(open(expected_file).read().splitlines()) ❻
    assert set(out.splitlines()) == expected ❼
```

❶ The function takes the name of the input file.

❷ The output file is the name of the input file plus *.out*.

❸ Make sure the output file exists.

❹ Run the program with the input file and capture the return value and output.

❺ Make sure the program reported a successful run.

❻ Read the expected output file, breaking the contents on lines and creating a set of the resulting strings.

❼ Break the output of the program on lines and create a set to compare to the expected results. Sets allow me to disregard the order of the lines.

This simplifies the tests. Note that the INPUT* and EMPTY variables are declared at the top of the module:

```
def test_ok1() -> None:
    run(INPUT1)

def test_ok2() -> None:
    run(INPUT2)

def test_mepty() -> None:
    run(EMPTY)
```

I would encourage you to spend some time reading the *_test.py* files for every program. I hope that you will integrate testing into your development workflow, and I'm sure you can find ample code to copy from my tests, which will save you time.

Going Further

The minimum (4) and maximum (12) values for the length of the sites are hardcoded in the program. Add command-line parameters to pass these as integer options using those default values. Change the code to use the given values, and add tests to ensure the correct sites are found for different ranges of these values.

Write a program that can identify English palindromes such as "A man, a plan, a canal—Panama!" Start by creating a new repository. Find several interesting palindromes to use in your tests. Be sure to provide phrases that are not palindromes and verify that your algorithm rejects those, too. Release your code to the internet, and reap the fame, glory, and profit of writing open source software.

Review

Key points from this chapter:

- You can reuse functions by placing them into a module and importing them as needed.
- The PYTHONPATH environment variable indicates directories which Python should search when looking for modules of code.

## CHAPTER 14

Finding Open Reading Frames

The ORF challenge (*https://oreil.ly/DPWXc*) is the last Rosalind problem I'll tackle in this book. The goal is to find all the possible open reading frames (ORFs) in a sequence of DNA. An ORF is a region of nucleotides between the start codon and the stop codon. The solution will consider both the forward and reverse complement as well as frameshifts. Although there are existing tools such as TransDecoder to find coding regions, writing a bespoke solution brings together many skills from previous chapters, including reading a FASTA file, creating the reverse complement of a sequence, using string slices, finding k-mers, using multiple for loops/iterations, translating DNA, and using regular expressions.

You will learn:

- How to truncate a sequence to a length evenly divisible by a codon size
- How to use the `str.find()` and `str.partition()` functions
- How to document a regular expression using code formatting, comments, and Python's implicit string concatenation

Getting Started

The code, tests, and solutions for this challenge are located in the *14_orf* directory. Start by copying the first solution to the program `orf.py`:

```
$ cd 14_orf/
$ cp solution1_iterate_set.py orf.py
```

If you request the usage, you'll see the program takes a single positional argument of a FASTA-formatted file of sequences:

```
$ ./orf.py -h
usage: orf.py [-h] FILE

Open Reading Frames

positional arguments:
  FILE        Input FASTA file

optional arguments:
  -h, --help  show this help message and exit
```

The first test input file has the same content as the example on the Rosalind page. Note that I've broken the sequence file here, but it's a single line in the input file:

```
$ cat tests/inputs/1.fa
>Rosalind_99
AGCCATGTAGCTAACTCAGGTTACATGGGGATGACCCCGCGACTTGGATTAGAGTCTCTTTTGGAATAAG\
CCTGAATGATCCGAGTAGCATCTCAG
```

Run the program with this input file and note the output. The order of the ORFs is not important:

```
$ ./orf.py tests/inputs/1.fa
M
MGMTPRLGLESLLE
MLLGSFRLIPKETLIQVAGSSPCNLS
MTPRLGLESLLE
```

Run the test suite to ensure the program passes the tests. When you are satisfied with how your program should work, start over:

```
$ new.py -fp 'Open Reading Frames' orf.py
Done, see new script "orf.py".
```

At this point, you probably need no help in defining a single positional file argument, but here is the code you can use:

```
class Args(NamedTuple):
    """ Command-line arguments """
    file: TextIO


def get_args() -> Args:
    """ Get command-line arguments """

    parser = argparse.ArgumentParser(
        description='Open Reading Frames',
        formatter_class=argparse.ArgumentDefaultsHelpFormatter)

    parser.add_argument('file', ❶
                        help='Input FASTA file',
                        metavar='FILE',
                        type=argparse.FileType('rt'))
```

```
    args = parser.parse_args()

    return Args(args.file)
```

❶ Define a positional argument that must be a readable text file.

Modify the `main()` to print the incoming filename:

```
def main() -> None:
    args = get_args()
    print(args.file.name)
```

Verify that the program prints the usage, rejects bad files, and prints the filename for a valid argument:

```
$ ./orf.py tests/inputs/1.fa
tests/inputs/1.fa
```

At this point, your program should pass the first three tests. Next, I'll talk about how to make the program find ORFs.

Translating Proteins Inside Each Frame

It might be helpful to write a bit of pseudocode to help sketch out what needs to happen:

```
def main() -> None:
    args = get_args()

    # Iterate through each DNA sequence in the file:
        # Transcribe the sequence from DNA to mRNA
        # Iterate using the forward and reverse complement of the mRNA:
            # Iterate through 0, 1, 2 for frames in this sequence:
                # Translate the mRNA frame into a protein sequence
                # Try to find the ORFs in this protein sequence
```

You can use a `for` loop to iterate through the input sequences using `Bio.SeqIO`:

```
def main() -> None:
    args = get_args()

    for rec in SeqIO.parse(args.file, 'fasta'):
        print(str(rec.seq))
```

Run the program to verify that this works:

```
$ ./orf.py tests/inputs/1.fa
AGCCATGTAGCTAACTCAGGTTACATGGGGATGACCCCGCGACTTGGATTAGAGTCTCTTTTGGA\
ATAAGCCTGAATGATCCGAGTAGCATCTCAG
```

I need to transcribe this to mRNA, which entails changing all the *T*s to *U*s. I'll let you use whatever solution from Chapter 2 you like so long as your program can now print this:

```
$ ./orf.py tests/inputs/1.fa
AGCCAUGUAGCUAACUCAGGUUACAUGGGGAUGACCCCGCGACUUGGAUUAGAGUCUCUUUUGGA\
AUAAGCCUGAAUGAUCCGAGUAGCAUCUCAG
```

Next, refer to Chapter 3 and have your program print both the forward and reverse complements of this sequence:

```
$ ./orf.py tests/inputs/1.fa
AGCCAUGUAGCUAACUCAGGUUACAUGGGGAUGACCCCGCGACUUGGAUUAGAGUCUCUUUUGGA\
AUAAGCCUGAAUGAUCCGAGUAGCAUCUCAG
CUGAGAUGCUACUCGGAUCAUUCAGGCUUAUUCCAAAAGAGACUCUAAUCCAAGUCGCGGGGUCA\
UCCCCAUGUAACCUGAGUUAGCUACAUGGCU
```

Refer to Chapter 7 to translate the forward and reverse complements to proteins:

```
$ ./orf.py tests/inputs/1.fa
SHVANSGYMGMTPRLGLESLLE*A*MIRVASQ
LRCYSDHSGLFQKRL*SKSRGHPHVT*VSYMA
```

Now, rather than reading each mRNA sequence from the beginning, implement frameshifts by reading them starting from the zeroth, first, and second characters, which you can do using a string slice. If you use Biopython to translate the mRNA slice, you may encounter the warning:

> *Partial codon, len(sequence) not a multiple of three. Explicitly trim the sequence or add trailing N before translation. This may become an error in the future.*

To fix this, I created a function to truncate a sequence to the nearest even division by a value:

```
def truncate(seq: str, k: int) -> str:
    """ Truncate a sequence to even division by k """

    return ''
```

Figure 14-1 shows the results of shifting through the string 0123456789 and truncating each result to a length that is evenly divisible by 3.

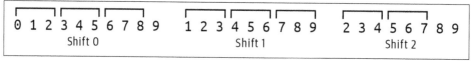

Figure 14-1. *Truncating the various frameshifts to a length that is evenly divisible by the codon size 3*

Here is a test you could use:

```
def test_truncate() -> None:
    """ Test truncate """

    seq = '0123456789'
```

```
assert truncate(seq, 3) == '012345678'
assert truncate(seq[1:], 3) == '123456789'
assert truncate(seq[2:], 3) == '234567'
```

Change your program to print the protein translations for the three shifts for both the forward and reverse complements of the mRNA. Be sure to print the entire translation, including all stop (*) codons, like so:

```
$ ./orf.py tests/inputs/1.fa
SHVANSGYMGMTPRLGLESLLE*A*MIRVASQ
AM*LTQVTWG*PRDLD*SLFWNKPE*SE*HL
PCS*LRLHGDDPATWIRVSFGISLNDPSSIS
LRCYSDHSGLFQKRL*SKSRGHPHVT*VSYMA
*DATRIIQAYSKRDSNPSRGVIPM*PELATW
EMLLGSFRLIPKETLIQVAGSSPCNLS*LHG
```

Finding the ORFs in a Protein Sequence

Now that the program can find all the protein sequences from each frameshift of the mRNA, it's time to look for the open reading frames in the proteins. Your code will need to consider every interval from each start codon to the first subsequent stop codon. The codon *AUG* is the most common start codon, and it codes for the amino acid methionine (*M*). There are three possible stop codons shown with the asterisk (*). For example, Figure 14-2 shows that the amino acid sequence *MAMAPR** contains two start codons and one stop codon and so has two possible proteins of *MAMAPR* and *MAPR*. Although it is common for tools to report only the longer sequence, the Rosalind challenge expects all possible sequences.

MAMAPR*

Figure 14-2. The protein sequence MAMAPR has two overlapping open reading frames*

I decided to write a function called find_orfs() that will accept an amino acid string and return a list of ORFs:

```
def find_orfs(aa: str) -> List[str]: ❶
    """ Find ORFs in AA sequence """

    return [] ❷
```

❶ The function accepts a string of amino acids and returns a list of possible protein strings.

❷ For now, return the empty list.

Here is a test for this function. If you can implement the find_orfs() that passes this test, then you should be able to pass the integration test:

```
def test_find_orfs() -> None:
    """ Test find_orfs """

    assert find_orfs('') == [] ❶
    assert find_orfs('M') == [] ❷
    assert find_orfs('*') == [] ❸
    assert find_orfs('M*') == ['M'] ❹
    assert find_orfs('MAMAPR*') == ['MAMAPR', 'MAPR'] ❺
    assert find_orfs('MAMAPR*M') == ['MAMAPR', 'MAPR'] ❻
    assert find_orfs('MAMAPR*MP*') == ['MAMAPR', 'MAPR', 'MP'] ❼
```

❶ The empty string should produce no ORFs.

❷ A single start codon with no stop codon should produce no ORFs.

❸ A single stop codon with no preceding start codon should produce no ORFs.

❹ The function should return the start codon even if there are no intervening bases before the stop codon.

❺ This sequence contains two ORFs.

❻ This sequence also contains only two ORFs.

❼ This sequence contains three putative ORFs in two separate sections.

Once you can find all the ORFs in each mRNA sequence, you should collect them into a distinct list. I suggest you use a set() for this. Though my solution prints the ORFs in sorted order, this is not a requirement for the test. The solution will bring together many of the skills you've already learned. The craft of writing longer and longer programs lies in composing smaller pieces that you understand and test. Keep plugging away at your program until you pass all the tests.

Solutions

I'll present three solutions to finding ORFs using two string functions and regular expressions.

Solution 1: Using the str.index() Function

To start, here is how I wrote the truncate() function that will assuage the Bio.Seq.translate() function when I try to translate the various frame-shifted mRNA sequences:

```
def truncate(seq: str, k: int) -> str:
    """ Truncate a sequence to even division by k """
```

```
length = len(seq) ❶
end = length - (length % k) ❷
return seq[:end] ❸
```

❶ Find the length of the sequence.

❷ The end of the desired subsequence is the length minus the length modulo k.

❸ Return the subsequence.

Next, here is one way to write the find_orfs() that uses the str.index() function to find each starting *M* codon followed by a * stop codon:

```
def find_orfs(aa: str) -> List[str]:
    orfs = [] ❶
    while 'M' in aa: ❷
        start = aa.index('M') ❸
        if '*' in aa[start + 1:]: ❹
            stop = aa.index('*', start + 1) ❺
            orfs.append(aa[start:stop]) ❻
            aa = aa[start + 1:] ❼
        else:
            break ❽

    return orfs
```

❶ Initialize a list to hold the ORFs.

❷ Create a loop to iterate while there are start codons present.

❸ Use str.index() to find the location of the start codon.

❹ See if the stop codon is present after the start codon's position.

❺ Get the index of the stop codon after the start codon.

❻ Use a string slice to grab the protein.

❼ Set the amino acid string to the index after the position of the start codon to find the next start codon.

❽ Leave the while loop if there is no stop codon.

Here is how I incorporate these ideas into the program:

```
def main() -> None:
    args = get_args()
    for rec in SeqIO.parse(args.file, 'fasta'): ❶
        rna = str(rec.seq).replace('T', 'U') ❷
```

```
            orfs = set() ❸

            for seq in [rna, Seq.reverse_complement(rna)]: ❹
                for i in range(3): ❺
                    if prot := Seq.translate(truncate(seq[i:], 3), to_stop=False): ❻
                        for orf in find_orfs(prot): ❼
                            orfs.add(orf) ❽

            print('\n'.join(sorted(orfs))) ❾
```

❶ Iterate through the input sequences.

❷ Transcribe the DNA sequence to mRNA.

❸ Create an empty set to hold all the ORFs.

❹ Iterate through the forward and reverse complement of the mRNA.

❺ Iterate through the frameshifts.

❻ Attempt to translate the truncated, frame-shifted mRNA into a protein sequence.

❼ Iterate through each ORF found in the protein sequence.

❽ Add the ORF to the set to maintain a unique list.

❾ Print the sorted ORFs.

Solution 2: Using the str.partition() Function

Here is another approach to writing the find_orfs() function that uses str.parti
tion(). This function breaks a string into the part before some substring, the substring, and the part after. For instance, the string *MAMAPR*MP** can be partitioned
on the stop codon (*):

```
>>> 'MAMAPR*MP*'.partition('*')
('MAMAPR', '*', 'MP*')
```

If the protein sequence does not contain a stop codon, the function returns the entire
sequence in the first position and empty strings for the others:

```
>>> 'M'.partition('*')
('M', '', '')
```

In this version, I use two infinite loops. The first tries to partition the given amino
acid sequence on the stop codon. If this is not successful, I exit the loop. Figure 14-3
shows that the protein sequence *MAMAPR*MP** contains two sections that have start
and end codons.

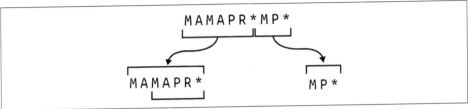

Figure 14-3. The protein sequence MAMAPR*MP* has three ORFs in two sections

The second loop checks the first partition to find all the subsequences starting with the *M* start codon. So in the partition *MAMAPR*, it finds the two sequences *MAMAPR* and *MAPR*. The code then truncates the amino acid sequence to the last partition, *MP**, to repeat the operation until all ORFs have been found:

```
def find_orfs(aa: str) -> List[str]:
    """ Find ORFs in AA sequence """

    orfs = []  ❶
    while True:  ❷
        first, middle, rest = aa.partition('*')  ❸
        if middle == '':  ❹
            break

        last = 0  ❺
        while True:  ❻
            start = first.find('M', last)  ❼
            if start == -1:  ❽
                break
            orfs.append(first[start:])  ❾
            last = start + 1  ❿
        aa = rest  ⓫

    return orfs  ⓬
```

❶ Initialize a list for the ORFs to return.

❷ Create the first infinite loop.

❸ Partition the amino acid sequence on the stop codon.

❹ The middle will be empty if the stop codon is not present, so break from the outer loop.

❺ Set a variable to remember the last position of a start codon.

❻ Create a second infinite loop.

❼ Use the str.find() method to locate the index of the start codon.

❽ The value -1 indicates that the start codon is not present, so leave the inner loop.

❾ Add the substring from the start index to the list of ORFs.

❿ Move the last known position to after the current start position.

⓫ Truncate the protein sequence to the last part of the initial partition.

⓬ Return the ORFs to the caller.

Solution 3: Using a Regular Expression

In this final solution, I'll once again point out that a regular expression is probably the most fitting solution to find a pattern of text. This pattern always starts with *M*, and I can use the re.findall() function to find the four *M*s in this protein sequence:

```
>>> import re
>>> re.findall('M', 'MAMAPR*MP*M')
['M', 'M', 'M', 'M']
```

The Rosalind challenge does not consider noncanonical start codons, so an ORF will always start with an *M* and extend to the first stop codon. In between these, there can be zero or more not-stop codons which I can represent using a negated character class of [^*] that excludes the stop codon followed by a * to indicate that there can be *zero or more* of the preceding pattern:

```
>>> re.findall('M[^*]*', 'MAMAPR*MP*M')
['MAMAPR', 'MP', 'M']
```

I need to add the stop codon * to this pattern. Because the literal asterisk is a meta-character, I must use a backslash to escape it:

```
>>> re.findall('M[^*]*\*', 'MAMAPR*MP*M')
['MAMAPR*', 'MP*']
```

I can also place the asterisk inside a character class where it has no meta meaning:

```
>>> re.findall('M[^*]*[*]', 'MAMAPR*MP*M')
['MAMAPR*', 'MP*']
```

Figure 14-4 shows this pattern using a finite state machine diagram.

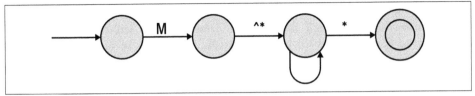

Figure 14-4. A finite state machine diagram of the regular expression to find an open reading frame

I can see that this pattern is close to working, but it's only finding two of the three ORFs because the first one overlaps the second one. As in Chapters 8 and 11, I can wrap the pattern in a positive look-ahead assertion. Further, I will use parentheses to create a capture group around the ORF up to the stop codon:

```
>>> re.findall('(?=(M[^*]*)[*])', 'MAMAPR*MP*M')
['MAMAPR', 'MAPR', 'MP']
```

Here is one version of the find_orfs() that uses this pattern:

```
def find_orfs(aa: str) -> List[str]:
    """ Find ORFs in AA sequence """

    return re.findall('(?=(M[^*]*)[*])', aa)
```

While this passes test_find_orfs(), this is a complicated regex that I will have to relearn every time I come back to it. An alternate way to write this is to place each functional piece of the regex on a separate line, followed by an end-of-line comment, and rely on Python's implicit string concatenation (first shown in Chapter 2) to join these into a single string. This is my preferred method to find the ORFs:

```
def find_orfs(aa: str) -> List[str]:
    """ Find ORFs in AA sequence """

    pattern = (      ❶
        '(?='        # start positive look-ahead to handle overlaps
        '('          # start a capture group
        'M'          # a literal M
        '[^*]*'      # zero or more of anything not the asterisk
        ')'          # end the capture group
        '[*]'        # a literal asterisk
        ')')         # end the look-ahead group

    return re.findall(pattern, aa)  ❷
```

❶ The parentheses group the following lines such that Python will automatically join the strings into a single string. Be sure there are no commas or Python will create a tuple.

❷ Use the pattern with the re.findall() function.

This is a longer function, but it will be much easier to understand the next time I see it. One downside is that yapf that I use to format my code will remove the vertical alignment of the comments, so I must manually format this section. Still, I think it's worth it to have more self-documenting code.

Going Further

Expand the program to process multiple input files, writing all the unique ORFs to an indicated output file.

Review

Key points from this chapter:

- The Bio.Seq.translate() function will print warnings if the input sequence is not evenly divisible by three, so I wrote a truncate() function to trim the protein.

- The str.find() and str.partition() functions each present ways to find subsequences in a string.

- A regular expression remains my preferred method to find a pattern in some text.

- A complicated regex can be written over multiple lines with comments so that Python will implicitly concatenate them into a single string.

## PART II

Other Programs

In the chapters in this part, I'll show you several programs I've written that capture patterns I have used repeatedly in bioinformatics. First, I'll show you how to write a program to find basic statistics from sequence files and format an output table. Next, I'll demonstrate how to select sequences using pattern matching on the header information followed by a program that will create artificial DNA sequences using data learned from training files. Then I'll show you a program that explores randomness to subsample sequences, and I'll finish by using Python to parse delimited text files both with and without header information. I hope you'll find patterns in these programs you can use when writing your own.

## CHAPTER 15

Seqmagique: Creating and Formatting Reports

Often in bioinformatics projects, you'll find yourself staring at a directory full of sequence files, probably in FASTA or FASTQ format. You'll probably want to start by getting an idea of the distribution of sequences in the files, such as how many are in each file and the average, minimum, and maximum lengths of the sequences. You need to know if any files are corrupted—maybe they didn't transfer completely from your sequencing center—or if any samples have far fewer reads, perhaps indicating a bad sequencing run that needs to be redone. In this chapter, I'll introduce some techniques for checking your sequence files using hashes and the Seqmagick tool (*https:// oreil.ly/VI9gr*). Then I'll write a small utility to mimic part of Seqmagick to illustrate how to create formatted text tables. This program serves as a template for any program that needs to process all the records in a given set of files and produce a table of summary statistics.

You will learn:

- How to install the `seqmagick` tool
- How to use MD5 hashes
- How to use `choices` in `argparse` to constrain arguments
- How to use the `numpy` module
- How to mock a filehandle
- How to use the `tabulate` and `rich` modules to format output tables

Using Seqmagick to Analyze Sequence Files

`seqmagick` is a useful command-line utility for handling sequence files. This should have been installed along with the other Python modules if you followed the setup instructions in the Preface. If not, you can install it with `pip`:

```
$ python3 -m pip install seqmagick
```

If you run **seqmagick --help**, you'll see the tool offers many options. I only want to focus on the `info` subcommand. I can run this on the test input FASTA files in the *15_seqmagique* directory like so:

```
$ cd 15_seqmagique
$ seqmagick info tests/inputs/*.fa
name                    alignment   min_len   max_len   avg_len   num_seqs
tests/inputs/1.fa       FALSE            50        50     50.00          1
tests/inputs/2.fa       FALSE            49        79     64.00          5
tests/inputs/empty.fa   FALSE             0         0      0.00          0
```

In this exercise, you will create a program called `seqmagique.py` (this should be pronounced with an exaggerated French accent) that will mimic this output. The point of the program is to provide a basic overview of the sequences in a given set of files so you can spot, for instance, a truncated or corrupted file.

Start by copying the solution to `seqmagique.py` and requesting the usage:

```
$ cp solution1.py seqmagique.py
$ ./seqmagique.py -h
usage: seqmagique.py [-h] [-t table] FILE [FILE ...]

Mimic seqmagick

positional arguments:
  FILE                  Input FASTA file(s) ❶

optional arguments:
  -h, --help            show this help message and exit
  -t table, --tablefmt table ❷
                        Tabulate table style (default: plain)
```

❶ The program accepts one or more input files which should be in FASTA format.

❷ This option controls the format of the output table.

Run this program on the same files and note that the output is almost identical, except that I have omitted the `alignment` column:

```
$ ./seqmagique.py tests/inputs/*.fa
name                    min_len   max_len   avg_len   num_seqs
tests/inputs/1.fa            50        50     50.00          1
```

```
tests/inputs/2.fa          49        79      64.00         5
tests/inputs/empty.fa       0         0       0.00         0
```

The `--tablefmt` option controls how the output table is formatted. This is the first program you'll write that constrains the value to a given list. To see this in action, use a bogus value like `blargh`:

```
$ ./seqmagique.py -t blargh tests/inputs/1.fa
usage: seqmagique.py [-h] [-t table] FILE [FILE ...]
seqmagique.py: error: argument -t/--tablefmt: invalid choice: 'blargh'
(choose from 'plain', 'simple', 'grid', 'pipe', 'orgtbl', 'rst',
 'mediawiki', 'latex', 'latex_raw', 'latex_booktabs')
```

Then try a different table format, such as `simple`:

```
$ ./seqmagique.py -t simple tests/inputs/*.fa
name                   min_len    max_len    avg_len      num_seqs
--------------------   --------   --------   --------   ----------
tests/inputs/1.fa           50         50      50.00            1
tests/inputs/2.fa           49         79      64.00            5
tests/inputs/empty.fa        0          0       0.00            0
```

Run the program with other table styles and then try the test suite. Next, I'll talk about getting data for our program to analyze.

Checking Files Using MD5 Hashes

The first step in most genomics projects will be transferring sequence files to some location where you can analyze them, and the first line of defense against data corruption is ensuring that the files were copied completely. The source of the files may be a sequencing center or a public repository like GenBank (*https://oreil.ly/2eaMj*) or the Sequence Read Archive (SRA) (*https://oreil.ly/kGNCv*). The files may arrive on a thumb drive, or you may download them from the internet. If the latter, you may find that your connection drops, causing some files to be truncated or corrupted. How can you find these types of errors?

One way to check that your files are complete is to compare the file sizes locally with those on the server. For instance, you can use the `ls -l` command to view the *long* listing of files where the file size, in bytes, is shown. For large sequence files, this is going to be a very large number, and you will have to manually compare the file sizes from the source to the destination, which is tedious and prone to error.

Another technique involves using a *hash* or *message digest* of the file, which is a signature of the file's contents generated by a one-way cryptographic algorithm that creates a unique output for every possible input. Although there are many tools you can use to create a hash, I'll focus on tools that use the MD5 algorithm. This algorithm was originally developed in the context of cryptography and security, but researchers have since identified numerous flaws that now make it suitable only for purposes such as verifying data integrity.

On macOS, I can use **md5** to generate a 128-bit hash value from the contents of the first test input file, like so:

```
$ md5 -r tests/inputs/1.fa
c383c386a44d83c37ae287f0aa5ae11d tests/inputs/1.fa
```

I can also use **openssl**:

```
$ openssl md5 tests/inputs/1.fa
MD5(tests/inputs/1.fa)= c383c386a44d83c37ae287f0aa5ae11d
```

On Linux, I use **md5sum**:

```
$ md5sum tests/inputs/1.fa
c383c386a44d83c37ae287f0aa5ae11d  tests/inputs/1.fa
```

As you can see, no matter the tool or platform, the hash value is the same for the same input file. If I change even one bit of the input file, a different hash value will be generated. Conversely, if I find another file that generates the same hash value, then the contents of the two files are identical. For instance, the *empty.fa* file is a zero-length file I created for testing, and it has the following hash value:

```
$ md5 -r tests/inputs/empty.fa
d41d8cd98f00b204e9800998ecf8427e tests/inputs/empty.fa
```

If I use the **touch foo** command to create another empty file, I'll find it has the same signature:

```
$ touch foo
$ md5 -r foo
d41d8cd98f00b204e9800998ecf8427e foo
```

It's common for data providers to create a file of the checksums so that you can verify that your copies of the data are complete. I created a *tests/inputs/checksums.md5* like so:

```
$ cd tests/inputs
$ md5 -r *.fa > checksums.md5
```

It has the following contents:

```
$ cat checksums.md5
c383c386a44d83c37ae287f0aa5ae11d 1.fa
863ebc53e28fdfe6689278e40992db9d 2.fa
d41d8cd98f00b204e9800998ecf8427e empty.fa
```

The md5sum tool has a --check option that I can use to automatically verify that the files match the checksums found in a given file. The macOS md5 tool does not have an option for this, but you can use **brew install md5sha1sum** to install an equivalent md5sum tool that can do this:

```
$ md5sum --check checksums.md5
1.fa: OK
2.fa: OK
empty.fa: OK
```

MD5 checksums present more complete and easier ways to verify data integrity than manually checking file sizes. Although file digests are not directly part of this exercise, I feel it's important to understand how to verify that you have complete and uncorrupted data before beginning any analyses.

Getting Started

You should work in the *15_seqmagique* directory for this exercise. I'll start the program as usual:

```
$ new.py -fp 'Mimic seqmagick' seqmagique.py
Done, see new script "seqmagique.py".
```

First I need to make the program accept one or more text files as positional parameters. I also want to create an option to control the output table format. Here is the code for that:

```
import argparse
from typing import NamedTuple, TextIO, List


class Args(NamedTuple):
    """ Command-line arguments """
    files: List[TextIO]
    tablefmt: str


def get_args() -> Args:
    """Get command-line arguments"""

    parser = argparse.ArgumentParser(
        description='Argparse Python script',
        formatter_class=argparse.ArgumentDefaultsHelpFormatter)

    parser.add_argument('file', ❶
                        metavar='FILE',
                        type=argparse.FileType('rt'),
                        nargs='+',
                        help='Input FASTA file(s)')

    parser.add_argument('-t',
                        '--tablefmt',
                        metavar='table',
                        type=str,
                        choices=[ ❷
                            'plain', 'simple', 'grid', 'pipe', 'orgtbl', 'rst',
```

```
                  'mediawiki', 'latex', 'latex_raw', 'latex_booktabs'
              ],
              default='plain',
              help='Tabulate table style')

    args = parser.parse_args()

    return Args(args.file, args.tablefmt)
```

❶ Define a positional parameter for one or more readable text files.

❷ Define an option that uses choices to constrain the argument to a value in the list, making sure to define a reasonable default value.

Using choices for the --tablefmt really saves you quite a bit of work in validating user input. As shown in "Using Seqmagick to Analyze Sequence Files" on page 290, a bad value for the table format option will trigger a useful error message.

Modify the main() function to print the input filenames:

```
def main() -> None:
    args = get_args()
    for fh in args.files:
        print(fh.name)
```

And verify that this works:

```
$ ./seqmagique.py tests/inputs/*.fa
tests/inputs/1.fa
tests/inputs/2.fa
tests/inputs/empty.fa
```

The goal is to iterate through each file and print the following:

name
    The filename

min_len
    The length of the shortest sequence

max_len
    The length of the longest sequence

avg_len
    The average/mean length of all the sequences

num_seqs
    The number of sequences

If you would like to have some real input files for the program, you can use the fastq-dump tool (*https://oreil.ly/Vmb0w*) from NCBI to download sequences from the study "Planktonic Microbial Communities from North Pacific Subtropical Gyre" (*https://oreil.ly/aAGUA*):

```
$ fastq-dump --split-3 SAMN00000013 ❶
```

❶ The --split-3 option will ensure that paired-end reads are correctly split into forward/reverse/unpaired reads. The SAMN00000013 string is the accession of one of the samples (*https://oreil.ly/kBCQU*) from the experiment.

Formatting Text Tables Using tabulate()

The output of the program will be a text table formatted using the tabulate() function from that module. Be sure to read the documentation:

```
>>> from tabulate import tabulate
>>> help(tabulate)
```

I need to define the headers for the table, and I decided to use the same ones as Seqmagick (minus the alignment column):

```
>>> hdr = ['name', 'min_len', 'max_len', 'avg_len', 'num_seqs']
```

The first test file, *tests/inputs/1.fa*, has just one sequence of 50 bases, so the columns for this are as follows:

```
>>> f1 = ['tests/inputs/1.fa', 50, 50, 50.00, 1]
```

The second test file, *tests/inputs/2.fa*, has five sequences ranging from 49 bases to 79 with an average length of 64 bases:

```
>>> f2 = ['tests/inputs/2.fa', 49, 79, 64.00, 5]
```

The tabulate() function expects the table data to be passed positionally as a list of lists, and I can specify the headers as a keyword argument:

```
>>> print(tabulate([f1, f2], headers=hdr))
name               min_len    max_len    avg_len    num_seqs
---------------    ---------  ---------  ---------  ----------
tests/inputs/1.fa       50         50         50           1
tests/inputs/2.fa       49         79         64           5
```

Alternatively, I can place the headers as the first row of data and indicate this is the location of the headers:

```
>>> print(tabulate([hdr, f1, f2], headers='firstrow'))
name               min_len    max_len    avg_len    num_seqs
---------------    ---------  ---------  ---------  ----------
tests/inputs/1.fa       50         50         50           1
tests/inputs/2.fa       49         79         64           5
```

Note that the default table style for the `tabulate()` function is `simple`, but the `plain` format is what I need to match Seqmagick's output. I can set this with the `tablefmt` option:

```
>>> print(tabulate([f1, f2], headers=hdr, tablefmt='plain'))
name               min_len    max_len    avg_len    num_seqs
tests/inputs/1.fa       50         50         50           1
tests/inputs/2.fa       49         79         64           5
```

One other thing to note is that the values in the `avg_len` column are being shown as integers but should be formatted as floating-point numbers to two decimal places. The `floatfmt` option controls this, using syntax similar to the f-string number formatting I've shown before:

```
>>> print(tabulate([f1, f2], headers=hdr, tablefmt='plain', floatfmt='.2f'))
name               min_len    max_len    avg_len    num_seqs
tests/inputs/1.fa       50         50      50.00           1
tests/inputs/2.fa       49         79      64.00           5
```

Your job is to process all the sequences in each file to find the statistics and print the final table. This should be enough for you to solve the problem. Don't read ahead until you can pass all the tests.

Solutions

I'll present two solutions that both show the file statistics but differ in the formatting of the output. The first solution uses the `tabulate()` function to create an ASCII text table and the second uses the `rich` module to create a fancier table sure to impress your labmates and principal investigator (PI).

Solution 1: Formatting with tabulate()

For my solution, I first decided to write a `process()` function that would handle each input file. Whenever I approach a problem that needs to handle some list of items, I prefer to focus on how to handle just one of the items. That is, rather than trying to find all the statistics for all the files, I first want to figure out how to find this information for just one file.

My function needs to return the filename and the four metrics: minimum/maximum/average sequence lengths, plus the number of sequences. Just as with the `Args` class, I like to create a type based on a `NamedTuple` for this so that I have a statically typed data structure that `mypy` can validate:

```
class FastaInfo(NamedTuple):
    """ FASTA file information """
    filename: str
    min_len: int
    max_len: int
```

```
    avg_len: float
    num_seqs: int
```

Now I can define a function that returns this data structure. Note that I'm using the numpy.mean() function to get the average length. The numpy module offers many powerful mathematical operations to handle numeric data and is especially useful for multidimensional arrays and linear algebra functions. When importing the dependencies, it's common to import the numpy module with the alias np:

```
import numpy as np
from tabulate import tabulate
from Bio import SeqIO
```

You can run **help(np)** in the REPL to read the documentation. Here's how I wrote this function:

```
def process(fh: TextIO) -> FastaInfo: ❶
    """ Process a file """

    if lengths := [len(rec.seq) for rec in SeqIO.parse(fh, 'fasta')]: ❷
        return FastaInfo(filename=fh.name, ❸
                         min_len=min(lengths), ❹
                         max_len=max(lengths), ❺
                         avg_len=round(float(np.mean(lengths)), 2), ❻
                         num_seqs=len(lengths)) ❼

    return FastaInfo(filename=fh.name, ❽
                     min_len=0,
                     max_len=0,
                     avg_len=0,
                     num_seqs=0)
```

❶ The function accepts a filehandle and returns a FastaInfo object.

❷ Use a list comprehension to read all the sequences from the filehandle. Use the len() function to return the length of each sequence.

❸ The name of the file is available through the fh.name attribute.

❹ The min() function will return the minimum value.

❺ The max() function will return the maximum value.

❻ The np.mean() function will return the mean from a list of values. The round() function is used to round this floating-point value to two significant digits.

❼ The number of sequences is the length of the list.

❽ If there are no sequences, return zeros for all the values.

As always, I want to write a unit test for this. While it's true that the integration tests I wrote cover this part of the program, I want to show how you can write a unit test for a function that reads a file. Rather than relying on actual files, I'll create a *mock* or fake filehandle.

The first test file looks like this:

```
$ cat tests/inputs/1.fa
>SEQ0
GGATAAAGCGAGAGGCTGGATCATGCACCAACTGCGTGCAACGAAGGAAT
```

I can use the `io.StringIO()` function to create an object that behaves like a filehandle:

```
>>> import io
>>> f1 = '>SEQ0\nGGATAAAGCGAGAGGCTGGATCATGCACCAACTGCGTGCAACGAAGGAAT\n' ❶
>>> fh = io.StringIO(f1) ❷
>>> for line in fh: ❸
...     print(line, end='') ❹
...
>SEQ0
GGATAAAGCGAGAGGCTGGATCATGCACCAACTGCGTGCAACGAAGGAAT
```

❶ This is the data from the first input file.

❷ Create a mock filehandle.

❸ Iterate through the lines of the mock filehandle.

❹ Print the line which has a newline (`\n`), so use `end=''` to leave off an additional newline.

There's a slight problem, though, because the `process()` function calls the `fh.name` attribute to get the input filename, which will raise an exception:

```
>>> fh.name
Traceback (most recent call last):
  File "<stdin>", line 1, in <module>
AttributeError: '_io.StringIO' object has no attribute 'name'
```

Luckily, there's another way to create a mock filehandle using Python's standard `unittest` module. While I favor the `pytest` module for almost everything I write, the `unittest` module has been around for a long time and is another capable framework for writing and running tests. In this case, I need to import the `unittest.mock.mock_open()` function. (*https://oreil.ly/EGvXh*) Here is how I can create a mock filehandle with the data from the first test file. I use `read_data` to define the data that will be returned by the `fh.read()` method:

```
>>> from unittest.mock import mock_open
>>> fh = mock_open(read_data=f1)()
```

```
>>> fh.read()
'>SEQ0\nGGATAAAGCGAGAGGCTGGATCATGCACCAACTGCGTGCAACGAAGGAAT\n'
```

In the context of testing, I don't care about the filename, only that this returns a string and does not throw an exception:

```
>>> fh.name
<MagicMock name='open().name' id='140349116126880'>
```

While I often place my unit tests in the same modules as the functions they test, in this instance, I'd rather put this into a separate unit.py module to keep the main program shorter. I wrote the test to handle an empty file, a file with one sequence, and a file with more than one sequence (which are also reflected in the three input test files). Presumably, if the function works for these three cases, it should work for all others:

```
from unittest.mock import mock_open  ❶
from seqmagique import process  ❷


def test_process() -> None:
    """ Test process """

    empty = process(mock_open(read_data='')())  ❸
    assert empty.min_len == 0
    assert empty.max_len == 0
    assert empty.avg_len == 0
    assert empty.num_seqs == 0

    one = process(mock_open(read_data='>SEQ0\nAAA')())  ❹
    assert one.min_len == 3
    assert one.max_len == 3
    assert one.avg_len == 3
    assert one.num_seqs == 1

    two = process(mock_open(read_data='>SEQ0\nAAA\n>SEQ1\nCCCC')())  ❺
    assert two.min_len == 3
    assert two.max_len == 4
    assert two.avg_len == 3.5
    assert two.num_seqs == 2
```

❶ Import the mock_open() function.

❷ Import the process() function I'm testing.

❸ A mock empty filehandle that should have zeros for all the values.

❹ A single sequence with three bases.

❺ A filehandle with two sequences of three and four bases.

Use **pytest** to run the tests:

```
$ pytest -xv unit.py
============================= test session starts =============================
...

unit.py::test_process PASSED                                            [100%]

============================== 1 passed in 2.55s ==============================
```

Where to Place Unit Tests

Note that the preceding unit.py module imports the process() function from the seqmagique.py module, so both modules need to be in the same directory. If I were to move unit.py to the *tests* directory, then pytest would break. I encourage you to try the following and note the errors.

You should get a notice like *ModuleNotFoundError: No module named seqmagique*:

```
$ cp unit.py tests
$ pytest -xv tests/unit.py
```

As noted in the documentation (*https://oreil.ly/um7N7*), I must invoke pytest as follows to add the current directory to sys.path so that seqmagique.py can be found:

```
$ python3 -m pytest -xv tests/unit.py
```

Placing unit.py in the same directory as the code it's testing is slightly more convenient as it means I can run the shorter pytest command, but grouping all the tests into the *tests* directory is tidier. I would normally prefer to have this module as *tests/unit_test.py* so that pytest will automatically discover it, and I would use a make target to run the longer invocation. Mostly I just want you to be aware of different ways to organize your code and tests.

Here is how I use my process() function in main():

```
def main() -> None:
    args = get_args()
    data = [process(fh) for fh in args.files] ❶
    hdr = ['name', 'min_len', 'max_len', 'avg_len', 'num_seqs'] ❷
    print(tabulate(data, tablefmt=args.tablefmt, headers=hdr, floatfmt='.2f')) ❸
```

❶ Process all the input files into a list of FastaInfo objects (tuples).

❷ Define the table headers.

❸ Use the tabulate() function to print a formatted output table.

To test this program, I run it with the following inputs:

- The empty file
- The file with one sequence
- The file with two sequences
- All the input files

To start, I run all these with the default table style. Then I need to verify that all 10 of the table styles are created correctly. Combining all the possible test inputs with all the table styles creates a high degree of *cyclomatic complexity*—the number of different ways that parameters can be combined.

To test this, I first need to manually verify that my program is working correctly. Then I need to generate sample outputs for each of the combinations I intend to test. I wrote the following bash script to create an *out* file for a given combination of an input file and possibly a table style:

```
$ cat mk-outs.sh
#!/usr/bin/env bash

PRG="./seqmagique.py"  ❶
DIR="./tests/inputs"   ❷
INPUT1="${DIR}/1.fa"   ❸
INPUT2="${DIR}/2.fa"
EMPTY="${DIR}/empty.fa"

$PRG $INPUT1 > "${INPUT1}.out"   ❹
$PRG $INPUT2 > "${INPUT2}.out"
$PRG $EMPTY > "${EMPTY}.out"
$PRG $INPUT1 $INPUT2 $EMPTY > "$DIR/all.fa.out"

STYLES="plain simple grid pipe orgtbl rst mediawiki latex latex_raw
  latex_booktabs"

for FILE in $INPUT1 $INPUT2; do  ❺
    for STYLE in $STYLES; do
        $PRG -t $STYLE $FILE > "$FILE.${STYLE}.out"
    done
done

echo Done.
```

❶ The program being tested.

❷ The directory for the input files.

❸ The input files.

❹ Run the program using the three input files and the default table style.

❺ Run the program with two of the input files and all the table styles.

The tests in *tests/seqmagique_test.py* will run the program with a given file and will compare the output to one of the *out* files in the *tests/inputs* directory. At the top of this module, I define the input and output files like so:

```
TEST1 = ('./tests/inputs/1.fa', './tests/inputs/1.fa.out')
```

I define a `run()` function in the module to run the program with the input file and compare the actual output to the expected output. This is a basic pattern you could copy for testing any program's output:

```
def run(input_file: str, expected_file: str) -> None:
    """ Runs on command-line input """

    expected = open(expected_file).read().rstrip() ❶
    rv, out = getstatusoutput(f'{RUN} {input_file}') ❷
    assert rv == 0 ❸
    assert out == expected ❹
```

❶ Read the expected output from the file.

❷ Run the program with the given input file using the default table style.

❸ Check that the return value is 0.

❹ Check that the output was the expected value.

I use it like so:

```
def test_input1() -> None:
    """ Runs on command-line input """

    run(*TEST1) ❶
```

❶ Splat the tuple to pass the two values positionally to the `run()` function.

The test suite also checks the table styles:

```
def test_styles() -> None:
    """ Test table styles """

    styles = [ ❶
        'plain', 'simple', 'grid', 'pipe', 'orgtbl', 'rst', 'mediawiki',
        'latex', 'latex_raw', 'latex_booktabs'
    ]

    for file in [TEST1[0], TEST2[0]]: ❷
        for style in styles: ❸
```

```
expected_file = file + '.' + style + '.out'  ❹
assert os.path.isfile(expected_file)  ❺
expected = open(expected_file).read().rstrip()  ❻
flag = '--tablefmt' if random.choice([0, 1]) else '-t'  ❼
rv, out = getstatusoutput(f'{RUN} {flag} {style} {file}')  ❽
assert rv == 0  ❾
assert out == expected
```

❶ Define a list of all possible styles.

❷ Use the two nonempty files.

❸ Iterate through each style.

❹ The output file is the name of the input file plus the style and the extension *.out*.

❺ Check that the file exists.

❻ Read the expected value from the file.

❼ Randomly choose the short or long flag to test.

❽ Run the program with the flag option, style, and file.

❾ Ensure that the program ran without error and produces the correct output.

If I make a change such that the program no longer creates the same output as before, these tests should catch it. This is a *regression* test, where I am comparing how a program works now to how it previously worked. That is, a failure to produce the same output would be considered a regression. While my test suite isn't completely exhaustive, it covers enough combinations that I feel confident the program is correct.

Solution 2: Formatting with rich

In this second solution I want to show a different way to create the output table, using the rich module to track the processing of the input files and make a fancier output table. Figure 15-1 shows how the output looks.

```
$ ./seqmagique_rich.py tests/inputs/*.fa
Working...  ━━━━━━━━━━━━━━━━━━━━━━━━━━━━━  100% 0:00:00
```

Name	Min. Len	Max. Len	Avg. Len	Num. Seqs
tests/inputs/1.fa	50	50	50.0	1
tests/inputs/2.fa	49	79	64.0	5
tests/inputs/empty.fa	0	0	0	0

Figure 15-1. The progress indicator and output table using the rich module are fancier

I still process the files in the same way, so the only difference is in creating the output. I first need to import the needed functions:

```python
from rich.console import Console
from rich.progress import track
from rich.table import Table, Column
```

Here is how I use these:

```python
def main() -> None:
    args = get_args()

    table = Table('Name', ❶
                  Column(header='Min. Len', justify='right'),
                  Column(header='Max. Len', justify='right'),
                  Column(header='Avg. Len', justify='right'),
                  Column(header='Num. Seqs', justify='right'),
                  header_style="bold black")

    for fh in track(args.file): ❷
        file = process(fh) ❸
        table.add_row(file.filename, str(file.min_len), str(file.max_len), ❹
                      str(file.avg_len), str(file.num_seqs))

    console = Console() ❺
    console.print(table)
```

❶ Create the table to hold the data. The Name column is a standard, left-justified string field. All the others need to be right-justified and require a custom Column object.

❷ Iterate through each filehandle using the track() function to create a progress bar for the user.

❸ Process the file to get the statistics.

❹ Add the file's statistics to the table. Note that all the values must be strings.

❺ Create a `Console` object, and use it to print the output.

Going Further

The `seqmagick` tool has many other useful options. Implement your own versions of as many as you can.

Review

Key points from this chapter:

- The `seqmagick` tool provides many ways to examine sequence files.
- There are many ways to verify that your input files are complete and not corrupted, from examining file sizes to using message digests such as MD5 hashes.
- The `choices` option for `argparse` parameters will force the user to select a value from a given list.
- The `tabulate` and `rich` modules can create text tables of data.
- The `numpy` module is useful for many mathematical operations.
- The `io.StringIO()` and `unittest.mock.mock_open()` functions offer two ways to mock a filehandle for testing.
- Regression testing verifies that a program continues to work as it did before.

FASTX grep: Creating a Utility Program to Select Sequences

A colleague asked me once to find all the RNA sequences in a FASTQ file that had a description or name containing the string *LSU* (for *long subunit* RNA). Although it's possible to solve this problem for FASTQ files by using the grep program[1] to find all the lines of a file matching some pattern, writing a solution in Python allows you to create a program that could be expanded to handle other formats, like FASTA, as well as to select records based on other criteria, such as length or GC content. Additionally, you can add options to change the output sequence format and introduce conveniences for the user like guessing the input file's format based on the file extension.

In this chapter, you will learn:

- About the structure of a FASTQ file
- How to perform a case-insensitive regular expression match
- About DWIM (Do What I Mean) and DRY (Don't Repeat Yourself) ideas in code
- How to use and and or operations to reduce Boolean values and bits

---

1 Some say this is short for *global regular expression print*.

Finding Lines in a File Using grep

The grep program can find all the lines in a file matching a given pattern. If I search for *LSU* in one of the FASTQ files, it finds two header lines containing this pattern:

```
$ grep LSU tests/inputs/lsu.fq
@ITSLSUmock2p.ITS_M01380:138:000000000-C9GKM:1:1101:14440:2042 2:N:0
@ITSLSUmock2p.ITS_M01384:138:000000000-C9GKM:1:1101:14440:2043 2:N:0
```

If the goal were only to find how many sequences contain this string, I could pipe this into wc (word count) to count the lines using the -l option:

```
$ grep LSU tests/inputs/lsu.fq | wc -l
       2
```

Since my goal is to extract the sequence records where the header contains the substring *LSU*, I have to do a little more work. As long as the input files are in FASTQ format, I can still use grep, but this requires a better understanding of the format.

The Structure of a FASTQ Record

The FASTQ sequence format is a common way to receive sequence data from a sequencer as it includes both the base calls and the quality scores for each base. That is, sequencers generally report both a base and a measure of certainty that the base is correct. Some sequencing technologies have trouble, for instance, with homopolymer runs, like a poly(*A*) run of many *A*s where the sequencer may be unable to count the correct number. Many sequencers also lose confidence in base calls as the reads grow longer. Quality scores are an important means for rejecting or truncating low-quality reads.

 Depending on the sequencer, some bases can be hard to distinguish, and the ambiguity may be reported using IUPAC codes I describe in Chapter 1, such as *R* for *A* or *G* or *N* for any base.

The FASTQ format is somewhat similar to the FASTA format used in many of the problems in the Rosalind challenges. As a reminder, FASTA records start with a > symbol followed by a header line that identifies the sequence and may contain metadata. The sequence itself follows, which may be one (possibly long) line of text or might be split over multiple lines. In contrast, FASTQ records must always be exactly four lines, as shown in Figure 16-1.

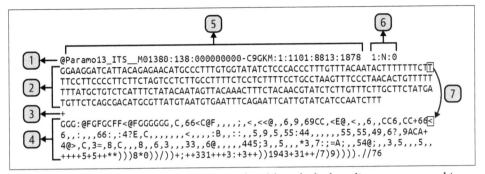

Figure 16-1. The elements of a FASTQ record—although the long lines are wrapped in this display, the actual record contains exactly four lines

Let's take a closer look at the contents of this figure:

1. The first line starts with the @ symbol and contains the header information.

2. The second line contains the sequence with no line breaks.

3. The third line begins with the + symbol. Often it will be only this symbol, but sometimes the header information may be repeated.

4. The fourth line contains the quality scores for each base in the sequence and also has no line breaks.

5. The sequence ID is all characters up to the first space.

6. Additional metadata may follow the ID and is included in the description.

7. Each base in the sequence has a partner in the quality line representing the confidence that this base is correct.

A FASTQ header has the same structure as the header in a FASTA record, with the exception that it starts with the @ sign instead of the >. The sequence identifier is usually all the characters after @ up to the first space. The second line containing the sequence cannot contain any line breaks, and each base in the sequence has a corresponding quality value in the fourth line. The quality scores on the fourth line use the ASCII values of the characters to encode the certainty of the base call. These scores are represented using the printable characters from the ASCII table first introduced in Chapter 3.

The first 32 values in the ASCII table are unprintable control characters and the space. The printable characters start at 33, with punctuation followed by numbers. The first letter, *A*, is not found until 65, and uppercase characters precede lowercase. The following is the output from the `asciitbl.py` program included in the repository that shows the ordinal values of the 128 values from the ASCII table:

```
$ ./asciitbl.py
  0 NA      26 NA       52 4       78 N      104 h
  1 NA      27 NA       53 5       79 O      105 i
  2 NA      28 NA       54 6       80 P      106 j
  3 NA      29 NA       55 7       81 Q      107 k
  4 NA      30 NA       56 8       82 R      108 l
  5 NA      31 NA       57 9       83 S      109 m
  6 NA      32 SPACE    58 :       84 T      110 n
  7 NA      33 !        59 ;       85 U      111 o
  8 NA      34 "        60 <       86 V      112 p
  9 NA      35 #        61 =       87 W      113 q
 10 NA      36 $        62 >       88 X      114 r
 11 NA      37 %        63 ?       89 Y      115 s
 12 NA      38 &        64 @       90 Z      116 t
 13 NA      39 '        65 A       91 [      117 u
 14 NA      40 (        66 B       92 \      118 v
 15 NA      41 )        67 C       93 ]      119 w
 16 NA      42 *        68 D       94 ^      120 x
 17 NA      43 +        69 E       95 _      121 y
 18 NA      44 ,        70 F       96 `      122 z
 19 NA      45 -        71 G       97 a      123 {
 20 NA      46 .        72 H       98 b      124 |
 21 NA      47 /        73 I       99 c      125 }
 22 NA      48 0        74 J      100 d      126 ~
 23 NA      49 1        75 K      101 e      127 DEL
 24 NA      50 2        76 L      102 f
 25 NA      51 3        77 M      103 g
```

Look at the quality line of the FASTQ record in Figure 16-1 and see how the characters change from higher values like uppercase letters at the beginning to lower values like punctuation and numbers toward the end. Note that the @ and + symbols on the fourth line represent possible quality values and so would not be metacharacters denoting the beginning of a record or the separator line. For this reason, FASTQ records can't use newlines to break the sequence (like in FASTA records) or quality lines: the symbols @ and + might end up as the first character on a line, making it impossible to find the start of a record. Combine this with the utterly useless third line that often consists of a single + symbol, and which sometimes needlessly recapitulates all the header information, and you see why biologists should never be allowed to define a file format.

There are multiple encoding standards using various ranges to represent the quality scores.

Because FASTQ records must be four lines long, I can use the -A|--after-context option for grep to specify the number of lines of trailing context after each match:

```
$ grep -A 4 LSU tests/inputs/lsu.fq | head -4
@ITSLSUmock2p.ITS_M01380:138:000000000-C9GKM:1:1101:14440:2042 2:N:0
CAAGTTACTTCCTCTAAATGACCAAGCCTAGTGTAGAACCATGTCGTCAGTGTCAGTCTGAGTGTAGATCT\
CGGTGGTCGCCGTATCATTAAAAAAAAAAATGTAATACTACTAGTAATTATTAATATTATAATTTTGTCTA\
TTAGCATCTTATTATAGATAGAAGATATTATTCATATTTCACTATCTTATACTGATATCAGCTTTATCAGA\
TCACACTCTAGTGAAGATTGTTCTTAACTGAAATTTCCTTCTTCATACAGACACATTAATCTTACCTA
+
EFGGGGGGGGGCGGGGGFCFFFGGGGGFGGGGGGGGGGGGFGGGGGGGFGFFFCFGGFFGGGGGGGGGFGGG\
GFGGGDG<FD@4@CFFGGGGCFFAFEFEG+,9,,,,99,,,5,,49,4,8,4,444,4,4,,,,,,,,,,,\
,,,8,,,,63,,,,,,,,376,3,,,,,,,8,,,,,,,,,,+++++++++++++3++25+++0+*+0+*0+*\
**))*0))1/+++***************.****.*******0*********/(,(/).)))1)).).).
```

This works as long as the substring of interest occurs only in the header, which is the first line of a record. If grep managed to find a match in any other line in the record, it would print that line plus the following three, yielding unusable garbage. Given that I would like to control exactly which parts of the record to search and the fact that the input files might be in FASTQ, FASTA, or any number of other formats, it quickly becomes evident that grep won't take me very far.

Getting Started

First, I'll show you how my solution works, and then I'll challenge you to implement your version. All the code and tests for this exercise are in the *16_fastx_grep* directory of the repository. Start by changing into this directory and copying the solution to fastx_grep.py:

```
$ cd 16_fastx_grep
$ cp solution.py fastx_grep.py
```

The grep usage shows that it accepts two positional arguments, a pattern and one or more files:

```
$ grep -h
usage: grep [-abcDEFGHhIiJLlmnOoqRSsUVvwxZ] [-A num] [-B num] [-C[num]]
            [-e pattern] [-f file] [--binary-files=value] [--color=when]
            [--context[=num]] [--directories=action] [--label] [--line-buffered]
            [--null] [pattern] [file ...]
```

Request help from the fastx_grep.py program and see that it has a similar interface that requires a pattern and one or more input files. Additionally, this program can parse different input file formats, produce various output formats, write the output to a file, and perform case-insensitive matching:

```
$ ./fastx_grep.py -h
usage: fastx_grep.py [-h] [-f str] [-O str] [-o FILE] [-i]
                     PATTERN FILE [FILE ...]

Grep through FASTX files

positional arguments:
```

```
    PATTERN                 Search pattern ❶
    FILE                    Input file(s) ❷

optional arguments:
  -h, --help              show this help message and exit
  -f str, --format str    Input file format (default: ) ❸
  -O str, --outfmt str    Output file format (default: ) ❹
  -o FILE, --outfile FILE
                          Output file (default: <_io.TextIOWrapper ❺
                          name='<stdout>' mode='w' encoding='utf-8'>)
  -i, --insensitive       Case-insensitive search (default: False) ❻
```

❶ The regular expression (pattern) is the first positional argument.

❷ One or more positional file arguments are required second.

❸ The input file format of the sequences, either *fasta* or *fastq*. The default is to guess from the file extension.

❹ The output file format, one of *fasta*, *fastq*, or *fasta-2line*. The default is to use the same as the input file.

❺ The output filename; the default is STDOUT.

❻ Whether to perform case-insensitive matches; the default is False.

This program has a more complicated set of arguments than many of the programs from Part I. As usual, I like to use a NamedTuple to model the options:

```
from typing import List, NamedTuple, TextIO


class Args(NamedTuple):
    """ Command-line arguments """
    pattern: str ❶
    files: List[TextIO] ❷
    input_format: str ❸
    output_format: str ❹
    outfile: TextIO ❺
    insensitive: bool ❻
```

❶ The regular expression to use.

❷ One or more input files.

❸ The format of the input file, such as FASTA or FASTQ.

❹ The format of the output file.

❺ The name of the output file.

❻ Whether to perform case-insensitive searching.

Here is how I define the program's parameters:

```python
def get_args() -> Args:
    """ Get command-line arguments """

    parser = argparse.ArgumentParser(
        description='Grep through FASTX files',
        formatter_class=argparse.ArgumentDefaultsHelpFormatter)

    parser.add_argument('pattern', ❶
                        metavar='PATTERN',
                        type=str,
                        help='Search pattern')

    parser.add_argument('file',
                        metavar='FILE',
                        nargs='+',
                        type=argparse.FileType('rt'), ❷
                        help='Input file(s)')

    parser.add_argument('-f',
                        '--format',
                        help='Input file format',
                        metavar='str',
                        choices=['fasta', 'fastq'], ❸
                        default='')

    parser.add_argument('-O',
                        '--outfmt',
                        help='Output file format',
                        metavar='str',
                        choices=['fasta', 'fastq', 'fasta-2line'], ❹
                        default='')

    parser.add_argument('-o',
                        '--outfile',
                        help='Output file',
                        type=argparse.FileType('wt'), ❺
                        metavar='FILE',
                        default=sys.stdout)

    parser.add_argument('-i', ❻
                        '--insensitive',
                        help='Case-insensitive search',
                        action='store_true')

    args = parser.parse_args()
```

```
    return Args(pattern=args.pattern,
                files=args.file,
                input_format=args.format,
                output_format=args.outfmt,
                outfile=args.outfile,
                insensitive=args.insensitive)
```

❶ The pattern will be a string.

❷ The inputs must be readable text files.

❸ Use `choices` to constrain the input values. The default will be guessed from the input file extension.

❹ Constrain values using `choices`; default to using the input format. The *fasta-2line* option will not break long sequences over multiple lines and so will use only two lines per record.

❺ The output file will be a writable text file. The default is STDOUT.

❻ A flag to indicate case-insensitive searching. The default is `False`.

If you run the following command to search for *LSU* in the *lsu.fq* test file, you should see eight lines of output representing two FASTQ records:

```
$ ./fastx_grep.py LSU tests/inputs/lsu.fq | wc -l
    8
```

If you search for lowercase *lsu*, however, you should see no output:

```
$ ./fastx_grep.py lsu tests/inputs/lsu.fq | wc -l
    0
```

Use the `-i|--insensitive` flag to perform a case-insensitive search:

```
$ ./fastx_grep.py -i lsu tests/inputs/lsu.fq  | wc -l
    8
```

You can use the `-o|--outfile` option to write the results to a file instead of STDOUT:

```
$ ./fastx_grep.py -o out.fq -i lsu tests/inputs/lsu.fq
$ wc -l out.fq
    8 out.fq
```

If you look at the *out.fq* file, you'll see it's in FASTQ format just like the original input. You can use the `-O|--outfmt` option to change this to something like FASTA and look at the output file to verify the format:

```
$ ./fastx_grep.py -O fasta -o out.fa -i lsu tests/inputs/lsu.fq
$ head -3 out.fa
>ITSLSUmock2p.ITS_M01380:138:000000000-C9GKM:1:1101:14440:2042 2:N:0
```

```
CAAGTTACTTCCTCTAAATGACCAAGCCTAGTGTAGAACCATGTCGTCAGTGTCAGTCTG
AGTGTAGATCTCGGTGGTCGCCGTATCATTAAAAAAAAAAATGTAATACTACTAGTAATT
```

Try using the *fasta-2line* output format to see how the long sequences are not broken over multiple lines. Note that the program also works on FASTA input without my having to indicate the file format because it is guessed from the *.fa* file extension:

```
$ ./fastx_grep.py -o out.fa -i lsu tests/inputs/lsu.fa
$ ../15_seqmagique/seqmagique.py out.fa
name       min_len    max_len    avg_len    num_seqs
out.fa         281        301     291.00           2
```

Run **pytest -v** to see all the tests for the program, which include guessing the file format, handling empty files, searching lowercase and uppercase input both with and without case-sensitivity, writing output files, and writing different output formats. When you think you understand all the options your program must handle, start over:

```
$ new.py -fp 'Grep through FASTX files' fastx_grep.py
Done, see new script "fastx_grep.py".
```

Guessing the File Format

If you look at *out.fa* created in the preceding section, you'll see it's in FASTA format, matching the input format, but I never indicated the input file format. The program intelligently checks the file extension of the input file and guesses at the format using the assumptions in Table 16-1. Similarly, if no output format is specified, then the input file format is assumed to be the desired output format. This is an example of the *DWIM* principle in software development: Do What I Mean.

Table 16-1. Common file extensions for FASTA/Q files

Extension	Format
.fasta	FASTA
.fa	FASTA
.fna	FASTA (nucleotides)
.faa	FASTA (amino acids)
.fq	FASTQ
.fastq	FASTQ

Your program will similarly need to guess the format of the input files. I created a guess_format() function that takes the name of a file and returns a string of either fasta or fastq. Here is a stub for the function:

```
def guess_format(filename: str) -> str:
    """ Guess format from extension """
```

```
    return ''
```

Here is the test I wrote. After defining the arguments, I would recommend you start with this function. Do not proceed until your code passes this:

```
def test_guess_format() -> None:
    """ Test guess_format """

    assert guess_format('/foo/bar.fa') == 'fasta'
    assert guess_format('/foo/bar.fna') == 'fasta'
    assert guess_format('/foo/bar.faa') == 'fasta'
    assert guess_format('/foo/bar.fasta') == 'fasta'
    assert guess_format('/foo/bar.fq') == 'fastq'
    assert guess_format('/foo/bar.fastq') == 'fastq'
    assert guess_format('/foo/bar.fx') == ''
```

It might help to sketch out how the program should work:

```
def main():
    get the program arguments

    for each input file:
        guess the input format or complain that it can't be guessed
        figure out the output format from the args or use the input format

        for each record in the input file:
            if the sequence ID or description matches the pattern:
                write the sequence to the output file in the output format
```

For instance, I can run the program on three input files by using the shell glob `*.f[aq]` to indicate all the files with an extension starting with the letter *f* and followed by either *a* or *q*:

```
$ ls tests/inputs/*.f[aq]
tests/inputs/empty.fa  tests/inputs/lsu.fa    tests/inputs/lsu.fq
```

This should write four sequences to the file *out.fa*:

```
$ ./fastx_grep.py -O fasta -o out.fa -i lsu tests/inputs/*.f[aq]
$ ../15_seqmagique/seqmagique.py out.fa
name      min_len    max_len    avg_len    num_seqs
out.fa        281        301     291.00           4
```

 This is a complex program that may take you quite a while to finish. There is value in your struggle, so just keep writing and running the tests, which you should also read to understand how to challenge your program.

Solution

In my experience, this is a realistically complicated program that captures many patterns I write often. It starts by validating and processing some number of input files. I'm a truly lazy programmer[2] who always wants to give as little information as possible to my programs, so I'm happy to write a little code to guess the file formats for me.

Guessing the File Format from the File Extension

I'll start with the function for guessing a file's format from the file extension:

```
def guess_format(filename: str) -> str:
    """ Guess format from extension """

    ext = re.sub('^[.]', '', os.path.splitext(filename)[1])  ❶

    return 'fasta' if re.match('f(ast|a|n)?a$', ext) else 'fastq' if re.match(  ❷
        'f(ast)?q$', ext) else ''
```

❶ Use the `os.path.splitext()` function to get the file extension and remove the leading dot.

❷ Return the string `fasta` if the extension matches one of the patterns for FASTA files from Table 16-1, `fastq` if it matches a FASTQ pattern, and the empty string otherwise.

Recall that `os.path.splitext()` will return both the root of the filename and the extension as a 2-tuple:

```
>>> import os
>>> os.path.splitext('/foo/bar.fna')
('/foo/bar', '.fna')
```

Since I'm only interested in the second part, I can use the _ to assign the first member of the tuple to a throwaway:

```
>>> _, ext = os.path.splitext('/foo/bar.fna')
>>> ext
'.fna'
```

Instead, I chose to index the tuple to select only the extension:

```
>>> ext = os.path.splitext('/foo/bar.fna')[1]
>>> ext
'.fna'
```

---

2 According to *Programming Perl* by Tom Christiansen et al. (O'Reilly, 2012), the three great virtues of a programmer are laziness, impatience, and hubris.

Since I don't want the leading dot, I could use a string slice to remove this, but this looks cryptic and unreadable to me:

```
>>> ext = os.path.splitext('/foo/bar.fna')[1][1:]
>>> ext
'fna'
```

Instead, I'd prefer to use the re.sub() function I first introduced in Chapter 2. The pattern I'm looking for is a literal dot at the beginning of the string. The caret ^ indicates the start of the string, and the . is a metacharacter that means one of anything. To show that I want a literal dot, I must either place a backslash in front of it like ^\. or place it inside a character class as in ^[.]:

```
>>> import re
>>> ext = re.sub('^[.]', '', os.path.splitext('/foo/bar.fna')[1]) ❶
>>> ext
'fna'
```

❶ Use re.sub() to remove a literal dot at the beginning of the file extension.

As shown in Table 16-1, there are four common extensions for FASTA files which I can represent using one compact regular expression. Recall that there are two functions in the re module for searching:

- re.match() finds a match from the beginning of a string.
- re.search() finds a match anywhere inside a string.

In this example, I'm using the re.match() function to ensure that the pattern (the first argument) is found at the beginning of the extension (the second argument):

```
>>> re.match('f(ast|a|n)?a$', ext)
<re.Match object; span=(0, 3), match='fna'>
```

To get the same results from re.search(), I would need to use a caret at the beginning to anchor the pattern to the start of the string:

```
>>> re.search('^f(ast|a|n)?a$', ext)
<re.Match object; span=(0, 3), match='fna'>
```

Figure 16-2 describes each part of the regular expression.

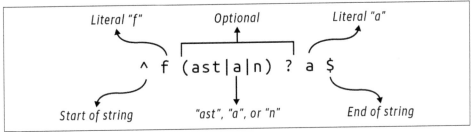

Figure 16-2. A regular expression for matching the four FASTA patterns

It may help to see this drawn as a finite state machine diagram, as shown in Figure 16-3.

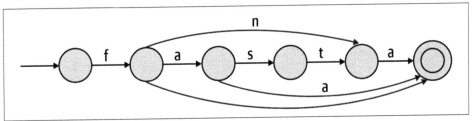

Figure 16-3. A finite state machine diagram for matching the four FASTA patterns

As there are only two patterns for FASTQ files, the pattern is somewhat simpler:

```
>>> re.search('^f(ast)?q$', 'fq')
<re.Match object; span=(0, 2), match='fq'>
>>> re.search('^f(ast)?q$', 'fastq')
<re.Match object; span=(0, 5), match='fastq'>
```

Figure 16-4 explains this regular expression.

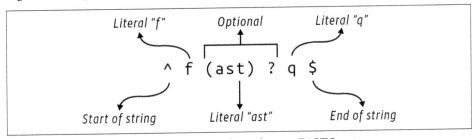

Figure 16-4. A regular expression for matching the two FASTQ patterns

Figure 16-5 shows the same idea expressed as a finite state machine.

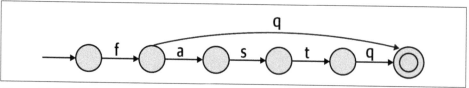

Figure 16-5. A finite state machine diagram for matching the two FASTQ patterns

I Love It When a Plan Comes Together

Following is how I wrote `main()` using the structure I introduced in the first part of the chapter:

```
def main() -> None:
    args = get_args()
    regex = re.compile(args.pattern, re.IGNORECASE if args.insensitive else 0) ❶

    for fh in args.files: ❷
        input_format = args.input_format or guess_format(fh.name) ❸

        if not input_format: ❹
            sys.exit(f'Please specify file format for "{fh.name}"')

        output_format = args.output_format or input_format ❺

        for rec in SeqIO.parse(fh, input_format): ❻
            if any(map(regex.search, [rec.id, rec.description])): ❼
                SeqIO.write(rec, args.outfile, output_format) ❽
```

❶ Compile a regular expression to find the given pattern.

❷ Iterate through the input files.

❸ Use the input format or guess it from the filename.

❹ Exit with an error if there is no input file format.

❺ Use the output format or use the input format.

❻ Iterate through each sequence in the file.

❼ See if either the sequence ID or description matches the pattern.

❽ If so, write the sequence to the output file.

There are several items I'd like to highlight, and I'll start with my use of `sys.exit()` to halt the program in the middle of processing the files if I'm unable to decide on the output file format. This is a value I don't necessarily expect from the user, and one that I'm hoping I can figure out when the program is running. If I can't, then I need to return an error message to the user and an exit value to the operating system to indicate a failure. I need the user to start over and correct the missing information before I can continue.

I also want to point out my use of the `any()` function, which has a corollary in the `all()` function. Both functions reduce a list of truthy values to a single Boolean value. The `all()` function will return `True` if *all* the values are truthy, and `False` otherwise:

```
>>> all([True, True, True])
True
>>> all([True, False, True])
False
```

While the `any()` function will return `True` if *any* of the values are truthy and `False` otherwise:

```
>>> any([True, False, True])
True
>>> any([False, False, False])
False
```

I use this with the compiled regular expression to search the record's ID and description fields. That regex is also using the `re.IGNORECASE` flag to turn on case-insensitive matching. To explain this, I'd like to go on a tangent into how Python combines Boolean values using and and or and bits using the respective bitwise operators & and |.

Combining Regular Expression Search Flags

By default, regular expressions are case-sensitive, but this program needs to handle both case-sensitive and case-insensitive searching. For example, if I search for lowercase *lsu* but the record header has only uppercase *LSU*, I would expect this to fail:

```
>>> import re
>>> type(re.search('lsu', 'This contains LSU'))
<class 'NoneType'>
```

One way to disregard case is to force both the search pattern and string to uppercase or lowercase:

```
>>> re.search('lsu'.upper(), 'This contains LSU'.upper())
<re.Match object; span=(14, 17), match='LSU'>
```

Another method is to provide an optional flag to the `re.search()` function:

```
>>> re.search('lsu', 'This contains LSU', re.IGNORECASE)
<re.Match object; span=(14, 17), match='LSU'>
```

This can be shortened to `re.I`:

```
>>> re.search('lsu', 'This contains LSU', re.I)
<re.Match object; span=(14, 17), match='LSU'>
```

In the program, I use this when I compile the regular expression:

```
regex = re.compile(args.pattern, re.IGNORECASE if args.insensitive else 0) ❶
```

❶ If `args.insensitive` is `True`, then use the `re.IGNORECASE` option when compiling the pattern; otherwise, use `0`, which means no options.

I first showed how to compile a regular expression in Chapter 11. The advantage is that Python only has to parse the pattern once, which usually makes your code run faster. Here I need to decide whether to change to case-insensitive matching using an optional flag. I can alter many aspects of the regular expression matching with other such flags, which can be combined using the bitwise or | operator. I think it's best to start with the documentation from `help(re)`:

```
Each function other than purge and escape can take an optional 'flags' argument
consisting of one or more of the following module constants, joined by "|".
A, L, and U are mutually exclusive.
    A  ASCII       For string patterns, make \w, \W, \b, \B, \d, \D
                   match the corresponding ASCII character categories
                   (rather than the whole Unicode categories, which is the
                   default).
                   For bytes patterns, this flag is the only available
                   behaviour and needn't be specified.
    I  IGNORECASE  Perform case-insensitive matching.
    L  LOCALE      Make \w, \W, \b, \B, dependent on the current locale.
    M  MULTILINE   "^" matches the beginning of lines (after a newline)
                   as well as the string.
                   "$" matches the end of lines (before a newline) as well
                   as the end of the string.
    S  DOTALL      "." matches any character at all, including the newline.
    X  VERBOSE     Ignore whitespace and comments for nicer looking RE's.
    U  UNICODE     For compatibility only. Ignored for string patterns (it
                   is the default), and forbidden for bytes patterns.
```

Looking closely, I can find that `re.IGNORECASE` is an enum (*https://oreil.ly/J6Wsy*) or *enumeration* of possible values:

```
>>> type(re.IGNORECASE)
<enum 'RegexFlag'>
```

According to the documentation (*https://oreil.ly/nONMy*), this is "a subclass of enum.IntFlag," which is described thusly (*https://oreil.ly/l1dyG*):

Base class for creating enumerated constants that can be combined using the bitwise operators without losing their IntFlag membership. IntFlag members are also subclasses of int.

This means that re.IGNORECASE is deep down an int, just like False is actually 0 and True is actually 1. I used a little detective work to figure out the integer values of the flags by adding 0:

```
>>> for flag in sorted([re.A, re.I, re.L, re.M, re.S, re.X, re.U]):
...     print(f'{flag:15} {flag + 0:5} {0 + flag:#011b}')
...
re.IGNORECASE     2 0b000000010
re.LOCALE         4 0b000000100
re.MULTILINE      8 0b000001000
re.DOTALL        16 0b000010000
re.UNICODE       32 0b000100000
re.VERBOSE       64 0b001000000
re.ASCII        256 0b100000000
```

Note how each value is a power of 2 so that each flag can be represented by a single, unique bit. This makes it possible to combine flags using the | operator mentioned in the documentation. To demonstrate, I can use the prefix 0b to represent a string of raw bytes. Here are the binary representations of the numbers 1 and 2. Note that each uses just a single bit set to 1:

```
>>> one = 0b001
>>> two = 0b010
```

If I use | to *or* the bits together, each of the three bits is combined using the truth table shown in Table 16-2.

Table 16-2. Truth table for or (|)

First	Second	Result
T	T	T
T	F	T
F	T	T
F	F	F

As shown in Figure 16-6, Python will look at each bit and select 1 if either bit is 1, and 0 only if both bits are 0, resulting in 0b011, which is the binary representation of the number 3 because the bits for positions 1 and 2 are both set:

```
>>> one | two
3
```

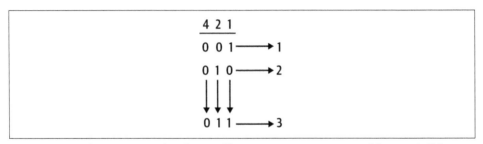

Figure 16-6. When or-ing each column of bits, a 1 in any position yields a 1; if all bits are 0, the result is 0

When using the & operator, Python will only yield a 1 when both bits are 1; otherwise, it will return 0, as shown in Table 16-3.

Table 16-3. Truth table for and (&)

First	Second	Result
T	T	T
T	F	F
F	T	F
F	F	F

Therefore, using & to combine one and two will result in the value 0b000, which is the binary representation of 0:

```
>>> one & two
0
```

I can use the | operator to join multiple regular expression bitflags. For example, re.IGNORECASE is 2, which is represented as 0b010, and re.LOCALE is 4, which is represented as 0b100. Bitwise or combines these as 0b110, which is the number 6:

```
>>> 0b010 | 0b100
6
```

I can verify that this is true:

```
>>> (re.IGNORECASE | re.LOCALE) == 6
True
```

To return to the re.compile() function, the default is to match case:

```
>>> regex = re.compile('lsu')
>>> type(regex.search('This contains LSU'))
<class 'NoneType'>
```

If the user wants case-insensitive searching, then I want to execute something like this:

```
>>> regex = re.compile('lsu', re.IGNORECASE)
>>> regex.search('This contains LSU')
<re.Match object; span=(14, 17), match='LSU'>
```

One way to avoid this would be to use an `if` statement:

```
regex = None
if args.insensitive:
    regex = re.compile(args.pattern, re.IGNORECASE)
else:
    regex = re.compile(args.pattern)
```

I dislike this solution as it violates the *DRY* principle: Don't Repeat Yourself. I can write an `if` expression to choose either the `re.IGNORECASE` flag or some default value that means *no flags*, which turns out to be the number 0:

```
regex = re.compile(args.pattern, re.IGNORECASE if args.insensitive else 0)
```

If I wanted to expand this program to include any of the additional search flags from the documentation, I could use | to combine them. Chapters 6 and 12 talk about the idea of *reducing* multiple values to a single value. For instance, I can use addition to reduce a list of numbers to their sum, or multiplication to create the product, and reduce a list of strings to a single value using the `str.join()` function. I can similarly use bitwise | to reduce all the regex flags:

```
>>> (re.A | re.I | re.L | re.M | re.S | re.X | re.U) + 0
382
```

Because these flags use unique bits, it's possible to find out exactly which flags were used to generate a particular value by using the & operator to determine if a given bit is on. For instance, earlier I showed how to combine the flags `re.IGNORECASE` and `re.LOCALE` using |:

```
>>> flags = re.IGNORECASE | re.LOCALE
```

To see if a given flag is present in the `flags` variable, I use &. It will be returned when I *and* it because only the 1 bits present in both values will be returned:

```
>>> flags & re.IGNORECASE
re.IGNORECASE
```

If I *and* a flag that's not present in the combined values, the result is 0:

```
>>> (flags & re.VERBOSE) + 0
0
```

That's a lot of information about combining bits. So, if you don't know, now you know.

Reducing Boolean Values

I'd like to bring this back to the any() function I used in this program. As with the bitwise combinations of integer values, I can similarly reduce multiple Boolean values. That is, here is the same information as in Table 16-2, using the or operator to combine Booleans:

```
>>> True or True
True
>>> True or False
True
>>> False or True
True
>>> False or False
False
```

This is the same as using any() with a list of Booleans. If *any* of the values is truthy, then the whole expression is True:

```
>>> any([True, True])
True
>>> any([True, False])
True
>>> any([False, True])
True
>>> any([False, False])
False
```

And here is the same data as in Table 16-3, using and to combine Booleans:

```
>>> True and True
True
>>> True and False
False
>>> False and True
False
>>> False and False
False
```

This is the same as using all(). Only if *all* of the values are truthy will the whole expression be True:

```
>>> all([True, True])
True
>>> all([True, False])
False
>>> all([False, True])
False
>>> all([False, False])
False
```

Here is the line of code where I use this idea:

```
if any(map(regex.search, [rec.id, rec.description])):
```

The `map()` function feeds each of the `rec.id` and `rec.description` values to the `regex.search()` function, resulting in a list of values that can be interpreted for their truthiness. If any of these is truthy—meaning I found a match in at least one of the fields—then `any()` will return `True` and the sequence should be written to the output file.

Going Further

Sometimes the sequence header contains key/value metadata like "Organism=Oryza sativa." Add an option to search these values. Be sure you add an input file example to the *tests/inputs* directory and the appropriate tests to *tests/fastx_grep_test.py*.

Expand the program to handle additional input sequence formats like GenBank, EMBL, and SwissProt. Again, be sure to add example files and tests to ensure that your program works.

Alter the program to select sequences with some minimum length and quality score.

Review

Key points from this chapter:

- The FASTQ file format requires each record to be represented by four lines: a header, the sequence, a separator, and the quality scores.
- Regular expression matches can accept flags that control, for example, whether to perform a case-insensitive match. By default, regexes are case-sensitive.
- To indicate multiple regex flags, use the | (*or*) bitwise operator to combine the integer values of the flags.
- Boolean values can be reduced using `and` and `or` operations as well as the `any()` and `all()` functions.
- The DWIM (Do What I Mean) aesthetic means you try to anticipate what your user would want a program to do naturally and intelligently.
- The DRY (Don't Repeat Yourself) principle means that you never duplicate the same idea in your code but rather isolate it to one locus or function.

## CHAPTER 17

DNA Synthesizer: Creating Synthetic Data with Markov Chains

A Markov chain is a model for representing a sequence of possibilities found in a given dataset. It is a machine learning (ML) algorithm because it discovers or learns patterns from input data. In this exercise, I'll show how to use Markov chains trained on a set of DNA sequences to generate novel DNA sequences.

In this exercise, you will:

- Read some number of input sequence files to find all the unique k-mers for a given *k*.
- Create a Markov chain using these k-mers to produce some number of novel sequences of lengths bounded by a minimum and maximum.
- Learn about generators.
- Use a random seed to replicate random selections.

Understanding Markov Chains

In Claude Shannon's "A Mathematical Theory of Communication" (*https://oreil.ly/ 8Gka4*) (1948), the author describes a *Markoff process* that is surprisingly similar to graphs and the finite state diagrams I've been using to illustrate regular expressions. Shannon describes this process as "a finite number of possible *states* of a system" and "a set of transition probabilities" that one state will lead to another.

For one example of a Markov process, Shannon describes a system for generating strings of text by randomly selecting from the 26 letters of the English alphabet and a space. In a "zero-order approximation," each character has an equal probability of being chosen. This process generates strings where letter combinations like *bz* and *qr* might appear as frequently as *st* and *qu*. An examination of actual English words, however, would show that the latter two are orders of magnitude more common than the first two:

```
$ for LETTERS in bz qr st qu
> do echo -n $LETTERS && grep $LETTERS /usr/share/dict/words | wc -l; done
bz      4
qr      1
st    21433
qu     3553
```

To more accurately model the possible transition from one letter to another, Shannon introduces a "first-order approximation... obtained by choosing successive letters independently but each letter having the same probability that it has in the natural language." For this model, I need to train the selection process on representative texts of English. Shannon notes that the letter *e* has a probability of 0.12, reflecting the frequency of its use in English words, whereas *w*, being much less frequently used, has a probability of 0.02, as shown in Figure 17-1.

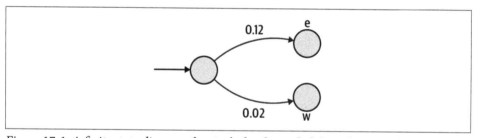

Figure 17-1. A finite state diagram that includes the probability of moving from any character in English to the letters "e" or "w"

Shannon goes on to describe a "second-order approximation" where subsequent letters are "chosen in accordance with the frequencies with which the various letters follow the first one." This relates to k-mers that I used several times in Part I. In linguistics, these are called *N-grams*. For instance, what possible 3-mers could be created given the 2-mer *th*? The letters *e* or *r* would be rather likely, whereas *z* would be impossible as no English word contains the sequence *thz*.

I can perform a rough estimation of how often I can find these patterns. I find approximately 236K English words using `wc -l` to count the lines of my system dictionary:

```
$ wc -l /usr/share/dict/words
  235886 /usr/share/dict/words
```

To find the frequency of the substrings, I need to account for the fact that some words may have the pattern twice. For instance, here are a few words that have more than one occurrence of the pattern *the*:

```
$ grep -E '.*the.*the.*' /usr/share/dict/words | head -3
diathermotherapy
enthelminthes
hyperthermesthesia
```

I can use `grep -io` to search in a case-insensitive fashion (`-i`) for the strings *thr* and *the*, while the `-o` flag tells `grep` to return *only* the matching strings, which will reveal all the matches in each word. I find that *thr* occurs 1,270 times, while *the* occurs 3,593 times:

```
$ grep -io thr /usr/share/dict/words | wc -l
    1270
$ grep -io the /usr/share/dict/words | wc -l
    3593
```

Dividing these numbers by the total number of words leads to a frequency of 0.005 for *thr* and 0.015 for *the*, as shown in Figure 17-2.

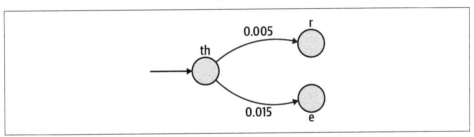

Figure 17-2. A finite state diagram showing the probability of moving from "th" to either an "r" or an "e"

I can apply these ideas to generate novel DNA sequences by reading some sample sequences and noting the ordering of the bases at the level of some k-mer like 10 base pairs (bp). It's important to note that different training texts will affect the model. For instance, English words and spellings have changed over time, so training on older English texts like *Beowulf* and *Canterbury Tales* will yield different results than articles from modern newspapers. This is the *learning* part of machine learning. Many ML algorithms are designed to find patterns from some sets of data to apply to another set. In the case of this program, the generated sequences will bear some resemblance in composition to the input sequences. Running the program with the human genome as training data will produce different results than using a viral metagenome from an oceanic hydrothermal flume.

Getting Started

You should work in the *17_synth* directory containing the inputs and tests for this program. Start by copying the solution to the program synth.py:

```
$ cd 17_synth
$ cp solution.py synth.py
```

This program has a large number of parameters. Run the help to see them:

```
$ ./synth.py -h
usage: synth.py [-h] [-o FILE] [-f format] [-n number] [-x max] [-m min]
                [-k kmer] [-s seed]
                FILE [FILE ...]

Create synthetic DNA using Markov chain

positional arguments:
  FILE                  Training file(s) ❶

optional arguments:
  -h, --help            show this help message and exit
  -o FILE, --outfile FILE
                        Output filename (default: out.fa) ❷
  -f format, --format format
                        Input file format (default: fasta) ❸
  -n number, --num number
                        Number of sequences to create (default: 100) ❹
  -x max, --max_len max
                        Maximum sequence length (default: 75) ❺
  -m min, --min_len min
                        Minimum sequence length (default: 50) ❻
  -k kmer, --kmer kmer  Size of kmers (default: 10) ❼
  -s seed, --seed seed  Random seed value (default: None) ❽
```

❶ The only required parameter is one or more input files.

❷ The output filename will default to *out.fa*.

❸ The input format should be either *fasta* or *fastq* and defaults to the first.

❹ The default number of sequences generated will be 100.

❺ The default maximum sequence length is 75 bp.

❻ The default minimum sequence length is 50 bp.

❼ The default k-mer length is 10 bp.

❽ The default random seed is the None value.

As usual, I create an `Args` class to represent these parameters. I use the following typing imports. Note the `Dict` is used later in the program:

```
from typing import NamedTuple, List, TextIO, Dict, Optional

class Args(NamedTuple):
    """ Command-line arguments """
    files: List[TextIO] ❶
    outfile: TextIO ❷
    file_format: str ❸
    num: int ❹
    min_len: int ❺
    max_len: int ❻
    k: int ❼
    seed: Optional[int] ❽
```

❶ The input `files` will be a list of open filehandles.

❷ The `outfile` will be an open filehandle.

❸ The `file_format` of the input files is a string.

❹ The number of sequences to generate (`num`) is an integer.

❺ The `min_len` is an integer.

❻ The `max_len` is an integer.

❼ The k for k-mer length is an integer.

❽ The random seed can either be the value `None` or an integer.

Here is how I define the program's parameters:

```
def get_args() -> Args:
    """ Get command-line arguments """

    parser = argparse.ArgumentParser(
        description='Create synthetic DNA using Markov chain',
        formatter_class=argparse.ArgumentDefaultsHelpFormatter)

    parser.add_argument('file',
                        help='Training file(s)',
                        metavar='FILE',
                        nargs='+',
                        type=argparse.FileType('rt')) ❶

    parser.add_argument('-o',
                        '--outfile',
                        help='Output filename',
```

```
                             metavar='FILE',
                             type=argparse.FileType('wt'), ❷
                             default='out.fa')

    parser.add_argument('-f',
                             '--format',
                             help='Input file format',
                             metavar='format',
                             type=str,
                             choices=['fasta', 'fastq'], ❸
                             default='fasta')

    parser.add_argument('-n',
                             '--num',
                             help='Number of sequences to create',
                             metavar='number',
                             type=int,
                             default=100) ❹

    parser.add_argument('-x',
                             '--max_len',
                             help='Maximum sequence length',
                             metavar='max',
                             type=int,
                             default=75) ❺

    parser.add_argument('-m',
                             '--min_len',
                             help='Minimum sequence length',
                             metavar='min',
                             type=int,
                             default=50) ❻

    parser.add_argument('-k',
                             '--kmer',
                             help='Size of kmers',
                             metavar='kmer',
                             type=int,
                             default=10) ❼

    parser.add_argument('-s',
                             '--seed',
                             help='Random seed value',
                             metavar='seed',
                             type=int,
                             default=None) ❽

    args = parser.parse_args()

    return Args(files=args.file,
                outfile=args.outfile,
                file_format=args.format,
```

```
            num=args.num,
            min_len=args.min_len,
            max_len=args.max_len,
            k=args.kmer,
            seed=args.seed)
```

❶ The `type` restricts the values to readable text files, and the `nargs` requires one or more values.

❷ The `type` restricts the value to a writable text file, and the default filename will be *out.fa*.

❸ The `choices` restrict the values to either `fasta` or `fastq`, and the default will be `fasta`.

❹ The `type` restricts this to a valid integer value, and the default is `100`.

❺ The `type` restricts this to a valid integer value, and the default is `75`.

❻ The `type` restricts this to a valid integer value, and the default is `50`.

❼ The `type` restricts this to a valid integer value, and the default is `10`.

❽ The `type` restricts this to a valid integer value, and the default is `None`.

It might seem a little odd that the `seed` has `type=int` but has a default of `None` because `None` is not an integer. What I'm saying is that if the user provides any value for the seed, it must be a valid integer; otherwise, the value will be `None`. This is also reflected in the `Args.seed` definition as an `Optional[int]`, which means the value can either be `int` or `None`. Note that this is equivalent to `typing.Union[int, None]`, the union of the `int` type and `None` value.

Understanding Random Seeds

There is an element of randomness to this program as you generate the sequences. I can start with Shannon's zero-order implementation where I select each base independently at random. I can use the `random.choice()` function to select one base:

```
>>> bases = list('ACGT')
>>> import random
>>> random.choice(bases)
'G'
```

If I wanted to generate a 10-bp sequence, I could use a list comprehension with the range() function, like this:

```
>>> [random.choice(bases) for _ in range(10)]
['G', 'T', 'A', 'A', 'C', 'T', 'C', 'T', 'C', 'T']
```

I could further select a random sequence length between some minimum and maximum length using the random.randint() function:

```
>>> [random.choice(bases) for _ in range(random.randint(10, 20))]
['G', 'T', 'C', 'A', 'C', 'C', 'A', 'G', 'C', 'A', 'G']
```

If you execute the preceding code on your computer, it's highly unlikely you will see the same output as shown. Fortunately, these selections are only pseudorandom as they are produced deterministically by a random number generator (RNG). Truly random, unreproducible choices would make testing this program impossible.

I can use a *seed* or initial value to force the pseudorandom selections to be predictable. If you read **help(random.seed)**, you'll see that the "supported seed types are None, int, float, str, bytes, and bytearray." For instance, I can seed using an integer:

```
>>> random.seed(1)
>>> [random.choice(bases) for _ in range(random.randint(10, 20))]
['A', 'G', 'A', 'T', 'T', 'T', 'T', 'C', 'A', 'T', 'A', 'T']
```

I can also use a string:

```
>>> random.seed('markov')
>>> [random.choice(bases) for _ in range(random.randint(10, 20))]
['G', 'A', 'G', 'C', 'T', 'A', 'A', 'C', 'G', 'T', 'C', 'C', 'C', 'G', 'G']
```

If you execute the preceding code, you should get the exact output shown. By default, the random seed is None, which you'll notice is the default for the program. This is the same as not setting the seed at all, so when the program runs with the default it will act in a pseudorandom manner. When testing, I can provide a value that will produce a known result to verify that the program works correctly.

Note that I have forced the user to provide an integer value. Although I find using integers to be convenient, you can seed using strings, numbers, or bytes when writing your own programs. Just remember that the integer 4 and the string '4' are two different values and will produce different results:

```
>>> random.seed(4) ❶
>>> [random.choice(bases) for _ in range(random.randint(10, 20))]
['G', 'A', 'T', 'T', 'C', 'A', 'A', 'A', 'T', 'G', 'A', 'C', 'G']
>>> random.seed('4') ❷
>>> [random.choice(bases) for _ in range(random.randint(10, 20))]
['G', 'A', 'T', 'C', 'G', 'G', 'A', 'G', 'A', 'C', 'C', 'A']
```

❶ Seed using the integer value 4.

❷ Seed using the string value '4'.

The random seed affects every call to random functions from that point forward. This creates a *global* change to your program, and so should be viewed with extreme caution. Typically, I will set the random seed in my program immediately after validating the arguments:

```
def main() -> None:
    args = get_args()
    random.seed(args.seed)
```

If the seed is the default value of None, this will not affect the random functions. If the user has provided a seed value, then all subsequent random calls will be affected.

Reading the Training Files

The first step in my program is to read the training files. Due to how I defined this argument with argparse, the process of validating the input files has been handled, and I know I will have a List[TextIO] which is a list of open filehandles. I will use Bio.SeqIO.parse(), as in previous chapters, to read the sequences.

From the training files, I want to produce a dictionary that describes the weighted possible bases that can follow each k-mer. I think it's helpful to use a type alias to define a couple of new types to describe this. First, I want a dictionary that maps a base like *T* to a floating-point value between 0 and 1 to describe the probability of choosing this base. I'll call it a WeightedChoice:

```
WeightedChoice = Dict[str, float]
```

For instance, in the sequence *ACGTACGC*, the 3-mer *ACG* is followed by either *T* or *C* with equal likelihood. I represent this like so:

```
>>> choices = {'T': 0.5, 'C': 0.5}
```

Next, I want a type that maps the k-mer *ACG* to the choices. I'll call this a Chain as it represents the Markov chain:

```
Chain = Dict[str, WeightedChoice]
```

It would look like this:

```
>>> weighted = {'ACG': {'T': 0.5, 'C': 0.5}}
```

Each k-mer from the sequences in the input file will have a dictionary of weighted options for use in selecting the next base. Here is how I use it to define a function to read the training files:

```
def read_training(fhs: List[TextIO], file_format: str, k: int) -> Chain: ❶
    """ Read training files, return dict of chains """

    pass ❷
```

❶ The function accepts a list of filehandles, the file format of the files, and the size of the k-mers to read. It returns the type Chain.

❷ Use pass to do nothing and return None for now.

Since k-mers figure prominently in this solution, you may want to use the find_kmers() function from Part I. As a reminder, for a function with this signature:

```
def find_kmers(seq: str, k: int) -> List[str]:
    """ Find k-mers in string """
```

I would use the following test:

```
def test_find_kmers() -> None:
    """ Test find_kmers """

    assert find_kmers('ACTG', 2) == ['AC', 'CT', 'TG']
    assert find_kmers('ACTG', 3) == ['ACT', 'CTG']
    assert find_kmers('ACTG', 4) == ['ACTG']
```

I think it's helpful to see exactly what goes into this function and what I expect it to return. In the *tests/unit_test.py* file, you'll find all the unit tests for this program. Here is the test for this function:

```
def test_read_training() -> None:  ❶
    """ Test read_training """

    f1 = io.StringIO('>1\nACGTACGC\n')  ❷
    assert read_training([f1], 'fasta', 4) == {  ❸
        'ACG': { 'T': 0.5, 'C': 0.5 },
        'CGT': { 'A': 1.0 },
        'GTA': { 'C': 1.0 },
        'TAC': { 'G': 1.0 }
    }

    f2 = io.StringIO('@1\nACGTACGC\n+\n!!!!!!!!!')  ❹
    assert read_training([f2], 'fastq', 5) == {  ❺
        'ACGT': { 'A': 1.0 },
        'CGTA': { 'C': 1.0 },
        'GTAC': { 'G': 1.0 },
        'TACG': { 'C': 1.0 }
    }
```

❶ The function takes no arguments and returns None.

❷ Define a mock filehandle containing a single sequence in FASTA format.

❸ Read the data in FASTA format and return the Markov chains for 4-mers.

❹ Define a mock filehandle containing a single sequence in FASTQ format.

❺ Read the data in FASTQ format and return the Markov chains for 5-mers.

To help you better understand k-mers, I've included a program called kmer_tiler.py that will show you the overlapping k-mers in a given sequence. The first test in the preceding function checks that the 3-mer *ACG* is followed by either *T* or *C* with equal probability to create the 4-mers *ACGT* and *ACGC*. Looking at the output from kmer_tiler.py, I can see these two possibilities:

```
$ ./kmer_tiler.py ACGTACGC -k 4
There are 5 4-mers in "ACGTACGC."
ACGTACGC
ACGT ❶
 CGTA
  GTAC
   TACG
    ACGC ❷
```

❶ *ACG* followed by *T*.

❷ *ACG* followed by *C*.

Using this information, I can create Shannon's second-order approximation. For instance, if I randomly select the 3-mer *ACG* to start generating a new sequence, I can add either *T* or *C* with equal probability. Given this training data, I could never append either *A* or *G*, as these patterns never occur.

This is a tricky function to write, so let me give you some pointers. First, you need to find all the k-mers in all the sequences in all the files. For each k-mer, you need to find all the possible endings for a sequence of length k - 1. That is, if k is 4, you first find all the 4-mers and then note how the leading 3-mer can be completed with the last base.

I used collections.Counter() and ended up with an interim data structure that looks like this:

```
{
    'ACG': Counter({'T': 1, 'C': 1}),
    'CGT': Counter({'A': 1}),
    'GTA': Counter({'C': 1}),
    'TAC': Counter({'G': 1})
}
```

Since the input files are all DNA sequences, each k-mer can have at most four possible choices. The key to the Markov chain is in giving these values weights, so next, I need to divide each option by the total number of options. In the case of *ACG*, there are two possible values each occurring once, so they each get a weight of 1/2 or 0.5. The data structure I return from this function looks like the following:

```
{
    'ACG': {'T': 0.5, 'C': 0.5},
    'CGT': {'A': 1.0},
    'GTA': {'C': 1.0},
    'TAC': {'G': 1.0}
}
```

I recommend you first focus on writing a function that passes this test.

Generating the Sequences

Next, I recommend you concentrate on using the Chain to generate new sequences.
Here is a stub for your function:

```
def gen_seq(chain: Chain, k: int, min_len: int, max_len: int) -> Optional[str]: ❶
    """ Generate a sequence """

    return '' ❷
```

❶ The function accepts the Chain, the size of the k-mers, and the minimum and
maximum sequence length. It might or might not return a new sequence as a
string, for reasons I'll explain shortly.

❷ For now, return the empty string.

 When stubbing a function, I interchange pass with returning some
dummy value. Here I use the empty string since the function
returns a str. The point is only to create a function that Python
parses and which I can use for testing. At this point, I *expect* the
function to fail.

Here is the test I wrote for this:

```
def test_gen_seq() -> None: ❶
    """ Test gen_seq """

    chain = { ❷
        'ACG': { 'T': 0.5, 'C': 0.5 },
        'CGT': { 'A': 1.0 },
        'GTA': { 'C': 1.0 },
        'TAC': { 'G': 1.0 }
    }

    state = random.getstate() ❸
    random.seed(1) ❹
    assert gen_seq(chain, k=4, min_len=6, max_len=12) == 'CGTACGTACG' ❺
    random.seed(2) ❻
    assert gen_seq(chain, k=4, min_len=5, max_len=10) == 'ACGTA' ❼
    random.setstate(state) ❽
```

❶ The function accepts no arguments and returns None.

❷ This is the data structure returned by the read_training() function.

❸ Save the current global state of the random module.

❹ Set the seed to a known value of 1.

❺ Verify that the proper sequence is generated.

❻ Set the seed to a different known value of 2.

❼ Verify that the proper sequence is generated.

❽ Restore the random module to any previous state.

 As noted before, calling random.seed() globally modifies the state of the random module. I use random.getstate() to save the current state before modifying and then restore that state when the testing is finished.

This is a tricky function to write, so I'll give you some direction. You will first randomly select the length of the sequence to generate, and the random.randint() function will do just that. Note that the upper and lower bounds are inclusive:

```
>>> min_len, max_len = 5, 10
>>> import random
>>> seq_len = random.randint(min_len, max_len)
>>> seq_len
9
```

Next, you should initialize the sequence using one of the keys from the Markov Chain structure. Note the need to coerce the list(chain.keys()) to avoid the error "dict_keys object is not subscriptable":

```
>>> chain = {
...     'ACG': { 'T': 0.5, 'C': 0.5 },
...     'CGT': { 'A': 1.0 },
...     'GTA': { 'C': 1.0 },
...     'TAC': { 'G': 1.0 }
... }
>>> seq = random.choice(list(chain.keys()))
>>> seq
'ACG'
```

I decided to set up a loop with the condition that the length of the sequence is less than the chosen sequence length. While inside the loop, I will keep appending bases.

To select each new base, I need to get the last k - 1 bases of the ever-growing sequence, which I can do using a list slice and negative indexing. Here's one pass through the loop:

```
>>> k = 4
>>> while len(seq) < seq_len:
...     prev = seq[-1 * (k - 1):]
...     print(prev)
...     break
...
ACG
```

If this previous value occurs in the given chain, then I can select the next base using the random.choices() function. If you read **help(random.choices)**, you will see that this function accepts a population from which to select, weights to consider when making the selection, and a k for the number of choices to return. The keys of the chain for a given k-mer are the population:

```
>>> opts = chain['ACG']
>>> pop = opts.keys()
>>> pop
dict_keys(['T', 'C'])
```

The values of the chain are the weights:

```
>>> weights = opts.values()
>>> weights
dict_values([0.5, 0.5])
```

Note the need to coerce the keys and values using list(), and that random.choices() always returns a list even when you ask for just one, so you'll need to select the first value:

```
>>> from random import choices
>>> next = choices(population=list(pop), weights=list(weights), k=1)
>>> next
['T']
```

I can append this to the sequence:

```
>>> seq += next[0]
>>> seq
'ACGT'
```

The loop repeats until either the sequence is the correct length or I select a previous value that does not exist in the chain. The next time through the loop, the prev 3-mer will be *CGT*, as these are the last three bases in seq. It happens that *CGT* is a key in the chain, but you may sometimes find that there is no way to continue the sequence because the next k-mer doesn't exist in the chain. In this case, you can exit your loop and return None from the function. This is why the gen_seq() function signature returns an `Optional[str]`; I don't want my function to return sequences that are too short. I recommend that you not move on until this function passes the unit test.

Structuring the Program

Once you can read the training files and generate a new sequence using the Markov chain algorithm, you are ready to print the new sequences to the output file. Here is a general outline of how my program works:

```
def main() -> None:
    args = get_args()
    random.seed(args.seed)
    chains = read_training(...)
    seqs = calls to gen_seq(...)
    print each sequence to the output file
    print the final status
```

Note that the program will only generate FASTA output, and each sequence should be numbered from 1 as the ID. That is, your output file should look something like this:

```
>1
GGATTAGATA
>2
AGTCAACG
```

The test suite is pretty large as there are so many options to check. I recommend you run **make test** or read the *Makefile* to see the longer command to ensure that you are properly running all the unit and integration tests.

Solution

I have just one solution for this program as it's complicated enough. I'll start with my function to read the training files, which requires you to import `defaultdict()` and `Counter()` from the `collections` module:

```
def read_training(fhs: List[TextIO], file_format: str, k: int) -> Chain:
    """ Read training files, return dict of chains """

    counts: Dict[str, Dict[str, int]] = defaultdict(Counter)  ❶
    for fh in fhs:  ❷
        for rec in SeqIO.parse(fh, file_format):  ❸
            for kmer in find_kmers(str(rec.seq), k):  ❹
                counts[kmer[:k - 1]][kmer[-1]] += 1  ❺

    def weight(freqs: Dict[str, int]) -> Dict[str, float]:  ❻
        total = sum(freqs.values())  ❼
        return {base: freq / total for base, freq in freqs.items()}  ❽

    return {kmer: weight(freqs) for kmer, freqs in counts.items()}  ❾
```

① Initialize a dictionary to hold the Markov chains.

② Iterate through each filehandle.

③ Iterate through each sequence in the filehandle.

④ Iterate through each k-mer in the sequence.

⑤ Use the prefix of the k-mer as the key into the Markov chain, and add to the count of the final base.

⑥ Define a function that will turn the counts into weighted values.

⑦ Find the total number of bases.

⑧ Divide the frequencies of each base by the total.

⑨ Use a dictionary comprehension to convert the raw counts into weights.

This uses the `find_kmers()` function from Part I, which is:

```python
def find_kmers(seq: str, k: int) -> List[str]:
    """ Find k-mers in string """

    n = len(seq) - k + 1 ①
    return [] if n < 1 else [seq[i:i + k] for i in range(n)] ②
```

① The number of k-mers is the length of the sequence minus k plus 1.

② Use a list comprehension to select all the k-mers from the sequence.

Here is how I wrote the `gen_seq()` function to generate a single sequence:

```python
def gen_seq(chain: Chain, k: int, min_len: int, max_len: int) -> Optional[str]:
    """ Generate a sequence """

    seq = random.choice(list(chain.keys())) ①
    seq_len = random.randint(min_len, max_len) ②

    while len(seq) < seq_len: ③
        prev = seq[-1 * (k - 1):] ④
        if choices := chain.get(prev): ⑤
            seq += random.choices(population=list(choices.keys()), ⑥
                                  weights=list(choices.values()),
                                  k=1)[0]
        else:
            break ⑦

    return seq if len(seq) >= min_len else None ⑧
```

❶ Initialize the sequence to a random choice from the keys of the chain.

❷ Select a length for the sequence.

❸ Execute a loop while the length of the sequence is less than the desired length.

❹ Select the last k - 1 bases.

❺ Attempt to get a list of choices for this k-mer.

❻ Randomly choose the next base using the weighted choices.

❼ If we cannot find this k-mer in the chain, exit the loop.

❽ Return the new sequence if it is long enough; otherwise return None.

To integrate all these, here is my main() function:

```
def main() -> None:
    args = get_args()
    random.seed(args.seed) ❶
    if chain := read_training(args.files, args.file_format, args.k): ❷
        seqs = (gen_seq(chain, args.k, args.min_len, args.max_len) ❸
                for _ in count())

        for i, seq in enumerate(filter(None, seqs), start=1): ❹
            print(f'>{i}\n{seq}', file=args.outfile) ❺
            if i == args.num: ❻
                break

        print(f'Done, see output in "{args.outfile.name}".') ❼
    else:
        sys.exit(f'No {args.k}-mers in input sequences.') ❽
```

❶ Set the random seed.

❷ Read the training files in the given format using the given size k. This may fail if the sequences are shorter than k.

❸ Create a generator to produce the sequences.

❹ Use filter() with a predicate of None to remove falsey elements from the seqs generator. Use enumerate() to iterate through the index positions and sequences starting at 1 instead of 0.

❺ Print the sequence in FASTA format using the index position as the ID.

❻ Break out of the loop if enough sequences have been generated.

❼ Print the final status.

❽ Let the user know why no sequences could be generated.

I'd like to take a moment to explain the generator in the preceding code. I use the range() function to generate the desired number of sequences. I could have used a list comprehension like so:

```
>>> from solution import gen_seq, read_training
>>> import io
>>> f1 = io.StringIO('>1\nACGTACGC\n')
>>> chain = read_training([f1], 'fasta', k=4)
>>> [gen_seq(chain, k=4, min_len=3, max_len=5) for _ in range(3)]
['CGTACG', 'CGTACG', 'TACGTA']
```

A list comprehension will force the creation of all the sequences before moving to the next line. If I were creating millions of sequences, the program would block here and likely use a large amount of memory to store all the sequences. If I replace the square brackets [] of the list comprehension with parentheses (), then it becomes a lazy generator:

```
>>> seqs = (gen_seq(chain, k=4, min_len=3, max_len=5) for _ in range(3))
>>> type(seqs)
<class 'generator'>
```

I can still treat this like a list by iterating over the values, but these values are only produced as needed. That means the line of code to create the generator executes almost immediately and moves on to the for loop. Additionally, the program only uses the memory needed to produce the next sequence.

One small problem with using range() and the number of sequences is that I know the gen_seq() function may sometimes return None to indicate that random choices lead down a chain that didn't produce a long enough sequence. I need the generator to run with no upper limit, and I'll write code to stop requesting sequences when enough have been generated. I can use itertools.count() to create an infinite sequence, and I use filter() with a predicate of None to remove falsey elements:

```
>>> seqs = ['ACGT', None, 'CCCGT']
>>> list(filter(None, seqs))
['ACGT', 'CCCGT']
```

I can run the final program to create an output file using the defaults:

```
$ ./synth.py tests/inputs/*
Done, see output in "out.fa".
```

And then I can use seqmagique.py from Chapter 15 to verify that it generated the correct number of sequences in the expected ranges:

```
$ ../15_seqmagique/seqmagique.py out.fa
name      min_len    max_len    avg_len    num_seqs
out.fa         50         75      63.56         100
```

Flippin' sweet.

Going Further

Add a sequence --type option to produce either DNA or RNA.

Expand the program to handle paired-end sequences where the forward and reverse reads are in two separate files.

Now that you understand Markov chains, you might be interested to see how they are used elsewhere in bioinformatics. For instance, the HMMER (*http://hmmer.org*) tool uses hidden Markov models to find homologs in sequence databases and to create sequence alignments.

Review

Key points from this chapter:

- Random seeds are used to replicate pseudorandom selections.
- Markov chains can be used to encode the probabilities that a node in a graph can move to another node or state.
- A list comprehension can be made into a lazy generator by replacing the square brackets with parentheses.

## CHAPTER 18

FASTX Sampler: Randomly Subsampling Sequence Files

Sequence datasets in genomics and metagenomics can get dauntingly large, requiring copious time and compute resources to analyze. Many sequencers can produce tens of millions of reads per sample, and many experiments involve tens to hundreds of samples, each with multiple technical replicates resulting in gigabytes to terabytes of data. Reducing the size of the input files by randomly subsampling sequences allows you to explore data more quickly. In this chapter, I will show how to use Python's random module to select some portion of the reads from FASTA/FASTQ sequence files.

You will learn about:

- Nondeterministic sampling

Getting Started

The code and tests for this exercise are in the *18_fastx_sampler* directory. Start by copying the solution for a program called sampler.py:

```
$ cd 18_fastx_sampler/
$ cp solution.py sampler.py
```

The FASTA input files for testing this program will be generated by the synth.py program you wrote in Chapter 17. If you didn't finish writing that program, be sure to copy the solution to that filename before executing make fasta to create three FASTA files with 1K, 10K, and 100K reads, each between 75 and 200 bp in length, with filenames of *n1k.fa*, *n10k.fa*, and *n100k.fa*, respectively. Use seqmagique.py to verify that the files are correct:

```
$ ../15_seqmagique/seqmagique.py tests/inputs/n1*
name                   min_len    max_len    avg_len    num_seqs
tests/inputs/n100k.fa       75        200     136.08      100000
tests/inputs/n10k.fa        75        200     136.13       10000
tests/inputs/n1k.fa         75        200     135.16        1000
```

Run `sampler.py` to select the default of 10% of the sequences from the smallest file. If you use a random seed of 1, you should get 95 reads:

```
$ ./sampler.py -s 1 tests/inputs/n1k.fa
Wrote 95 sequences from 1 file to directory "out"
```

The results can be found in a file called *n1k.fa* in an output directory named *out*. One way to verify this is to use **grep -c** to count how many times it finds the symbol > at the start of each record:

```
$ grep -c '>' out/n1k.fa
95
```

Note that there is a pernicious error waiting for you if you happen to forget the quotes around >:

```
$ grep -c > out/n1k.fa
usage: grep [-abcDEFGHhIiJLlmnOoqRSsUVvwxZ] [-A num] [-B num] [-C[num]]
        [-e pattern] [-f file] [--binary-files=value] [--color=when]
        [--context[=num]] [--directories=action] [--label] [--line-buffered]
        [--null] [pattern] [file ...]
```

Wait, what just happened? Remember that > is the `bash` operator to *redirect* the STDOUT from one program into a file. In the preceding command, I ran `grep` without enough arguments and redirected the output into *out/n1k.fa*. The output you see is the usage that is printed to STDERR. Nothing was printed to STDOUT, so this null output overwrote the *out/n1k.fa* file and it is now empty:

```
$ wc out/n1k.fa
       0       0       0 out/n1k.fa
```

I point this out because I have lost several sequence files due to this gem. The data has been lost permanently, so I must rerun the earlier command to regenerate the file. After doing that, I recommend instead that you use `seqmagique.py` to verify the contents:

```
$ ../15_seqmagique/seqmagique.py out/n1k.fa
name            min_len    max_len    avg_len    num_seqs
out/n1k.fa           75        200     128.42          95
```

Reviewing the Program Parameters

This is a fairly complex program with many options. Run the `sampler.py` program to request help. Notice that the only required arguments are the input files, as all the options are set to reasonable defaults:

```
$ ./sampler.py -h
usage: sampler.py [-h] [-f format] [-p reads] [-m max] [-s seed] [-o DIR]
                  FILE [FILE ...]

Probabilistically subset FASTA files

positional arguments:
  FILE                  Input FASTA/Q file(s) ❶

optional arguments:
  -h, --help            show this help message and exit
  -f format, --format format
                        Input file format (default: fasta) ❷
  -p reads, --percent reads
                        Percent of reads (default: 0.1) ❸
  -m max, --max max     Maximum number of reads (default: 0) ❹
  -s seed, --seed seed  Random seed value (default: None) ❺
  -o DIR, --outdir DIR  Output directory (default: out) ❻
```

❶ One or more *FASTA* or *FASTQ* files are required.

❷ The default sequence format of the input files is *FASTA*.

❸ By default, the program will select 10% of the reads.

❹ This option will stop sampling when a given maximum is reached.

❺ This option sets the random seed to reproduce the selections.

❻ The default directory for the output files is *out*.

As with previous programs, the program will reject invalid or unreadable input files, and the random seed argument must be an integer value. The -p|--percent option should be a floating-point value between 0 and 1 (not inclusive), and the program will reject anything outside of that range. I manually validate this argument and use parser.error(), as in Chapters 4 and 9:

```
$ ./sampler.py -p 3 tests/inputs/n1k.fa
usage: sampler.py [-h] [-f format] [-p reads] [-m max] [-s seed] [-o DIR]
                  FILE [FILE ...]
sampler.py: error: --percent "3.0" must be between 0 and 1
```

The -f|--format option will only accept the values fasta or fastq and will default to the first. I use the choices option with argparse, as in Chapters 15 and 16, to automatically reject unwanted values. For instance, the program will reject a value of fastb:

```
$ ./sampler.py -f fastb tests/inputs/n1k.fa
usage: sampler.py [-h] [-f format] [-p reads] [-m max] [-s seed] [-o DIR]
                  FILE [FILE ...]
```

```
sampler.py: error: argument -f/--format: invalid choice:
'fastb' (choose from 'fasta', 'fastq')
```

Finally, the -m|--max option defaults to 0, meaning that the program will sample about --percent of the reads with no upper limit. Practically speaking, you may have input files with tens of millions of reads but you only want at most 100K from each. Use this option to halt sampling when the desired number is reached. For instance, I can use -m 30 to stop sampling at 30 reads:

```
$ ./sampler.py -m 30 -s 1 tests/inputs/n1k.fa
  1: n1k.fa
Wrote 30 sequences from 1 file to directory "out"
```

When you think you understand how the program should work, start over with your version:

```
$ new.py -fp 'Probabilistically subset FASTA files' sampler.py
Done, see new script "sampler.py".
```

Defining the Parameters

The arguments to the program contain many different data types, which I represent with the following class:

```
class Args(NamedTuple):
    """ Command-line arguments """
    files: List[TextIO] ❶
    file_format: str ❷
    percent: float ❸
    max_reads: int ❹
    seed: Optional[int] ❺
    outdir: str ❻
```

❶ The files are a list of open filehandles.

❷ The input file format is a string.

❸ The percentage of reads is a floating-point value.

❹ The maximum number of reads is an integer.

❺ The random seed value is an optional integer.

❻ The output directory name is a string.

Here is how I define the arguments with argparse:

```
def get_args() -> Args:
    parser = argparse.ArgumentParser(
        description='Probabilistically subset FASTA files',
```

```
        formatter_class=argparse.ArgumentDefaultsHelpFormatter)

parser.add_argument('file',
                    metavar='FILE',
                    type=argparse.FileType('r'), ❶
                    nargs='+',
                    help='Input FASTA/Q file(s)')

parser.add_argument('-f',
                    '--format',
                    help='Input file format',
                    metavar='format',
                    type=str,
                    choices=['fasta', 'fastq'], ❷
                    default='fasta')

parser.add_argument('-p',
                    '--percent',
                    help='Percent of reads',
                    metavar='reads',
                    type=float, ❸
                    default=.1)

parser.add_argument('-m',
                    '--max',
                    help='Maximum number of reads',
                    metavar='max',
                    type=int,
                    default=0) ❹

parser.add_argument('-s',
                    '--seed',
                    help='Random seed value',
                    metavar='seed',
                    type=int,
                    default=None) ❺

parser.add_argument('-o',
                    '--outdir',
                    help='Output directory',
                    metavar='DIR',
                    type=str,
                    default='out') ❻

args = parser.parse_args()

if not 0 < args.percent < 1: ❼
    parser.error(f'--percent "{args.percent}" must be between 0 and 1')

if not os.path.isdir(args.outdir): ❽
    os.makedirs(args.outdir)
```

```
      return Args(files=args.file, ❾
                file_format=args.format,
                percent=args.percent,
                max_reads=args.max,
                seed=args.seed,
                outdir=args.outdir)
```

❶ Define the file inputs as one or more readable text files.

❷ Use choices to restrict the file formats and default to fasta.

❸ The percent argument is a floating-point value with a default of 10%.

❹ The maximum number of reads should be an integer with a default of 0.

❺ The random seed value is optional but should be a valid integer if present.

❻ The output directory is a string with a default value of out.

❼ Verify that the percentage is between 0 and 1.

❽ Create the output directory if it does not exist.

❾ Return the Args object.

Note that, while the program will accept both FASTA and FASTQ inputs, it should only write FASTA-formatted output files.

Nondeterministic Sampling

Since there is no inherent ordering to sequences, one might be tempted to take the first however many sequences indicated by the user. For instance, I could use head to select some number of lines from each file. This would work on FASTQ files as long as this number is a multiple of four, but such an approach could create an invalid file for most any other sequence format, like multiline FASTA or Swiss-Prot.

I've shown several programs that read and select sequences from files, so I could repurpose one of those to select records until the desired number is reached. When I was first asked to write this program by my boss, I did exactly this. The output was unusable, however, because I had not realized that the input was a synthetic dataset created by a colleague to simulate a *metagenome*, which is an environmental sample comprised of unknown organisms. The input file was created by concatenating various reads from known genomes so that, for instance, the first 10K reads were from a bacteria, the next 10K from another bacteria, the next 10K from representative archaea, the next 10K from viruses, and so on. Taking just the first N records failed to include the diversity of the input. It was not only a boring program to write, but worse, it always generated the same output and so could not be used to generate different subsamples. This is an example of a *deterministic* program because the outputs were always the same given the same inputs.

Since I needed to find a way to randomly select some percentage of the reads, my first thought was to count the reads so I could figure out how many would be, for instance, 10%. To do this, I stored all the sequences in a list, used `len()` to figure out how many were present, and then randomly selected 10% of the numbers in this range. While such an approach might work for very small input files, I hope you can see how this fails to scale in any meaningful way. It's not uncommon to encounter input files with tens of millions of reads. Keeping all that data in something like a Python list could easily require more memory than is available on a machine.

Eventually, I settled on a solution that reads one sequence at a time and randomly decides whether to choose or reject it. That is, for each sequence I randomly select a number from a *continuous uniform distribution* between 0 and 1, meaning that all values in this range are equally likely to be selected. If that number is less than or equal to the given percentage, I select the read. This approach only ever holds one sequence record in memory at a time, and so should scale at least linearly or *O(n)*.

To demonstrate the selection process, I'll import the `random` module and select a number between 0 and 1 using the `random.random()` function:

```
>>> import random
>>> random.random()
0.465289867914331
```

It's unlikely that you'll get the same number as I do. We have to agree on a seed to both produce the same value. Use the integer 1, and you should get this number:

```
>>> random.seed(1)
>>> random.random()
0.13436424411240122
```

The `random.random()` function uses a uniform distribution. The `random` module can also sample from other distributions, like normal or Gaussian. Consult `help(random)` to see these other functions and how to use them.

As I iterate through the sequences in each file, I use this function to select a number. If the number is less than or equal to the selected percentage, I want to write the sequence to an output file. That is, the `random.random()` function ought to produce a number less than or equal to `.10` about 10% of the time. In this way, I'm using a *nondeterministic* approach to sampling because the selected reads will be different each time I run the program (assuming I do not set a random seed). This allows me to generate many different subsamples from the same input file, which could prove useful in generating technical replicates for analyses.

Structuring the Program

You may feel somewhat overwhelmed by the complexity of this program, so I'll present pseudocode you may find helpful:

```
set the random seed
iterate through each input file
    set the output filename to output directory plus the input file's basename
    open the output filehandle
    initialize a counter for how many records have been taken

    iterate through each record of the input file
        select a random number between 0 and 1
        if this number is less than or equal to the percent
            write the sequence in FASTA format to the output filehandle
            increment the counter for taking records

        if there is a max number of records and the number taken is equal
            leave the loop

    close the output filehandle

print how many sequences were taken from how many files and the output location
```

I would encourage you to run the solution.py program on one file and then on several files and try to reverse-engineer the output. Keep running the test suite to ensure your program is on track.

I hope you can see how similar in structure this program is to many previous programs that process some number of input files and create some output. For example, in Chapter 2 you processed files of DNA sequences to produce files of RNA sequences in an output directory. In Chapter 15, you processed sequence files to produce a summary table of statistics. In Chapter 16, you processed sequence files to select those records matching a pattern and wrote the selected sequences to an output file. The programs in Chapters 2 and 16 most closely resemble what you need to do here, so I recommend you borrow from those solutions.

Solutions

I want to share two versions of a solution. The first works to solve the exact problem as described. The second solution goes beyond the original requirements because I want to show you how to overcome two common problems you may face in processing large bioinformatics datasets, namely opening too many filehandles and reading compressed files.

Solution 1: Reading Regular Files

If you are dealing with a limited number of uncompressed input files, the following solution is appropriate:

```
def main() -> None:
    args = get_args()
    random.seed(args.seed) ❶

    total_num = 0 ❷
    for i, fh in enumerate(args.files, start=1): ❸
        basename = os.path.basename(fh.name)
        out_file = os.path.join(args.outdir, basename)
        print(f'{i:3}: {basename}') ❹

        out_fh = open(out_file, 'wt') ❺
        num_taken = 0 ❻

        for rec in SeqIO.parse(fh, args.file_format): ❼
            if random.random() <= args.percent: ❽
                num_taken += 1
                SeqIO.write(rec, out_fh, 'fasta')

            if args.max_reads and num_taken == args.max_reads: ❾
                break

        out_fh.close() ❿
        total_num += num_taken

    num_files = len(args.files) ⓫
    print(f'Wrote {total_num:,} sequence{"" if total_num == 1 else "s"} '
          f'from {num_files:,} file{"" if num_files == 1 else "s"} '
          f'to directory "{args.outdir}"')
```

❶ Set the random seed, if present. The default None value will be the same as not setting the seed.

❷ Initialize a variable to note the total number of sequences selected.

❸ Iterate through each input filehandle.

❹ Construct the output filename by joining the output directory with the file's basename.

❺ Open the output filehandle for writing text.

❻ Initialize a variable to note how many sequences have been taken from this file.

❼ Iterate through each sequence record in the input file.

❽ If the record is randomly selected, increment the counter and write the sequence to the output file.

❾ If there is a maximum limit defined and the number of records selected is equal to this, exit the inner `for` loop.

❿ Close the output filehandle and increment the number of total records taken.

⓫ Note the number of files processed and inform the user of the final status.

Solution 2: Reading a Large Number of Compressed Files

The original problem does not involve reading compressed files, but you will often find that data will be stored in this way to save bandwidth in transferring data and disk space in storing it. Python can directly read files compressed with tools like `zip` and `gzip`, so it's not necessary to uncompress the input files before processing.

Additionally, if you are processing hundreds to thousands of input files, you will find that using `type=argparse.FileType()` will cause your program to fail because you may exceed the maximum number of open files your operating system will allow. In that case, you should declare `Args.files` as `List[str]` and create the parameter like this:

```
parser.add_argument('file',
                    metavar='FILE',
                    type=str, ❶
                    nargs='+',
                    help='Input FASTA/Q file(s)')
```

❶ Set the parameter type to one or more string values.

This means you will have to validate the input files yourself, which you can do in the `get_args()` function, like so:

```
if bad_files := [file for file in args.file if not os.path.isfile(file)]: ❶
    parser.error(f'Invalid file: {", ".join(bad_files)}') ❷
```

❶ Find all the arguments that are not valid files.

❷ Use `parser.error()` to report the bad inputs.

The `main()` processing needs to change slightly as now `args.files` will be a list of strings. You will need to open the filehandles yourself with `open()`, and this is the crucial change to your program needed to handle compressed files. I will use a simple heuristic that examines the file extension for `.gz` to determine if a file is zipped and will instead use the `gzip.open()` function to open it:

```
def main() -> None:
    args = get_args()
    random.seed(args.seed)

    total_num = 0
    for i, file in enumerate(args.files, start=1):  ❶
        basename = os.path.basename(file)
        out_file = os.path.join(args.outdir, basename)
        print(f'{i:3}: {basename}')

        ext = os.path.splitext(basename)[1]  ❷
        fh = gzip.open(file, 'rt') if ext == '.gz' else open(file, 'rt')  ❸
        out_fh = open(out_file, 'wt')
        num_taken = 0

        for rec in SeqIO.parse(fh, args.file_format):
            if random.random() <= args.percent:
                num_taken += 1
                SeqIO.write(rec, out_fh, 'fasta')

            if args.max_reads and num_taken == args.max_reads:
                break

        out_fh.close()
        total_num += num_taken

    num_files = len(args.files)
    print(f'Wrote {total_num:,} sequence{"" if total_num == 1 else "s"} '
          f'from {num_files:,} file{"" if num_files == 1 else "s"} '
          f'to directory "{args.outdir}".')
```

❶ args.files is now a list of strings, not filehandles.

❷ Get the file extension.

❸ If the file extension is .gz, use gzip.open() to open the file; otherwise, use the normal open() function.

Finally, there are times when nargs='+' will also not work. For one project, I had to download over 350,000 XML files. Passing all these as arguments will lead to an "Argument list too long" error from the command line itself. My workaround is to accept the directory names as arguments:

```
parser.add_argument('-d',
                    '--dir',
                    metavar='DIR',
                    type=str,
                    nargs='+',
                    help='Input directories of FASTA/Q file(s)')
```

I then use Python to recursively search the directories for files. For this code, I added `from pathlib import Path` so that I could use the `Path.rglob()` function:

```
files = []
for dirname in args.dir:
    if os.path.isdir(dirname):
        files.extend(list(Path(dirname).rglob('*')))

if not files:
    parser.error('Found no files')

return Args(files=files,
            file_format=args.format,
            percent=args.percent,
            max_reads=args.max,
            seed=args.seed,
            outdir=args.outdir)
```

The program can continue as before because Python does not have a problem storing several hundred thousand items in a list.

Going Further

This program always produces FASTA output. Add an `--outfmt` output format option so that you can specify the output format. Consider detecting the input file format and writing the output format in the same way as you did in Chapter 16. Be sure to add the appropriate tests to verify that your program works.

Review

- The > record marker in a FASTA file is also the redirect operator in `bash`, so care must be taken to quote this value on the command line.

- Deterministic approaches always produce the same output for the given input. Nondeterministic approaches produce different outputs for the same inputs.

- The `random` module has functions to select numbers from various distributions, such as the uniform and normal distributions.

## CHAPTER 19

Blastomatic: Parsing Delimited Text Files

Delimited text files are a standard way to encode columnar data. You are likely familiar with spreadsheets like Microsoft Excel or Google Sheets, where each worksheet may hold a dataset with columns across the top and records running down. You can export this data to a text file where the columns of data are *delimited*, or separated by a character. Quite often the delimiter is a comma, and the file will have an extension of *.csv*. This format is called *CSV*, for *comma-separated values*. When the delimiter is a tab, the extension may be *.tab*, *.txt*, or *.tsv* for *tab-separated values*. The first line of the file usually will contain the names of the columns. Notably, this is not the case with the tabular output from BLAST (Basic Local Alignment Search Tool), one of the most popular tools in bioinformatics used to compare sequences. In this chapter, I will show you how to parse this output and merge the BLAST results with metadata from another delimited text file using the csv and pandas modules.

In this exercise, you will learn:

- How to use csvkit and csvchk to view delimited text files
- How to use the csv and pandas modules to parse delimited text files

Introduction to BLAST

The BLAST program is one of the most ubiquitous tools in bioinformatics for determining sequence similarity. In Chapter 6, I showed how the Hamming distance between two sequences is one measure of similarity and compared this to the concept of alignment. Whereas the Hamming distance compares both sequences starting from the beginning, an alignment with BLAST starts wherever both sequences begin to overlap and will allow for insertions, deletions, and mismatches to find the longest possible areas of similarity.

I'll show you the National Center for Biotechnology (NCBI) BLAST web interface, but you can use `blastn` if you have BLAST installed locally. I will compare 100 sequences from the Global Ocean Sampling Expedition (GOS) (*https://oreil.ly/POkOV*) to a sequence database at NCBI. GOS is one of the earliest metagenomic studies, dating from the early 2000s when Dr. Craig Venter funded a two-year expedition to collect and analyze ocean samples from around the globe. It's a *metagenomic* project because the genetic material was taken directly from an environmental sample. The purpose of using BLAST is to compare the unknown GOS sequences to known sequences at NCBI to determine their possible taxonomic classification.

I used the FASTX sampler from Chapter 18 to randomly select the 100 input sequences in *tests/inputs/gos.fa*:

```
$ ../15_seqmagique/seqmagique.py tests/inputs/gos.fa
name                    min_len    max_len    avg_len     num_seqs
tests/inputs/gos.fa         216       1212    1051.48          100
```

I used the NCBI BLAST tool (*https://oreil.ly/gXErw*) to compare these sequences to the *nr/nt* (nonredundant nucleotide) database using the `blastn` program to compare nucleotides. The results page allows me to select the detailed results for each of the 100 sequences. As shown in Figure 19-1, the first sequence has four *hits* or matches to known sequences. The first and best hit is about 93% identical over 99% of its length to a portion of the genome of *Candidatus Pelagibacter* (*https://oreil.ly/qywN2*), a marine bacteria of the SAR11 clade. Given that the GOS query sequence came from the ocean, this seems a likely match.

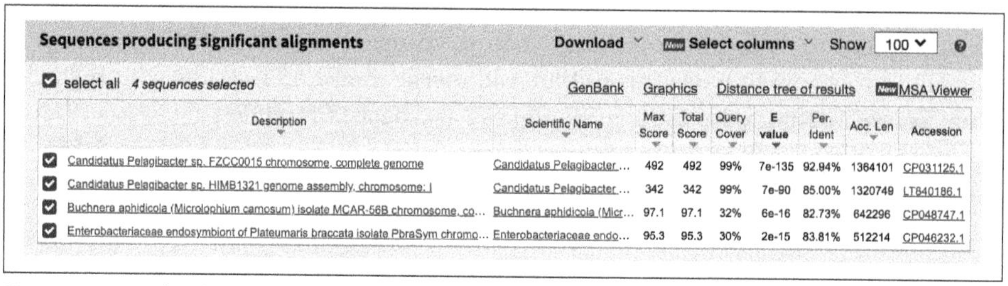

Sequences producing significant alignments
Download
Select columns
Show 100
select all 4 sequences selected
GenBank
Graphics
Distance tree of results
MSA Viewer
Description
Scientific Name
Max Score
Total Score
Query Cover
E value
Per. Ident
Acc. Len
Accession
Candidatus Pelagibacter sp. FZCC0015 chromosome, complete genome | Candidatus Pelagibacter... | 492 | 492 | 99% | 7e-135 | 92.94% | 1364101 | CP031125.1
Candidatus Pelagibacter sp. HIMB1321 genome assembly, chromosome: I | Candidatus Pelagibacter... | 342 | 342 | 99% | 7e-90 | 85.00% | 1320749 | LT840186.1
Buchnera aphidicola (Microlophium carnosum) isolate MCAR-56B chromosome, co... | Buchnera aphidicola (Micr... | 97.1 | 97.1 | 32% | 6e-16 | 82.73% | 642296 | CP048747.1
Enterobacteriaceae endosymbiont of Plateumaris braccata isolate PbraSym chromo... | Enterobacteriaceae endo... | 95.3 | 95.3 | 30% | 2e-15 | 83.81% | 512214 | CP046232.1

Figure 19-1. The first GOS sequence has four possible matches from nr/nt

Figure 19-2 shows how similar the query sequence is to a region of the *Candidatus Pelagibacter* genome. Notice how the alignment allows for single-nucleotide variations (SNVs) and gaps caused by deletions or insertions between the sequences. If you want to challenge yourself, try writing a sequence aligner. You can see an example in Figure 19-2.

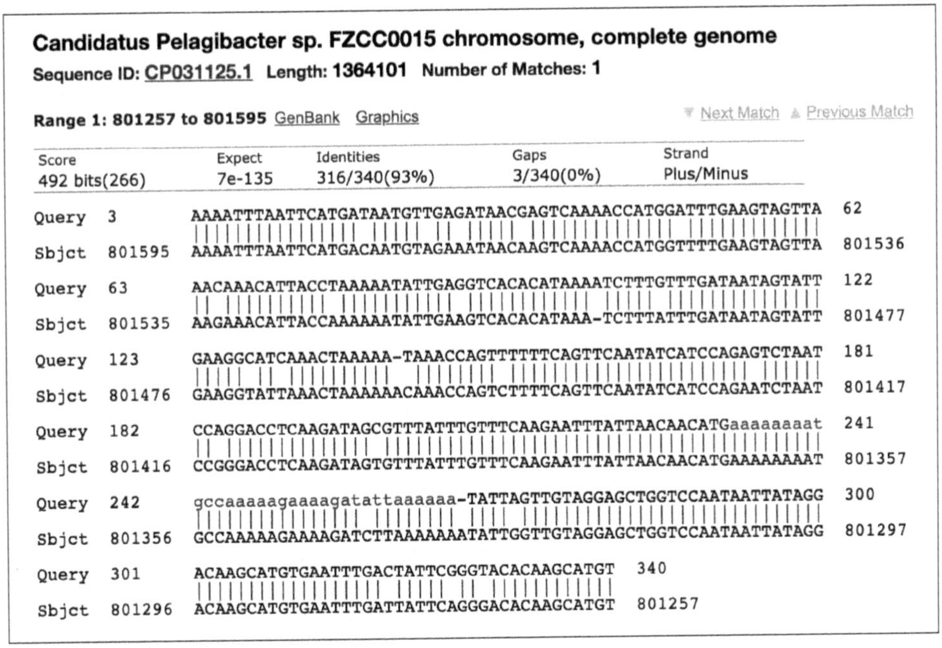

```
Candidatus Pelagibacter sp. FZCC0015 chromosome, complete genome
Sequence ID: CP031125.1  Length: 1364101  Number of Matches: 1

Range 1: 801257 to 801595 GenBank  Graphics          ▼ Next Match ▲ Previous Match

Score             Expect    Identities      Gaps        Strand
492 bits(266)     7e-135    316/340(93%)    3/340(0%)   Plus/Minus

Query  3      AAAATTTAATTCATGATAATGTTGAGATAACGAGTCAAAACCATGGATTTGAAGTAGTTA  62
              |||||||||||||| ||||  |||| | |||||||||||||||||| ||  |||||||||
Sbjct  801595 AAAATTTAATTCATGACAATGTAGAAATAACAAGTCAAAACCATGGTTTTGAAGTAGTTA  801536

Query  63     AACAAACATTACCTAAAAATATTGAGGTCACACATAAAATCTTTGTTTGATAATAGTATT  122
              || |||||||||| |||||||||||||| |||||||||  |||| ||||||||||||||||
Sbjct  801535 AAGAAACATTACCAAAAAATATTGAAGTCACACATAAA-TCTTTATTTGATAATAGTATT  801477

Query  123    GAAGGCATCAAACTAAAAA-TAAACCAGTTTTTTCAGTTCAATATCATCCAGAGTCTAAT  181
              ||||| || || |||||||| |||||||||  |||||||||||||||||||||||| ||||
Sbjct  801476 GAAGGTATTAAACTAAAAAAACAAACCAGTCTTTTCAGTTCAATATCATCCAGAATCTAAT  801417

Query  182    CCAGGACCTCAAGATAGCGTTTATTTGTTTCAAGAATTTATTAACAACATGaaaaaaaat  241
              || |||||||||||| || ||||||||||||||||||| |||||||||||||||  |||||
Sbjct  801416 CCGGGACCTCAAGATAGTGTTTATTTGTTTCAAGAATTTATTAACAACATGAAAAAAAAT  801357

Query  242    gccaaaaagaaaagatattaaaaaa-TATTAGTTGTAGGAGCTGGTCCAATAATTATAGG  300
              |||||||||||||||| ||||| ||| |||| ||||||||||||||||||||||||||||||
Sbjct  801356 GCCAAAAAGAAAAGATCTTAAAAAAATATTGGTTGTAGGAGCTGGTCCAATAATTATAGG  801297

Query  301    ACAAGCATGTGAATTTGACTATTCGGGTACACAAGCATGT  340
              |||||||||||||||||| |||||| |||||||||||||||
Sbjct  801296 ACAAGCATGTGAATTTGATTATTCAGGGACACAAGCATGT  801257
```

Figure 19-2. The alignment of the top BLAST hit

As interesting as it is to explore each individual hit, I want to download a table of all the hits. There is a Download All menu with 11 download formats. I chose the "Hit table(csv)" format and split this data into *hits1.csv* and *hits2.csv* in the *tests/inputs* directory:

```
$ wc -l tests/inputs/hits*.csv
    500 tests/inputs/hits1.csv
    255 tests/inputs/hits2.csv
    755 total
```

If you open these files with a text editor, you'll see they contain comma-separated values. You can also open a file with a spreadsheet program like Excel to see the data in columnar format, and you may notice that the columns are not named. If you were on a remote machine like a cluster node, you would likely not have access to a graphical program like Excel to inspect the results. Further, Excel is limited to about 1 million rows and 16,000 columns. In real-world bioinformatics, it's pretty easy to exceed both of those values, so I'll show you command-line tools you can use to look at delimited text files.

Using csvkit and csvchk

First, I'd like to introduce the csvkit module, "a suite of command-line tools for converting to and working with CSV." The *requirements.txt* file for the repo lists this as a dependency, so it's probably installed. If not, you can use this command to install it:

```
$ python3 -m pip install csvkit
```

This will install several useful utilities, and I encourage you to read the documentation (*https://oreil.ly/QDAn2*) to learn about them. I want to highlight csvlook, which "renders a CSV file in the console as a Markdown-compatible, fixed-width table." Run **csvlook --help** to view the usage and notice there is an -H|--no-header-row option to view files that have no header row. The following command will display the first three rows of the hits table. Depending on the size of your screen, this might be unreadable:

```
$ csvlook -H --max-rows 3 tests/inputs/hits1.csv
```

The csvchk program (*https://oreil.ly/T2QSo*) will transpose a wide record like this to a tall one vertically oriented with the column names on the left rather than across the top. This, too, should have been installed with other module dependencies, but you can use pip to install it if needed:

```
$ python3 -m pip install csvchk
```

If you read the usage, you'll see that this tool also has an -N|--noheaders option. Use csvchk to inspect the first record in the same hits file:

```
$ csvchk -N tests/inputs/hits1.csv
// ****** Record 1 ****** //
Field1  : CAM_READ_0234442157
Field2  : CP031125.1
Field3  : 92.941
Field4  : 340
Field5  : 21
Field6  : 3
Field7  : 3
Field8  : 340
Field9  : 801595
Field10 : 801257
Field11 : 6.81e-135
Field12 : 492
```

The output files you can download from NCBI BLAST match the output formats from the command-line versions of the BLAST programs, like blastn for comparing nucleotides, blastp for comparing proteins, etc. The help documentation for blastn includes an -outfmt option to specify the output format using a number between 0 and 18. The preceding output file format is the "Tabular" option 6:

```
*** Formatting options
-outfmt <String>
  alignment view options:
    0 = Pairwise,
    1 = Query-anchored showing identities,
    2 = Query-anchored no identities,
    3 = Flat query-anchored showing identities,
    4 = Flat query-anchored no identities,
    5 = BLAST XML,
    6 = Tabular,
    7 = Tabular with comment lines,
    8 = Seqalign (Text ASN.1),
    9 = Seqalign (Binary ASN.1),
   10 = Comma-separated values,
   11 = BLAST archive (ASN.1),
   12 = Seqalign (JSON),
   13 = Multiple-file BLAST JSON,
   14 = Multiple-file BLAST XML2,
   15 = Single-file BLAST JSON,
   16 = Single-file BLAST XML2,
   17 = Sequence Alignment/Map (SAM),
   18 = Organism Report
```

You may find yourself wondering why the tabular output file does not contain the column headers. If you read through all the formatting options, you may notice that output format 7 is "Tabular with comment lines," and you may ask yourself: Is this the option that will include the column names? Dear reader, you will be sorely disappointed to learn it does not.[1] Option 7 is the same as the "Hits table(text)" option on the NCBI BLAST page. Download and open that file to see that it contains metadata about the search as unstructured text on lines that begin with the # character. Because so many languages (including Python) use this as a comment character to indicate a line that should be ignored, it's common to say that the metadata is *commented out*, and many delimited text parsers will skip these lines.

So what are the column names? I must parse through hundreds of lines of the blastn usage to find that "Options 6, 7, 10 and 17 can be additionally configured" to include any of 53 optional fields. If the fields are not specified, then the default fields are as follows:

- qaccver: Query sequence accession/ID
- saccver: Subject sequence accession/ID
- pident: Percentage of identical matches
- length: Alignment length

---

1 You may say to yourself, "My God! What have they done?"

- mismatch: Number of mismatches
- gapopen: Number of gap openings
- qstart: Start of alignment in query
- qend: End of alignment in query
- sstart: Start of alignment in subject
- send: End of alignment in subject
- evalue: Expect value
- bitscore: Bit score

If you look again at the usage for csvchk, you'll find there is an option to name the -f|--fieldnames for the record. Following is how I could view the first record from a hits file and specify column names:

```
$ csvchk -f 'qseqid,sseqid,pident,length,mismatch,gapopen,qstart,qend,\
  sstart,send,evalue,bitscore' tests/inputs/hits1.csv
// ****** Record 1 ****** //
qseqid    : CAM_READ_0234442157
sseqid    : CP031125.1
pident    : 92.941
length    : 340
mismatch  : 21
gapopen   : 3
qstart    : 3
qend      : 340
sstart    : 801595
send      : 801257
evalue    : 6.81e-135
bitscore  : 492
```

This is a much more useful output. If you like this command, you can create an alias called blstchk in bash, like so:

```
alias blstchk='csvchk -f "qseqid,sseqid,pident,length,mismatch,gapopen,\
  qstart,qend,sstart,send,evalue,bitscore"'
```

 Most shells allow you to define aliases like this in a file that is read each time you start a new shell. In bash, you could add this line to a file in your $HOME directory, like *.bash_profile*, *.bashrc*, or *.profile*. Other shells have similar properties. Aliases are a handy way to create global shortcuts for common commands. If you wish to create a command shortcut inside a particular project or directory, consider using a target in a *Makefile*.

Here is how I use the `blstchk` command:

```
$ blstchk tests/inputs/hits1.csv
// ****** Record 1 ****** //
qseqid   : CAM_READ_0234442157
sseqid   : CP031125.1
pident   : 92.941
length   : 340
mismatch : 21
gapopen  : 3
qstart   : 3
qend     : 340
sstart   : 801595
send     : 801257
evalue   : 6.81e-135
bitscore : 492
```

The goal of the program in this chapter is to link the BLAST hits to the depth and location of the GOS sequences found in the file *tests/inputs/meta.csv*. I will use the `-g|--grep` option to `csvchk` to find the preceding query sequence, *CAM_READ_0234442157*:

```
$ csvchk -g CAM_READ_0234442157 tests/inputs/meta.csv
// ****** Record 1 ****** //
seq_id     : CAM_READ_0234442157
sample_acc : CAM_SMPL_GS112
date       : 8/8/05
depth      : 4573
salinity   : 32.5
temp       : 26.6
lat_lon    : -8.50525,80.375583
```

The BLAST results can be joined to the metadata where the former's `qseqid` is equal to the latter's `seq_id`. There is a command-line tool called `join` that will do exactly this. The inputs must both be sorted, and I will use the `-t` option to indicate that the comma is the field delimiter. By default, `join` assumes the first column in each file is the common value, which is true here. The output is a comma-separated union of the fields from both files:

```
$ cd tests/inputs/
$ join -t , <(sort hits1.csv) <(sort meta.csv) | csvchk -s "," -N - ❶
// ****** Record 1 ****** //
Field1 : CAM_READ_0234442157
Field2 : CP046232.1
Field3 : 83.810
Field4 : 105
Field5 : 12
Field6 : 5
Field7 : 239
Field8 : 340
Field9 : 212245
```

```
Field10 : 212143
Field11 : 2.24e-15
Field12 : 95.3
Field13 : CAM_SMPL_GS112
Field14 : 8/8/05
Field15 : 4573
Field16 : 32.5
Field17 : 26.6
Field18 : -8.50525,80.375583
```

❶ The two positional inputs to join use shell redirection < to read in the results of sorting the two input files. The output from join is piped to csvchk.

Although it's good to know how to use join, this output is not particularly useful because it does not have the column headers. (Also, the point is to learn how to do this in Python.) How might you add headers to this information? Would you cobble together some shell commands in a bash script or a *Makefile* target, or would you write a Python program? Let's keep moving, shall we? Next, I'll show you how the program should work and the output it will create.

Getting Started

All the code and tests for this exercise can be found in the *19_blastomatic* directory of the repository. Change to this directory and copy the second solution to the program blastomatic.py:

```
$ cd 19_blastomatic/
$ cp solution2_dict_writer.py blastomatic.py
```

The program will accept the BLAST hits and the metadata file and will produce an output file showing the sequence ID, the percent identity match, the depth, and the latitude and longitude of the sample. Optionally, the output can be filtered by the percent identity. Request help from the program to see the options:

```
$ ./blastomatic.py -h
usage: blastomatic.py [-h] -b FILE -a FILE [-o FILE] [-d DELIM] [-p PCTID]

Annotate BLAST output

optional arguments:
  -h, --help            show this help message and exit
  -b FILE, --blasthits FILE
                        BLAST -outfmt 6 (default: None) ❶
  -a FILE, --annotations FILE
                        Annotations file (default: None) ❷
  -o FILE, --outfile FILE
                        Output file (default: out.csv) ❸
  -d DELIM, --delimiter DELIM
                        Output field delimiter (default: ) ❹
```

```
    -p PCTID, --pctid PCTID
                        Minimum percent identity (default: 0.0) ❺
```

❶ The tabular output file from a BLAST search in -outfmt 6.

❷ An annotations file with metadata about the sequences.

❸ The name of the output file, which defaults to *out.csv*.

❹ The output file delimiter, which defaults to a guess based on the output file extension.

❺ The minimum percent identity, which defaults to 0.

If I run the program using the first hits file, it will write 500 sequences to the output file *out.csv*:

```
$ ./blastomatic.py -b tests/inputs/hits1.csv -a tests/inputs/meta.csv
Exported 500 to "out.csv".
```

I can use csvlook with the --max-rows option to view the first two rows of the table:

```
$ csvlook --max-rows 2 out.csv
| qseqid            | pident | depth | lat_lon          |
| ----------------- | ------ | ----- | ---------------- |
| CAM_READ_0234442157 | 92.941 | 4,573 | -8.50525,80.375583 |
| CAM_READ_0234442157 | 85.000 | 4,573 | -8.50525,80.375583 |
| ...               |        |  ...  |  ...  | ...        |
```

Or I can use csvchk with -l|--limit to do the same:

```
$ csvchk --limit 2 out.csv
// ****** Record 1 ****** //
qseqid  : CAM_READ_0234442157
pident  : 92.941
depth   : 4573
lat_lon : -8.50525,80.375583
// ****** Record 2 ****** //
qseqid  : CAM_READ_0234442157
pident  : 85.000
depth   : 4573
lat_lon : -8.50525,80.375583
```

If I want to only export hits with a percent identity greater than or equal to 90%, I can use the -p|--pctid option to find that only 190 records are found:

```
$ ./blastomatic.py -b tests/inputs/hits1.csv -a tests/inputs/meta.csv -p 90
Exported 190 to "out.csv".
```

I can peek at the file to see that it appears to have selected the correct data:

```
$ csvlook --max-rows 4 out.csv
| qseqid                | pident | depth | lat_lon        |
```

```
| ---------------------- | ------ | ----- | -------------------- |
| CAM_READ_0234442157    | 92.941 | 4,573 | -8.50525,80.375583   |
| JCVI_READ_1091145027519 | 97.368 |     2 | 44.137222,-63.644444 |
| JCVI_READ_1091145742680 | 98.714 |    64 | 44.690277,-63.637222 |
| JCVI_READ_1091145742680 | 91.869 |    64 | 44.690277,-63.637222 |
| ...                    |  ...   | ... | ...                  |
```

Using awk and cut with Delimited Text

If I only wanted to inspect the second column of data, I could use the cut tool, which has the -f option to select the second field (1-based counting) and the -d option to indicate the comma as the delimiter:

```
$ cut -f2 -d, out.csv
```

Alternatively, I could use awk with the -F option to indicate the comma as the field separator and the instruction {print $2} (also 1-based) to indicate that it should print the second field:

```
$ awk -F, '{print $2}' out.csv
```

The quotes in the awk command must be delimited by single quotes in bash because double quotes would try to interpolate the variable $2. Ah, the vagaries of shell quoting.

If I additionally wanted to verify that all the values in the second column were indeed greater than or equal to 90, I could use awk with cut. Note I need to add 0 to the first field to force awk to treat the value as an integer:

```
$ cut -f2 -d, out.csv | awk '$1 + 0 >= 90' | wc -l
    190
```

Or I could use awk to only print the values of column 2 when they are greater than or equal to 90:

```
$ awk -F"," '$2 + 0 >= 90 {print $2}' out.csv | wc -l
    190
```

Or, since Perl was one of my first true loves, here's a Perl one-liner:

```
$ perl -F"," -ane '$F[1] >= 90 && print($F[1], "\n")' out.csv | wc -l
    190
```

It is well worth your time to learn how to use these small, limited tools like join, paste, comm, awk, sed, grep, cut, sort, and even Perl. Sufficient proficiency with these tools and Unix pipes can often obviate writing longer Python programs.

The `blastomatic.py` program defaults to writing the output to the comma-separated file *out.csv*. You can use the `-d|--delimiter` option to specify a different delimiter and the `-o|--outfile` option to specify a different file. Note that the delimiter will be guessed from the extension of the output filename if it is not specified. The extension *.csv* will be taken to mean commas, and otherwise tabs will be used.

Run **make test** to see the full test suite. When you think you understand how the program should work, start anew:

```
$ new.py -fp 'Annotate BLAST output' blastomatic.py
Done, see new script "blastomatic.py".
```

Defining the Arguments

Here is the class I used to define my arguments:

```
class Args(NamedTuple):
    """ Command-line arguments """
    hits: TextIO ❶
    annotations: TextIO ❷
    outfile: TextIO ❸
    delimiter: str ❹
    pctid: float ❺
```

❶ The BLAST hits file will be an open filehandle.

❷ The metadata file will be an open filehandle.

❸ The output file will be an open filehandle.

❹ The output file delimiter will be a string.

❺ The percent identity will be a floating-point number.

Here is how I parse and validate the arguments:

```
def get_args():
    """ Get command-line arguments """

    parser = argparse.ArgumentParser(
        description='Annotate BLAST output',
        formatter_class=argparse.ArgumentDefaultsHelpFormatter)

    parser.add_argument('-b',
                        '--blasthits',
                        metavar='FILE',
                        type=argparse.FileType('rt'), ❶
                        help='BLAST -outfmt 6',
                        required=True)
```

```
parser.add_argument('-a',
                     '--annotations',
                     help='Annotations file',
                     metavar='FILE',
                     type=argparse.FileType('rt'), ❷
                     required=True)

parser.add_argument('-o',
                     '--outfile',
                     help='Output file',
                     metavar='FILE',
                     type=argparse.FileType('wt'), ❸
                     default='out.csv')

parser.add_argument('-d',
                     '--delimiter',
                     help='Output field delimiter', ❹
                     metavar='DELIM',
                     type=str,
                     default='')

parser.add_argument('-p',
                     '--pctid',
                     help='Minimum percent identity', ❺
                     metavar='PCTID',
                     type=float,
                     default=0.)

args = parser.parse_args()

return Args(hits=args.blasthits, ❻
            annotations=args.annotations,
            outfile=args.outfile,
            delimiter=args.delimiter or guess_delimiter(args.outfile.name), ❼
            pctid=args.pctid)
```

❶ The BLAST file must be a readable text file.

❷ The metadata file must be a readable text file.

❸ The output file must be a writable text file.

❹ The output field delimiter is a string that defaults to the empty string I will guess from the output filename.

❺ The minimum percent identity should be a floating-point number that defaults to 0.

❻ Create the `Args` object. Note that the fields of `Args` do not need to match the parameter names.

❼ I wrote a function to guess the delimiter from the output filename.

 This program has two required file arguments: the BLAST hits, and the annotations. I don't want to make these positional arguments because then my user would have to remember the order. It's better to have these as named options, but then they become optional, which I don't want either. To overcome this, I use `required=True` for both the file parameters to ensure the user supplies them.

You might like to start with the `guess_delimiter()` function. Here is the test I wrote:

```
def test_guess_delimiter() -> None:
    """ Test guess_delimiter """

    assert guess_delimiter('/foo/bar.csv') == ','
    assert guess_delimiter('/foo/bar.txt') == '\t'
    assert guess_delimiter('/foo/bar.tsv') == '\t'
    assert guess_delimiter('/foo/bar.tab') == '\t'
    assert guess_delimiter('') == '\t'
```

Start your `main()` with some minimal code that will work:

```
def main() -> None:
    args = get_args()
    print('hits', args.hits.name)
    print('meta', args.annotations.name)
```

Verify that this works:

```
$ ./blastomatic.py -a tests/inputs/meta.csv -b tests/inputs/hits1.csv
hits tests/inputs/hits1.csv
meta tests/inputs/meta.csv
```

At this point, you should be able to pass several tests when you run **make test**. Next, I'll show you how to parse the delimited text files.

Parsing Delimited Text Files Using the csv Module

Python has a `csv` module that will handle delimited text files easily, but I would first like to show you exactly what it's doing so you can appreciate how much effort it saves. To begin, I will open the metadata file and read the headers from the first line. I can call the `fh.readline()` method on a filehandle to read one line of text. This will still have the newline attached, so I call `str.rstrip()` to remove any whitespace from the right side of the string. Finally, I call `str.split(',')` to break the line on the delimiting comma:

```
>>> fh = open('tests/inputs/meta.csv')
>>> headers = fh.readline().rstrip().split(',')
>>> headers
['seq_id', 'sample_acc', 'date', 'depth', 'salinity', 'temp', 'lat_lon']
```

So far, so good. I'll try parsing the next line of data:

```
>>> line = fh.readline()
>>> data = line.split(',')
>>> data
['JCVI_READ_1092301105055', 'JCVI_SMPL_1103283000037', '2/11/04', '1.6', '',
 '25.4', '"-0.5938889', '-91.06944"']
```

Can you see a problem here? I have split the lat_lon field, which contains a comma, into two values, giving me eight values for seven fields:

```
>>> len(headers), len(data)
(7, 8)
```

Using str.split() will not work because it fails to consider when the separator is part of the field value. That is, when the field separator is enclosed in quotes, it's not a field separator. Notice that the lat_lon value is properly quoted:

```
>>> line[50:]
'11/04,1.6,,25.4,"-0.5938889,-91.06944"\n\'
```

One way to correctly parse this line uses the pyparsing module:

```
>>> import pyparsing as pp
>>> data = pp.commaSeparatedList.parseString(line).asList()
>>> data
['JCVI_READ_1092301105055', 'JCVI_SMPL_1103283000037', '2/11/04', '1.6',
 '', '25.4', '"-0.5938889,-91.06944"']
```

That's close, but the lat_lon field still has the quotes around it. I can use a regular expression to remove them:

```
>>> import re
>>> data = list(map(lambda s: re.sub(r'^"|"$', '', s), data))  ❶
>>> data
['JCVI_READ_1092301105055', 'JCVI_SMPL_1103283000037', '2/11/04', '1.6', '',
 '25.4', '-0.5938889,-91.06944']
```

❶ This regular expression replaces a quote anchored to either the beginning or the end of a string with the empty string.

Now that I have a list of the headers and a list of the data for a given record, I could create a dictionary by zipping these together. I've used the zip() function in Chapters 6 and 13 to join two lists into a list of tuples. Because zip() is a lazy function, I must use the list() function in the REPL to force the evaluation:

```
>>> from pprint import pprint
>>> pprint(list(zip(headers, data)))
```

```
[('seq_id', 'JCVI_READ_1092301105055'),
 ('sample_acc', 'JCVI_SMPL_1103283000037'),
 ('date', '2/11/04'),
 ('depth', '1.6'),
 ('salinity', ''),
 ('temp', '25.4'),
 ('lat_lon', '-0.5938889,-91.06944')]
```

I can change the `list()` function to `dict()` to turn this into a dictionary:

```
>>> pprint(dict(zip(headers, data)))
{'date': '2/11/04',
 'depth': '1.6',
 'lat_lon': '-0.5938889,-91.06944',
 'salinity': '',
 'sample_acc': 'JCVI_SMPL_1103283000037',
 'seq_id': 'JCVI_READ_1092301105055',
 'temp': '25.4'}
```

I could iterate through each line of the file and create a dictionary of the records by zipping the headers and data. That would work just fine, but all this work has already been done for me in the `csv` module. Following is how I can parse the same file into a list of dictionaries using `csv.DictReader()`. By default, it will use the comma as the delimiter:

```
>>> import csv
>>> reader = csv.DictReader(open('tests/inputs/meta.csv'))
>>> for rec in reader:
...     pprint(rec)
...     break
...
{'date': '2/11/04',
 'depth': '1.6',
 'lat_lon': '-0.5938889,-91.06944',
 'salinity': '',
 'sample_acc': 'JCVI_SMPL_1103283000037',
 'seq_id': 'JCVI_READ_1092301105055',
 'temp': '25.4'}
```

That's much easier. Here's how I might use this to create a dictionary of all the annotations keyed on the sequence ID. Be sure to add `from pprint import pprint` for this:

```
def main():
    args = get_args()
    annots_reader = csv.DictReader(args.annotations, delimiter=',')  ❶
    annots = {row['seq_id']: row for row in annots_reader}  ❷
    pprint(annots)
```

❶ Use `csv.DictReader()` to parse the CSV data in the annotations filehandle.

❷ Use a dictionary comprehension to create a dictionary keyed on the seq_id field from each record.

Run this with the input files and see if you get a reasonable-looking data structure. Here I'll redirect STDOUT to a file called *out* and use head to inspect it:

```
$ ./blastomatic.py -a tests/inputs/meta.csv -b tests/inputs/hits1.csv > out
$ head out
{'CAM_READ_0231669837': {'date': '8/4/05',
                         'depth': '7',
                         'lat_lon': '-12.092617,96.881733',
                         'salinity': '32.4',
                         'sample_acc': 'CAM_SMPL_GS108',
                         'seq_id': 'CAM_READ_0231669837',
                         'temp': '25.8'},
 'CAM_READ_0231670003': {'date': '8/4/05',
                         'depth': '7',
                         'lat_lon': '-12.092617,96.881733',
```

Before I move on to reading the BLAST hits, I'd like to open the output filehandle. The format of the output file should be another delimited text file. By default it will be a CSV file, but the user may choose something else, like a tab separator. The first line of the file should be the headers, so I'll immediately write those:

```
def main():
    args = get_args()
    annots_reader = csv.DictReader(args.annotations, delimiter=',')
    annots = {row['seq_id']: row for row in annots_reader}

    headers = ['qseqid', 'pident', 'depth', 'lat_lon'] ❶
    args.outfile.write(args.delimiter.join(headers) + '\n') ❷
```

❶ These are the output file's column names.

❷ args.outfile is a filehandle opened for writing text. Write the headers joined on the args.delimiter string. Be sure to add a newline.

Alternatively, you could use print() with a file argument:

```
print(args.delimiter.join(headers), file=args.outfile)
```

Next, I'll cycle through the BLAST hits. It's necessary to supply the fieldnames to csv.DictReader() since the first line of the file is missing the column names:

```
def main():
    args = get_args()
    annots_reader = csv.DictReader(args.annotations, delimiter=',')
    annots = {row['seq_id']: row for row in annots_reader}

    headers = ['qseqid', 'pident', 'depth', 'lat_lon']
    args.outfile.write(args.delimiter.join(headers) + '\n')
```

```
hits = csv.DictReader(args.hits, ❶
                      delimiter=',',
                      fieldnames=[
                          'qseqid', 'sseqid', 'pident', 'length',
                          'mismatch', 'gapopen', 'qstart', 'qend',
                          'sstart', 'send', 'evalue', 'bitscore'
                      ])

for hit in hits: ❷
    if float(hit.get('pident', -1)) < args.pctid: ❸
        continue
    print(hit.get('qseqid')) ❹
```

❶ Parse the BLAST CSV file.

❷ Iterate through each BLAST hit.

❸ Skip those hits where the percent ID is less than the minimum. Use the float() function to convert the text to a floating-point value.

❹ Print the query sequence ID.

Run this version of the program with a minimum percent ID of 90, and verify that you get 190 hits from the first file:

```
$ ./blastomatic.py -a tests/inputs/meta.csv -b tests/inputs/hits1.csv -p 90 \
  | wc -l
  190
```

If the BLAST hit's qseqid value is found as a seq_id in the metadata file, then print to the output file the sequence ID, the percent ID from the BLAST hit, and the depth and latitude/longitude values from the metadata file. That should be enough to get you rolling on this program. Be sure to run the tests to verify that your program is correct.

Parsing Delimited Text Files Using the pandas Module

The pandas module presents another effective way to read a delimited file. This module, along with NumPy, is one of the foundational Python libraries used in data science. I'll use the pd.read_csv() function, which closely resembles the read_csv() function from the R programming language, if you are familiar with that. Note that the function can read text delimited by any delimiter you specify using a sep field separator, but the default is a comma.

Normally the delimiter is a single character, but it's possible to split text using a string. If you do this, you may encounter the warning "ParserWarning: Falling back to the *python* engine because the *c* engine does not support regex separators (separators > 1 char and different from \s+ are interpreted as regex); you can avoid this warning by specifying engine=*python*."

It's common to import pandas with the alias pd:

```
>>> import pandas as pd
>>> meta = pd.read_csv('tests/inputs/meta.csv')
```

Much of pandas is based on ideas from R. A pandas dataframe is a two-dimensional object that holds all the columns and rows of the metadata file in a single object, just like a dataframe in R. That is, the reader in the previous example is an interface used to sequentially retrieve each of the records, but the pandas dataframe is a full representation of all the data from the file. As such, the size of a dataframe will be limited to the amount of memory on your computer. Just as I've warned about using fh.read() to read an entire file into memory, you must be judicious about which files can be practically read using pandas. If you must process millions of rows of delimited text in gigabyte-sized files, I would recommend using cvs.DictReader() to process one record at a time.

If you evaluate the meta object in the REPL, a sample of the table will be shown. You can see that pandas used the first row of the file for the column headers. As indicated by ellipses, some of the columns have been elided due to the constrained width of the screen:

```
>>> meta
                   seq_id  ...                lat_lon
0   JCVI_READ_1092301105055  ...  -0.5938889,-91.06944
1   JCVI_READ_1092351051817  ...  -0.5938889,-91.06944
2   JCVI_READ_1092301096881  ...  -0.5938889,-91.06944
3   JCVI_READ_1093017101914  ...  -0.5938889,-91.06944
4   JCVI_READ_1092342065252  ...   9.164444,-79.83611
..                      ...  ...                   ...
95  JCVI_READ_1091145742670  ...  44.690277,-63.637222
96  JCVI_READ_1091145742680  ...  44.690277,-63.637222
97  JCVI_READ_1091150268218  ...  44.690277,-63.637222
98  JCVI_READ_1095964929867  ...  -1.9738889,-95.014725
99  JCVI_READ_1095994150021  ...  -1.9738889,-95.014725

[100 rows x 7 columns]
```

To find the number of rows and columns in a dataframe, inspect the meta.shape attribute. Note that this is not followed by parentheses because it is not a method call. This dataframe has 100 rows and 7 columns:

```
>>> meta.shape
(100, 7)
```

I can inspect the `meta.columns` attribute for the column names:

```
>>> meta.columns
Index(['seq_id', 'sample_acc', 'date', 'depth', 'salinity', 'temp', 'lat_lon'],
dtype='object')
```

One advantage to dataframes is that you can query all the values from a column using a syntax that looks like accessing a field in a dictionary. Here I'll select the salinity values, and note that pandas has converted the values from text to floating-point values, with missing values represented, with NaN (not a number):

```
>>> meta['salinity']
0      NaN
1      NaN
2      NaN
3      NaN
4      0.1
      ...
95     30.2
96     30.2
97     30.2
98     NaN
99     NaN
Name: salinity, Length: 100, dtype: float64
```

I can find the rows with a salinity greater than 50 using syntax almost identical to that in R. This returns an array of Boolean values based on the predicate *salinity is greater than 50*:

```
>>> meta['salinity'] > 50
0      False
1      False
2      False
3      False
4      False
      ...
95     False
96     False
97     False
98     False
99     False
Name: salinity, Length: 100, dtype: bool
```

I can use these Booleans values as a mask to only select the rows where this condition is `True`:

```
>>> meta[meta['salinity'] > 50]
                    seq_id  ...              lat_lon
23  JCVI_READ_1092351234516  ...   -1.2283334,-90.42917
24  JCVI_READ_1092402566200  ...   -1.2283334,-90.42917
```

```
25   JCVI_READ_1092402515846   ...   -1.2283334,-90.42917

[3 rows x 7 columns]
```

The result is a new dataframe, so I could then look at the salinity values that were found:

```
>>> meta[meta['salinity'] > 50]['salinity']
23    63.4
24    63.4
25    63.4
Name: salinity, dtype: float64
```

If you read the BLAST hits file with pandas, you will need to supply the column names as you did in the previous example:

```
>>> cols = ['qseqid', 'sseqid', 'pident', 'length', 'mismatch', 'gapopen',
'qstart', 'qend', 'sstart', 'send', 'evalue', 'bitscore']
>>> hits = pd.read_csv('tests/inputs/hits1.csv', names=cols)
>>> hits
                     qseqid        sseqid  ...        evalue  bitscore
0          CAM_READ_0234442157  CP031125.1  ...  6.810000e-135     492.0
1          CAM_READ_0234442157  LT840186.1  ...   7.260000e-90     342.0
2          CAM_READ_0234442157  CP048747.1  ...   6.240000e-16      97.1
3          CAM_READ_0234442157  CP046232.1  ...   2.240000e-15      95.3
4      JCVI_READ_1095946186912  CP038852.1  ...   0.000000e+00    1158.0
..                        ...         ...  ...           ...       ...
495    JCVI_READ_1095403503430  EU805356.1  ...   0.000000e+00    1834.0
496    JCVI_READ_1095403503430  EU804987.1  ...   0.000000e+00    1834.0
497    JCVI_READ_1095403503430  EU804799.1  ...   0.000000e+00    1834.0
498    JCVI_READ_1095403503430  EU804695.1  ...   0.000000e+00    1834.0
499    JCVI_READ_1095403503430  EU804645.1  ...   0.000000e+00    1834.0

[500 rows x 12 columns]
```

One element of this program is to select only those hits with a percent ID greater than or equal to some minimum. pandas will automatically convert the pident column to a floating-point value. Here I will select those hits with a percent ID greater than or equal to 90:

```
>>> wanted = hits[hits['pident'] >= 90]
>>> wanted
                     qseqid        sseqid  ...        evalue  bitscore
0          CAM_READ_0234442157  CP031125.1  ...  6.810000e-135     492.0
12     JCVI_READ_1091145027519  CP058306.1  ...   6.240000e-06      65.8
13     JCVI_READ_1091145742680  CP000084.1  ...   0.000000e+00    1925.0
14     JCVI_READ_1091145742680  CP038852.1  ...   0.000000e+00    1487.0
111    JCVI_READ_1091145742680  CP022043.2  ...   1.320000e-07      71.3
..                        ...         ...  ...           ...       ...
495    JCVI_READ_1095403503430  EU805356.1  ...   0.000000e+00    1834.0
496    JCVI_READ_1095403503430  EU804987.1  ...   0.000000e+00    1834.0
497    JCVI_READ_1095403503430  EU804799.1  ...   0.000000e+00    1834.0
498    JCVI_READ_1095403503430  EU804695.1  ...   0.000000e+00    1834.0
```

```
    499  JCVI_READ_1095403503430  EU804645.1  ...  0.000000e+00  1834.0

[190 rows x 12 columns]
```

To iterate over the rows in a dataframe, use the `wanted.iterrows()` method. Note that this works like the `enumerate()` function in that it returns a tuple of the row index and the row value:

```
>>> for i, hit in wanted.iterrows():
...     print(hit)
...     break
...
qseqid       CAM_READ_0234442157
sseqid                 CP031125.1
pident                     92.941
length                        340
mismatch                       21
gapopen                         3
qstart                          3
qend                          340
sstart                     801595
send                       801257
evalue                      0.000
bitscore                  492.000
Name: 0, dtype: object
```

To print a single field from a record in the dataframe, you can treat the record like a dictionary using field access through square brackets or by using the familiar `dict.get()` method. As with dictionaries, the first method will create an exception if you misspell a field name, while the latter method will quietly return None:

```
>>> for i, hit in wanted.iterrows():
...     print(hit['qseqid'], hit.get('pident'), hit.get('nope'))
...     break
...
CAM_READ_0234442157 92.941 None
```

As in the previous example, I recommend you first read the metadata and then iterate through the BLAST hits. You can look up the metadata from the `meta` dataframe by searching over the `seq_id` field. The sequence IDs in the metadata file are unique, so you should only find at most one:

```
>>> seqs = meta[meta['seq_id'] == 'CAM_READ_0234442157']
>>> seqs
                 seq_id       sample_acc  ...  temp         lat_lon
91  CAM_READ_0234442157  CAM_SMPL_GS112  ...  26.6  -8.50525,80.375583

[1 rows x 7 columns]
```

You can either iterate over the matches or use the `iloc` accessor to get the first (zeroth) record:

```
>>> seqs.iloc[0]
seq_id          CAM_READ_0234442157
sample_acc          CAM_SMPL_GS112
date                       8/8/05
depth                     4573.0
salinity                    32.5
temp                        26.6
lat_lon        -8.50525,80.375583
Name: 91, dtype: object
```

If you fail to find any matches, you will get an empty dataframe:

```
>>> seqs = meta[meta['seq_id'] == 'X']
>>> seqs
Empty DataFrame
Columns: [seq_id, sample_acc, date, depth, salinity, temp, lat_lon]
Index: []
```

You can inspect the seqs.empty attribute to see if it's empty:

```
>>> seqs.empty
True
```

or inspect the rows value from seqs.shape:

```
>>> seqs.shape[0]
0
```

Dataframes can also write their values to a file using the to_csv() method. As with read_csv(), you can specify any sep field separator, and the default is the comma. Note that by default pandas will include the row index as the first field of the output file. I generally use index=False to omit this. For example, I'll save the metadata records with a salinity greater than 50 to the *salty.csv* file with one line of code:

```
>>> meta[meta['salinity'] > 50].to_csv('salty.csv', index=False)
```

I can verify that the data was written using csvchk or csvlook:

```
$ csvchk salty.csv
// ****** Record 1 ****** //
seq_id      : JCVI_READ_1092351234516
sample_acc : JCVI_SMPL_1103283000038
date        : 2/19/04
depth       : 0.2
salinity   : 63.4
temp        : 37.6
lat_lon     : -1.2283334,-90.42917
```

A thorough review of pandas is well beyond the scope of this book, but this should be enough for you to figure out a solution. If you would like to learn more, I recommend *Python for Data Analysis* by Wes McKinney (O'Reilly, 2017) and *Python Data Science Handbook* by Jake VanderPlas (O'Reilly, 2016).

Solutions

I have four solutions, two using the csv module and two using pandas. All of the solutions use the same guess_delimiter() function, which I wrote like this:

```python
def guess_delimiter(filename: str) -> str:
    """ Guess the field separator from the file extension """

    ext = os.path.splitext(filename)[1] ❶
    return ',' if ext == '.csv' else '\t' ❷
```

❶ Select the file extension from os.path.splitext().

❷ Return a comma if the file extension is *.csv* and the tab character otherwise.

Solution 1: Manually Joining the Tables Using Dictionaries

This version closely follows all the suggestions from earlier in the chapter:

```python
def main():
    args = get_args()
    annots_reader = csv.DictReader(args.annotations, delimiter=',') ❶
    annots = {row['seq_id']: row for row in annots_reader} ❷

    headers = ['qseqid', 'pident', 'depth', 'lat_lon'] ❸
    args.outfile.write(args.delimiter.join(headers) + '\n') ❹

    hits = csv.DictReader(args.hits, ❺
                          delimiter=',',
                          fieldnames=[
                              'qseqid', 'sseqid', 'pident', 'length',
                              'mismatch', 'gapopen', 'qstart', 'qend',
                              'sstart', 'send', 'evalue', 'bitscore'
                          ])

    num_written = 0 ❻
    for hit in hits: ❼
        if float(hit.get('pident', -1)) < args.pctid: ❽
            continue

        if seq_id := hit.get('qseqid'): ❾
            if seq := annots.get(seq_id): ❿
                num_written += 1 ⓫
                args.outfile.write(
                    args.delimiter.join( ⓬
                        map(lambda s: f'"{s}"', [
                            seq_id,
                            hit.get('pident'),
                            seq.get('depth'),
                            seq.get('lat_lon')
                        ])) + '\n')
```

```
    args.outfile.close() ⑬
    print(f'Exported {num_written:,} to "{args.outfile.name}".') ⑭
```

❶ Create a parser for the annotations file.

❷ Read all the annotations into a dictionary keyed on the sequence ID.

❸ Define the headers of the output file.

❹ Write the headers to the output file.

❺ Create a parser for the BLAST hits.

❻ Initialize a counter for the number of records written.

❼ Iterate through the BLAST hits.

❽ Skip records with a percent ID less than the minimum.

❾ Attempt to get the BLAST query sequence ID.

❿ Attempt to find this sequence ID in the annotations.

⓫ If found, increment the counter and write the output values.

⓬ Quote all the fields to ensure the delimiter is protected.

⓭ Close the output file.

⓮ Print a final status to the user. The comma in the formatting for num_written will
add a thousands separator to the number.

Solution 2: Writing the Output File with csv.DictWriter()

This next solution differs from the first only in that I use csv.DictWriter() to write
the output file. I generally prefer to use this method as it will handle, for instance,
properly quoting fields that contain the field separator:

```
def main():
    args = get_args()
    annots_reader = csv.DictReader(args.annotations, delimiter=',')
    annots = {row['seq_id']: row for row in annots_reader}

    writer = csv.DictWriter( ❶
        args.outfile,
        fieldnames=['qseqid', 'pident', 'depth', 'lat_lon'],
```

```
            delimiter=args.delimiter)
        writer.writeheader() ❷

    hits = csv.DictReader(args.hits,
                          delimiter=',',
                          fieldnames=[
                              'qseqid', 'sseqid', 'pident', 'length',
                              'mismatch', 'gapopen', 'qstart', 'qend',
                              'sstart', 'send', 'evalue', 'bitscore'
                          ])

    num_written = 0
    for hit in hits:
        if float(hit.get('pident', -1)) < args.pctid:
            continue

        if seq_id := hit.get('qseqid'):
            if seq := annots.get(seq_id):
                num_written += 1
                writer.writerow({ ❸
                    'qseqid': seq_id,
                    'pident': hit.get('pident'),
                    'depth': seq.get('depth'),
                    'lat_lon': seq.get('lat_lon'),
                })

    print(f'Exported {num_written:,} to "{args.outfile.name}".') ❹
```

❶ Create a writer object to create the delimited text output file.

❷ Write the header row to the output file.

❸ Write a row of data, passing in a dictionary with the same keys as the `fieldnames` defined for the writer.

❹ The formatting instruction {:,} will cause the number to be printed with thousands separators.

Solution 3: Reading and Writing Files Using pandas

The pandas version is a little simpler in some ways and a little more complicated in others. I chose to store all the output records in a Python list and instantiate a new dataframe from that to write the output file:

```
def main():
    args = get_args()
    annots = pd.read_csv(args.annotations, sep=',') ❶
    hits = pd.read_csv(args.hits, ❷
                       sep=',',
                       names=[
```

```
                        'qseqid', 'sseqid', 'pident', 'length', 'mismatch',
                        'gapopen', 'qstart', 'qend', 'sstart', 'send',
                        'evalue', 'bitscore'
                ])

    data = [] ❸
    for _, hit in hits[hits['pident'] >= args.pctid].iterrows(): ❹
        meta = annots[annots['seq_id'] == hit['qseqid']] ❺
        if not meta.empty: ❻
            for _, seq in meta.iterrows(): ❼
                data.append({ ❽
                    'qseqid': hit['qseqid'],
                    'pident': hit['pident'],
                    'depth': seq['depth'],
                    'lat_lon': seq['lat_lon'],
                })

    df = pd.DataFrame.from_records(data=data) ❾
    df.to_csv(args.outfile, index=False, sep=args.delimiter) ❿

    print(f'Exported {len(data):,} to "{args.outfile.name}".') ⓫
```

❶ Read the metadata file into a dataframe.

❷ Read the BLAST hits into a dataframe.

❸ Initialize a list for the output data.

❹ Select all the BLAST hits with a percent ID greater than or equal to the minimum percent.

❺ Select the metadata for the given query sequence ID.

❻ Verify that the metadata is not empty.

❼ Iterate over the metadata records (even though there should only be one).

❽ Store a new dictionary with the output data.

❾ Create a new dataframe from the output data.

❿ Write the dataframe to the output file, omitting the dataframe index values.

⓫ Print the status to the console.

Solution 4: Joining Files Using pandas

In this last solution, I use pandas to join the metadata and BLAST dataframes, much like the `join` program I illustrated earlier in the chapter:

```
def main():
    args = get_args()
    annots = pd.read_csv(args.annotations, sep=',', index_col='seq_id') ❶
    hits = pd.read_csv(args.hits,
                       sep=',',
                       index_col='qseqid', ❷
                       names=[
                           'qseqid', 'sseqid', 'pident', 'length', 'mismatch',
                           'gapopen', 'qstart', 'qend', 'sstart', 'send',
                           'evalue', 'bitscore'
                       ])

    joined = hits[hits['pident'] >= args.pctid].join(annots, how='inner') ❸

    joined.to_csv(args.outfile, ❹
                  index=True,
                  index_label='qseqid',
                  columns=['pident', 'depth', 'lat_lon'],
                  sep=args.delimiter)

    print(f'Exported {joined.shape[0]:,} to "{args.outfile.name}".')
```

❶ Read the annotations file and set the index column to `seq_id`.

❷ Read the BLAST hits and set the index column to `qseqid`.

❸ Select the BLAST hits with the desired percent ID, and perform an inner join to the annotations using the index columns.

❹ Write the desired columns of the `joined` dataframe to the output file using the indicated delimiter. Include the index and name it `qseqid`.

The join operation is quite complex, so let me take a moment to explain this. First, each dataframe must have a unique index, which by default is the row index:

```
>>> import pandas as pd
>>> annots = pd.read_csv('tests/inputs/meta.csv')
>>> annots.index
RangeIndex(start=0, stop=100, step=1)
```

Instead, I want pandas to use the `seq_id` column as the index, which I indicate with the `index_col` argument:

```
>>> annots = pd.read_csv('tests/inputs/meta.csv', index_col='seq_id')
```

I can also indicate the zeroth field:

```
>>> annots = pd.read_csv('tests/inputs/meta.csv', index_col=0)
```

Now the index is set to the seq_id:

```
>>> annots.index[:10]
Index(['JCVI_READ_1092301105055', 'JCVI_READ_1092351051817',
       'JCVI_READ_1092301096881', 'JCVI_READ_1093017101914',
       'JCVI_READ_1092342065252', 'JCVI_READ_1092256406745',
       'JCVI_READ_1092258001174', 'JCVI_READ_1092959499253',
       'JCVI_READ_1092959656555', 'JCVI_READ_1092959499263'],
      dtype='object', name='seq_id')
```

Similarly, I want the BLAST hits to be indexed on the query sequence ID:

```
>>> cols = ['qseqid', 'sseqid', 'pident', 'length', 'mismatch', 'gapopen',
    'qstart', 'qend', 'sstart', 'send', 'evalue', 'bitscore']
>>> hits = pd.read_csv('tests/inputs/hits1.csv', names=cols, index_col='qseqid')
>>> hits.index[:10]
Index(['CAM_READ_0234442157', 'CAM_READ_0234442157', 'CAM_READ_0234442157',
       'CAM_READ_0234442157', 'JCVI_READ_1095946186912',
       'JCVI_READ_1095946186912', 'JCVI_READ_1095946186912',
       'JCVI_READ_1095946186912', 'JCVI_READ_1095946186912',
       'JCVI_READ_1091145027519'],
      dtype='object', name='qseqid')
```

I can select the BLAST hits with pident greater than or equal to the minimum. For instance, I find 190 rows with a value of 90:

```
>>> wanted = hits[hits['pident'] >= 90]
>>> wanted.shape
(190, 11)
```

The resulting dataframe is still indexed on the qseqid column, so I can join it to the annotations where the index values (the sequence IDs) are in common. By default, pandas will perform a *left join*, selecting all the rows from the first or *left* dataframe and substituting null values for rows that have no mate in the right dataframe. A *right join* is the opposite of a left join, selecting all the records from the *right* dataframe regardless of matches to the left. Since I only want the hits that have annotations, I use an *inner join*. Figure 19-3 demonstrates the joins using Venn diagrams.

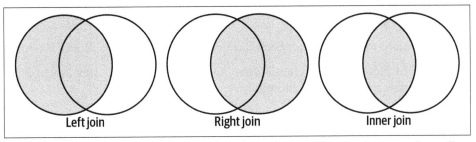

| Left join | Right join | Inner join |

Figure 19-3. A left join selects all the records from the left table, a right joins selects all the records from the right table, and an inner join selects only those records present in both

The join operation creates a new dataframe with the columns of both dataframes, just like the join tool I showed in "Using csvkit and csvchk" on page 364:

```
>>> joined = wanted.join(annots, how='inner')
>>> joined
                          sseqid   pident  ...  temp           lat_lon
CAM_READ_0234442157       CP031125.1  92.941  ...  26.6   -8.50525,80.375583
JCVI_READ_1091120852400   CP012541.1 100.000  ...  25.0    24.488333,-83.07
JCVI_READ_1091141680691   MN693562.1  90.852  ...  27.7  10.716389,-80.25445
JCVI_READ_1091141680691   MN693445.1  90.645  ...  27.7  10.716389,-80.25445
JCVI_READ_1091141680691   MN693445.1  91.935  ...  27.7  10.716389,-80.25445
...                             ...     ...  ...   ...                  ...
JCVI_READ_1095913058159   CP000437.1  94.737  ...   9.4  41.485832,-71.35111
JCVI_READ_1095913058159   AM286280.1  92.683  ...   9.4  41.485832,-71.35111
JCVI_READ_1095913058159   DQ682149.1  94.737  ...   9.4  41.485832,-71.35111
JCVI_READ_1095913058159   AM233362.1  94.737  ...   9.4  41.485832,-71.35111
JCVI_READ_1095913058159   AY871946.1  94.737  ...   9.4  41.485832,-71.35111

[190 rows x 17 columns]
```

Another way to write this is to use the `pd.merge()` function, which will default to an inner join. I must indicate which columns to use for the joins from the left and right dataframes, which in this case are the indexes:

```
>>> joined = pd.merge(wanted, annots, left_index=True, right_index=True)
```

I can use the `joined.to_csv()` method to write the dataframe to the output file. Note that the common sequence IDs are the index, which has no column name. I want the index included in the output file, so I use `index=True` and `index_name='qseqid'` so that the file matches the expected output:

```
>>> out_fh = open('out.csv', 'wt')
>>> joined.to_csv(out_fh, index=True, index_label='qseqid',
columns=['pident', 'depth', 'lat_lon'], sep=',')
```

Going Further

Add the options to filter by other fields like temperature, salinity, or BLAST e-value.

Default to including all the columns from both files in the output file, and add an option to select a subset of the columns.

Review

Key points from this chapter:

- Shell aliases can be used to create shortcuts for common commands.
- Delimited text files do not always have column headers. This is the case with BLAST's tabular output formats.
- The csv and pandas modules can read and write delimited text files.
- Datasets can be joined on common columns using the join command-line tool or in Python by using common keys from dictionaries or common indexes in pandas dataframes.
- pandas is a good choice for reading delimited files if you need access to all the data in memory—for example, if you need to perform statistical analysis of the data or want to quickly access all the values for a column. If you need to parse very large delimited files and can process records independently, then use the csv module for better performance.

Documenting Commands and Creating Workflows with make

The make program was created in 1976 to help build executable programs from source code files. Though it was originally developed to assist with programming in the C language, it is not limited to that language or even to the task of compiling code. According to the manual, one "can use it to describe any task where some files must be updated automatically from others whenever the others change." The make program has evolved far beyond its role as a build tool to become a workflow system.

Makefiles Are Recipes

When you run the make command, it looks for a file called *Makefile* (or *makefile*) in the current working directory. This file contains recipes that describe discrete actions that combine to create some output. Think of how a recipe for a lemon meringue pie has steps that need to be completed in a particular order and combination. For instance, I need to separately create the crust, filling, and meringue and then put them together and bake them before I can enjoy a tasty treat. I can visualize this with something called a *string diagram*, as illustrated in Figure A-1.

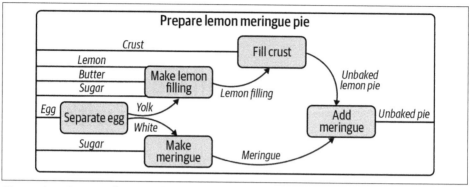

Figure A-1. A string diagram describing how to make a pie, adapted from Brendan Fong and David Spivak, An Invitation to Applied Category Theory (Seven Sketches in Compositionality), *Cambridge University Press, 2019*

It's not important if you make the pie crust the day before and keep it chilled, and the same might hold true for the filling, but it's certainly true that the crust needs to go into the dish first, followed by the filling and finally the meringue. An actual recipe might refer to generic recipes for crust and meringue elsewhere and list only the steps for the lemon filling and baking instructions.

I can write a *Makefile* to mock up these ideas. I'll use shell scripts to pretend I'm assembling the various ingredients into some output files like *crust.txt* and *filling.txt*. In the *app01_makefiles/pie* directory, I've written a *combine.sh* script that expects a filename and a list of "ingredients" to put into the file:

```
$ cd app01_makefiles/pie/
$ ./combine.sh
usage: combine.sh FILE ingredients
```

I can pretend to make the crust like this:

```
$ ./combine.sh crust.txt flour butter water
```

There is now a *crust.txt* file with the following contents:

```
$ cat crust.txt
Will combine flour butter water
```

It's common but not necessary for a recipe in a *Makefile* to create an output file. Note in this example that the clean target removes files:

```
all: crust.txt filling.txt meringue.txt ❶
        ./combine.sh pie.txt crust.txt filling.txt meringue.txt ❷
        ./cook.sh pie.txt 375 45

filling.txt:                                ❸
        ./combine.sh filling.txt lemon butter sugar
```

```
meringue.txt:                            ❹
        ./combine.sh meringue.txt eggwhites sugar

crust.txt:                               ❺
        ./combine.sh crust.txt flour butter water

clean:                                   ❻
        rm -f crust.txt meringue.txt filling.txt pie.txt
```

❶ This defines a target called `all`. The first target will be the one that is run when no target is specified. Convention holds that the `all` target will run *all* the targets necessary to accomplish some default goal, like building a piece of software. Here I want to create the *pie.txt* file from the component files and "cook" it. The name `all` is not as important as the fact that it is defined first. The target name is followed by a colon and then any dependencies that must be satisfied before running this target.

❷ The `all` target has two commands to run. Each command is indented with a Tab character.

❸ This is the `filling.txt` target. The goal of this target is to create the file called *filling.txt*. It's common but not necessary to use the output filename as the target name. This target has just one command, which is to combine the ingredients for the filling.

❹ This is the `meringue.txt` target, and it combines the egg whites and sugar.

❺ This is the `crust.txt` target that combines flour, butter, and water.

❻ It's common to have a `clean` target to remove any files that were created in the normal course of building.

As you can see in the preceding example, the target has a name followed by a colon. Any dependent actions can be listed after the colon in the order you wish them to be run. The actions for a target must be indented with a Tab character, as shown in Figure A-2, and you are allowed to define as many commands as you like.

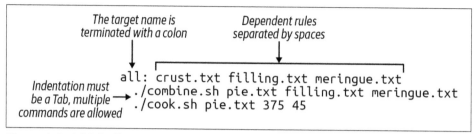

Figure A-2. A Makefile target is terminated by a colon and optionally followed by dependencies; all the target's actions must be indented with a single tab character

Running a Specific Target

Each action in a *Makefile* is called a *target*, *rule*, or *recipe*. The order of the targets is not important beyond the first target being the default. Targets, like the functions in Python programs, can reference other targets defined earlier or later in the file.

To run a specific target, I run **make target** to have make run the commands for a given recipe:

```
$ make filling.txt
./combine.sh filling.txt lemon butter sugar
```

And now there is a file called *filling.txt*:

```
$ cat filling.txt
Will combine lemon butter sugar
```

If I try to run this target again, I'll be told there's nothing to do because the file already exists:

```
$ make filling.txt
make: 'filling.txt' is up to date.
```

One of the reasons for the existence of make is precisely not to do extra work to create files unless some underlying source has changed. In the course of building software or running a pipeline, it may not be necessary to generate some output unless the inputs have changed, such as the source code being modified. To force make to run the *filling.txt* target, I can either remove that file or run **make clean** to remove any of the files that have been created:

```
$ make clean
rm -f crust.txt meringue.txt filling.txt pie.txt
```

Running with No Target

If you run the make command with no arguments, it will automatically run the first target. This is the main reason to place the all target (or something like it) first. Be careful not to put something destructive like a `clean` target first, as you might end up accidentally running it and removing valuable data.

Here's the output when I run `make` with the preceding *Makefile*:

```
$ make                                                          ❶
./combine.sh crust.txt flour butter water                      ❷
./combine.sh filling.txt lemon butter sugar                    ❸
./combine.sh meringue.txt eggwhites sugar                      ❹
./combine.sh pie.txt crust.txt filling.txt meringue.txt        ❺
./cook.sh pie.txt 375 45                                        ❻
Will cook "pie.txt" at 375 degrees for 45 minutes.
```

❶ I run `make` with no arguments. It looks for the first target in a file called *Makefile* in the current working directory.

❷ The *crust.txt* recipe is run first. Because I didn't specify a target, `make` runs the `all` target which is defined first, and this target lists *crust.txt* as the first dependency.

❸ Next, the *filling.txt* target is run.

❹ This is followed by the *meringue.txt*.

❺ Next I assemble *pie.txt*.

❻ And then I "cook" the pie at 375 degrees for 45 minutes.

If I run `make` again, I'll see that the intermediate steps to produce the *crust.txt*, *filling.txt*, and *meringue.txt* files are skipped because they already exist:

```
$ make
./combine.sh pie.txt crust.txt filling.txt meringue.txt
./cook.sh pie.txt 375 45
Will cook "pie.txt" at 375 degrees for 45 minutes.
```

If I want to force them to be recreated I can run `make clean && make`, where the `&&` is a logical *and* that will only run the second command if the first command succeeds:

```
$ make clean && make
rm -f crust.txt meringue.txt filling.txt pie.txt
./combine.sh crust.txt flour butter water
./combine.sh filling.txt lemon butter sugar
./combine.sh meringue.txt eggwhites sugar
./combine.sh pie.txt crust.txt filling.txt meringue.txt
./cook.sh pie.txt 375 45
Will cook "pie.txt" at 375 degrees for 45 minutes.
```

Makefiles Create DAGs

Each target can specify other targets as prerequisites or dependencies that must be completed first. These actions create a graph structure with a starting point and paths through targets to finally create some output file(s). The path described for any target should be a *directed* (from a start to a stop) *acyclic* (having no cycles or infinite loops) *graph*, or DAG, as shown in Figure A-3.

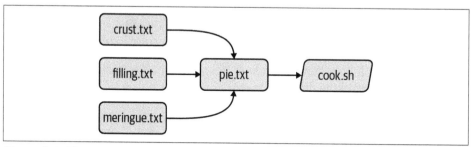

Figure A-3. The targets may join together to describe a directed acyclic graph of actions to produce some result

Many analysis pipelines are just that—a graph of some input, like a FASTA sequence file, and some transformations (trimming, filtering, comparisons) into some output, like BLAST hits, gene predictions, or functional annotations. You would be surprised at just how far make can be abused to document your work and even create fully functional analysis pipelines.

Using make to Compile a C Program

I believe it helps to use make for its intended purpose at least once in your life to understand why it exists. I'll take a moment to write and compile a "Hello, World" example in the C language. In the *app01_makefiles/c-hello* directory, you will find a simple C program that will print "Hello, World!" Here is the *hello.c* source code:

```
#include <stdio.h>          ❶
int main() {                ❷
    printf("Hello, World!\n"); ❸
    return 0;               ❹
}                           ❺
```

❶ Like in bash, the # character introduces comments in the C language, but this is a special comment that allows external modules of code to be used. Here, I want to use the printf (print-format) function, so I need to include the standard I/O (input/output) module, called stdio. I only need to include the "header" file, stdio.h, to get at the function definitions in that module. This is a standard module, and the C compiler will look in various locations for any included files to find it. There may be times when you are unable to compile C (or C++ programs) from source code because some header file cannot be found. For example, the gzip library is often used to de/compress data, but it is not always installed in a library form that other programs may include in this way. Therefore, you will have to download and install the libgz program, being sure to install the headers into the proper include directories. Note that package managers like apt-get and yum often have -dev or -devel packages that you have to install to get these headers; that is, you'll need to install both libgz and libgz-dev or whatnot.

❷ This is the start of a function declaration in C. The function name (main) is preceded by its return type (int). The parameters to the function are listed inside the parentheses after its name. In this case there are none, so the parentheses are empty. The opening curly brace ({) shows the start of the code that belongs to the function. Note that C will automatically execute the main() function, and every C program must have a main() function where the program starts.

❸ The printf() function will print the given string to the command line. This function is defined in the stdio library, which is why I need to #include the header file above.

❹ return will exit the function and return the value 0. Since this is the return value for the main() function, this will be the exit value for the entire program. The value 0 indicates that the program ran normally—think "zero errors." Any non-zero value would indicate a failure.

❺ The closing curly brace (}) is the mate for the one on line 2 and marks the end of the main() function.

To turn that into an executable program you will need to have a C compiler on your machine. For instance, I can use gcc, the GNU C compiler:

```
$ gcc hello.c
```

That will create a file called *a.out*, which is an executable file. On my Macintosh, this is what file will report:

```
$ file a.out
a.out: Mach-O 64-bit executable arm64
```

And I can execute that:

```
$ ./a.out
Hello, World!
```

I don't like the name *a.out*, though, so I can use the -o option to name the output file *hello*:

```
$ gcc -o hello hello.c
```

Run the resulting *hello* executable. You should see the same output.

Rather than typing `gcc -o hello hello.c` every time I modify *hello.c*, I can put that in a *Makefile*:

```
hello:
        gcc -o hello hello.c
```

And now I can run **make hello** or just **make** if this is the first target:

```
$ make
gcc -o hello hello.c
```

If I run **make** again, nothing happens because the *hello.c* file hasn't changed:

```
$ make
make: 'hello' is up to date.
```

What happens if I alter the *hello.c* code to print "Hola" instead of "Hello," and then try running make again?

```
$ make
make: 'hello' is up to date.
```

I can force make to run the targets using the -B option:

```
$ make -B
gcc -o hello hello.c
```

And now the new program has been compiled:

```
$ ./hello
Hola, World!
```

This is a trivial example, and you may be wondering how this saves time. A real-world project in C or any language would likely have multiple *.c* files with headers (*.h* files) describing their functions so that they could be used by other *.c* files. The C compiler would need to turn each *.c* file into a *.o* (*out*) file and then link them together into a single executable. Imagine you have dozens of *.c* files, and you change one line of code in one file. Do you want to type dozens of commands to recompile and link all your code? Of course not. You would build a tool to automate those actions for you.

I can add targets to the *Makefile* that don't generate new files. It's common to have a clean target that will clean up files and directories that I no longer need. Here I can create a clean target to remove the *hello* executable:

```
clean:
        rm -f hello
```

If I want to be sure that the executable is always removed before running the hello target, I can add it as a dependency:

```
hello: clean
        gcc -o hello hello.c
```

It's good to document for `make` that this is a *phony* target because the result of the target is not a newly created file. I use the `.PHONY:` target and list all the phonies. Here is the complete *Makefile* now:

```
$ cat Makefile
.PHONY: clean

hello: clean
        gcc -o hello hello.c

clean:
        rm -f hello
```

If you run **make** in the *c-hello* directory with the preceding *Makefile*, you should see this:

```
$ make
rm -f hello
gcc -o hello hello.c
```

And there should now be a *hello* executable in your directory that you can run:

```
$ ./hello
Hello, World!
```

Notice that the `clean` target can be listed as a dependency to the `hello` target even *before* the target itself is mentioned. `make` will read the entire file and then use the dependencies to resolve the graph. If you were to put `foo` as an additional dependency to `hello` and then run **make** again, you would see this:

```
$ make
make: *** No rule to make target 'foo', needed by 'hello'.  Stop.
```

A *Makefile* allows me to write independent groups of actions that are ordered by their dependencies. They are like *functions* in a higher-level language. I have essentially written a program whose output is another program.

I'd encourage you to run **cat hello** to view the contents of the *hello* file. It's mostly binary information that will look like gibberish, but you will probably be able to make out some plain English, too. You can also use **strings hello** to extract just the strings of text.

Using make for a Shortcut

Let's look at how I can abuse a *Makefile* to create shortcuts for commands. In the *app01_makefiles/hello* directory, you will find the following *Makefile*:

```
$ cat Makefile
.PHONY: hello            ❶

hello:                   ❷
        echo "Hello, World!" ❸
```

❶ Since the `hello` target doesn't produce a file, I list it as a phony target.

❷ This is the `hello` target. The name of the target should be composed only of letters and numbers, should have no spaces before it, and is followed by a colon (`:`).

❸ The command(s) to run for the `hello` target are listed on lines that are indented with a tab character.

I can execute this with **make**:

```
$ make
echo "Hello, World!"
Hello, World!
```

I often use a *Makefile* to remember how to invoke a command with various arguments. That is, I might write an analysis pipeline and then document how to run the program on various datasets with all their parameters. In this way, I'm documenting my work in a way that I can immediately reproduce by running the target.

Defining Variables

Here is an example of a *Makefile* I wrote to document how I used the Centrifuge program for making taxonomic assignments to short reads:

```
INDEX_DIR = /data/centrifuge-indexes        ❶

clean_paired:
    rm -rf $(HOME)/work/data/centrifuge/paired-out

paired: clean_paired                        ❷
    ./run_centrifuge.py \                   ❸
    -q $(HOME)/work/data/centrifuge/paired \ ❹
    -I $(INDEX_DIR) \                       ❺
    -i 'p_compressed+h+v' \
    -x "9606, 32630" \
    -o $(HOME)/work/data/centrifuge/paired-out \
    -T "C/Fe Cycling"
```

❶ Here I define the variable `INDEX_DIR` and assign a value. Note that there must be spaces on either side of the =. I prefer ALL_CAPS for my variable names, but this is my personal preference.

---

❷ Run the `clean_paired` target prior to running this target. This ensures that there is no leftover output from a previous run.

❸ This action is long, so I used backslashes (\) as on the command line to indicate that the command continues to the next line.

❹ To have `make` *deference* or use the value of the $HOME environment variable, use the syntax $(HOME).

❺ $(INDEX_DIR) refers to the variable defined at the top.

Writing a Workflow

In the *app01_makefiles/yeast* directory is an example of how to write a workflow as make targets. The goal is to download the yeast genome and characterize various gene types as "Dubious," "Uncharacterized," "Verified," and so on. This is accomplished with a collection of command-line tools such as wget, grep, and awk, combined with a custom shell script called *download.sh*, all pieced together and run in order by make:

```
.PHONY: all fasta features test clean

FEATURES = http://downloads.yeastgenome.org/curation/$\
    chromosomal_feature/
    SGD_features.tab

all: fasta genome chr-count chr-size features gene-count verified-genes \
    uncharacterized-genes gene-types terminated-genes test

clean:
        find . \( -name \*gene\* -o -name chr-\* \) -exec rm {} \;
        rm -rf fasta SGD_features.tab

fasta:
        ./download.sh

genome: fasta
        (cd fasta && cat *.fsa > genome.fa)

chr-count: genome
        grep -e '^>' "fasta/genome.fa" | grep 'chromosome' | wc -l > chr-count

chr-size: genome
        grep -ve '^>' "fasta/genome.fa" | wc -c > chr-size

features:
        wget -nc $(FEATURES)

gene-count: features
```

```
        cut -f 2 SGD_features.tab | grep ORF | wc -l > gene-count

verified-genes: features
        awk -F"\t" '$$3 == "Verified" {print}' SGD_features.tab | \
                wc -l > verified-genes

uncharacterized-genes: features
        awk -F"\t" '$$2 == "ORF" && $$3 == "Uncharacterized" {print $$2}' \
                SGD_features.tab | wc -l > uncharacterized-genes

gene-types: features
        awk -F"\t" '{print $$3}' SGD_features.tab | sort | uniq -c > gene-types

terminated-genes:
        grep -o '/G=[^ ]*' palinsreg.txt | cut -d = -f 2 | \
                sort -u > terminated-genes

test:
        pytest -xv ./test.py
```

I won't bother commenting on all the commands. Mostly I want to demonstrate how far I can abuse a *Makefile* to create a workflow. Not only have I documented all the steps, but they are *runnable* with nothing more than the command **make**. Absent using make, I'd have to write a shell script to accomplish this or, more likely, move to a more powerful language like Python. The resulting program written in either language would probably be longer, buggier, and more difficult to understand. Sometimes, all you need is a *Makefile* and some shell commands.

Other Workflow Managers

As you bump up against the limitations of make, you may choose to move to a workflow manager. There are many to choose from. For example:

- Snakemake extends the basic concepts of make with Python.

- The Common Workflow Language (CWL) defines workflows and parameters in a configuration file (in YAML), and you use tools like cwltool or cwl-runner (both implemented in Python) to execute the workflow with another configuration file that describes the arguments.

- The Workflow Description Language (WDL) takes a similar approach to describe workflows and arguments and can be run with the Cromwell engine.

- Pegasus allows you to use Python code to describe a workflow that then is written to an XML file, which is the input for the engine that will run your code.

- Nextflow is similar in that you use a full programming language called Groovy (a subset of Java) to write a workflow that can be run by the Nextflow engine.

All of these systems follow the same basic ideas as make, so understanding how make works and how to write the pieces of your workflow and how they interact is the basis for any larger analysis workflow you may create.

Further Reading

Here are some other resources you can use to learn about make:

- The GNU Make Manual (*https://oreil.ly/D9daZ*)
- *The GNU Make Book* by John Graham-Cumming (No Starch Press, 2015)
- *Managing Projects with GNU Make* (*https://oreil.ly/D8Oyk*) by Robert Mecklenburg (O'Reilly, 2004)

Understanding $PATH and Installing Command-Line Programs

PATH is an environment variable that defines the directories that will be searched for a given command. That is, if I type foo and there's no built-in command, shell function, command alias, or program anywhere in my PATH that the shell can execute as **foo**, I'll be told this command cannot be found:

```
$ foo
-bash: foo: command not found
```

In Windows PowerShell, I can inspect the PATH with **echo $env:Path**, whereas on Unix platforms I use the command **echo $PATH**. Both paths are printed as a long string with no spaces, listing all the directory names separated by semicolons on Windows or by colons on Unix. If the operating system didn't have some concept of a path, it would have to search *every directory* on the machine for a given command. This could take minutes to hours, so it makes sense to restrict the searching to just a few directories.

Following is my path on my Macintosh. Note that I have to put a dollar sign ($) in front of the name to tell my shell (bash) that this is a variable and not the literal string PATH. To make this more readable, I'll use Perl to replace the colons with newlines. Note that this command will only work on a Unix command line where Perl is installed:

```
$ echo $PATH | perl -pe 's/:/\n/g' ❶
/Users/kyclark/.local/bin ❷
/Library/Frameworks/Python.framework/Versions/3.9/bin ❸
/usr/local/bin ❹
/usr/bin ❺
/bin
```

```
/usr/sbin
/sbin
```

❶ The Perl substitute (s//) command replaces the first pattern (:) with the second (\n) globally (g).

❷ This is a custom directory I usually create for installing my own programs.

❸ This is where Python installed itself.

❹ This is a standard directory for user-installed software.

❺ The rest are more standard directories for finding programs.

 The directories will be searched in the order they are defined, so the order can be quite important. For instance, the Python path is listed before system paths so that when I type **python3** it will use the version found in my local Python directory *before* one that might have been preinstalled on my system.

Notice that all the directory names in my PATH end in *bin*. This is short for *binaries* and comes from the fact that many programs exist in a binary form. For example, the source code for a C program is written in a pseudo-English language that is compiled into a machine-readable executable file. The contents of this file are binary-encoded instructions that the operating system can execute.

Python programs, by contrast, are usually installed as their source code files, which are executed by Python at runtime. If you want to globally install one of your Python programs, I suggest you copy it to one of the directories that are already listed in your PATH. For instance, */usr/local/bin* is a typical directory for *local* installations of software by the user. It's such a common directory that it's normally present in the PATH. If you are working on your personal machine, like a laptop, where you have administrator privileges, you should be able to write new files into this location.

For instance, if I wanted to be able to run the dna.py program from Chapter 1 without providing the full path to the source code, I could copy it to a location in my PATH:

```
$ cp 01_dna/dna.py /usr/local/bin
```

You may not have sufficient permissions to do this, however. Unix systems were designed from the beginning to be *multitenant* operating systems, meaning that they support many different people using the system concurrently. It's important to keep users from writing and deleting files they shouldn't, and so the OS may prevent you from writing dna.py to a directory that you don't own. If, for instance, you are working on a shared high-performance computing (HPC) system at a university, you certainly won't have such privileges.

When you cannot install into system directories, it's easiest to create a location in your HOME directory for such files. On my laptop, this is my HOME directory:

```
$ echo $HOME
/Users/kyclark
```

On almost all my systems, I create a *$HOME/.local* directory for installing programs. Most shells interpret the tilde (~) as HOME:

```
$ mkdir ~/.local
```

 By convention, files and directories that have names starting with a dot are normally hidden by the ls command. You can use ls -a to list *all* the contents of a directory. You may notice many other *dotfiles* that are used by various programs to persist options and program state. I like to call this *.local* so I won't normally see it in my directory listing.

Creating a directory in your HOME for software installations is especially useful when compiling programs from source, a very common operation in bioinformatics. Most installations of this sort begin by using a `configure` program to gather information about your system, such as the location of your C compiler and such. This program almost always has a `--prefix` option that I'll set to this directory:

```
$ ./configure --prefix=$HOME/.local
```

The resulting installation will put the binary compiled files into *$HOME/.local/bin*. It might also install header files and manual pages and other supporting data into other directories in *$HOME/.local*.

Wherever you decide to install local programs, you'll need to ensure that your PATH is updated to search in that directory in addition to the others. I tend to use the bash shell, and one of the dotfiles in my HOME is a file called *.bashrc* (or sometimes *.bash_profile* or even *.profile*). I can add this line to put my custom directory first in the PATH:

```
export PATH=$HOME/.local/bin:$PATH
```

You may need something slightly different if you are using a different shell. Recently macOS started using zsh (Z shell) as the default shell, or your HPC system might use another shell. They all have the idea of PATH and all allow you to customize this in some manner. On Windows, you can use this command to append the directory to your path:

```
> $env:Path += ";~/.local/bin"
```

Here is how I can make the directory and copy the program:

```
$ mkdir -p ~/.local/bin
$ cp 01_dna/dna.py ~/.local/bin
```

I should now be able to execute **dna.py** from any location on a Unix machine:

```
$ dna.py
usage: dna.py [-h] DNA
dna.py: error: the following arguments are required: DNA
```

Windows shells like cmd.exe and PowerShell don't read and execute the shebang like Unix shells, so you are required to include the command **python.exe** or **python3.exe** before the program name:

```
> python.exe C:\Users\kyclark\.local\bin\dna.py
usage: dna.py [-h] DNA
dna.py: error: the following arguments are required: DNA
```

Be sure python.exe --version shows that you are using version 3 and not version 2. You may need to install the latest version of Python. I have only shown Windows commands using python.exe, assuming this means Python 3, but you may need to use python3.exe, depending on your system.

Epilogue

The tools we use have a profound (and devious!) influence on our thinking habits, and, therefore, on our thinking abilities.

—Edsger Dijkstra

This book was inspired by the Rosalind problems, which I've spent years revisiting, first as I was trying to understand more about biology and then as I learned new programming languages. I initially attempted them with Perl, and since have tried using JavaScript, Haskell, Python, and Rust, with varying degrees of success. I would challenge you to also write solutions using any other languages you know.

I've tried to show you patterns in Python you can reuse in your own programs. More than anything, I hope I've demonstrated that types and tests and various formatting and linting tools can drastically improve the programs you write.

Index

Symbols
. (any character), 106
& (bitwise and), 324
| (bitwise or), 323
(comment), 365, 396
== (compare for equality), 27, 37
    eq() functional version, 270
{} (curly brackets for string formatting), 32, 119
- (dash)
    flag names, 48
    parameters, 48
.<filename> (dotfiles), 193, 194, 407
$? (exit status variable), 85
+= (increment), 27
&& (logical and), 395
?=(<pattern>) (look-ahead assertion), 184
% (modulo), 249
__name__, 164
$# (number of arguments), 232
!= (operator.ne), 152
| (pipe), 113
< (redirect input), 113
> (redirect output), 192, 350
() (regex capture groups), 184, 285
-> (return value), 17
#! (shebang), 7, 232
* (splat a tuple), 33, 151
_ (throwaway variable), 90, 91
:= (variable assignment and testing), 133, 220
* (zero or more characters), 106

A
alias in bash shell, 366
alias in Python

NumPy imported as np, 297
    pandas imported as pd, 378
all() function for boolean, 321, 326
amino acid–codon table, 157, 251-256
and, bitwise (&), 324
and, logical (&&), 395
anonymous functions, 107
any() function for boolean, 321, 326
argparse
    about, 7
    FileType() to validate File arguments, 49
    -h and --help flags, 7
    using, 7-10
arguments
    $# (number of arguments), 232
    blastomatic program, 371
    download directory optional argument, 229
    get_args() called by main(), 7
    HOFs taking other functions as, 99
    make with no arguments, 395
    named tuples representing, 16-18
    new.py for new Python programs, 5
        argparse, 7-10
    one or more values required, 48, 49
    optional, 6, 46
        positional as, 114
    positional, 6, 46
        as optional, 114
    sequence file subsampling, 350
        parameters defined, 352
    testing for no arguments, 22
    tetranucleotide frequency, 5-10
    type error, 9
    valid file as, 113

About the Author

Ken Youens-Clark has been programming for about 25 years. After a wandering undergraduate education at the University of North Texas that started in music and ended in English literature, he learned programming on the job using various and sundry languages. Eventually he ended up in a bioinformatics lab and thought it seemed way cooler than anything he'd done before, so he stuck with that. Ken lives in Tucson, AZ, where he earned his MS in biosystems engineering in 2019 from the University of Arizona. When he's not coding, he enjoys cooking, playing music, riding bicycles, and being with his wife and three very interesting children.

Colophon

The animal on the cover of *Mastering Python for Bioinformatics* is the Mojave Rattlesnake (*Crotalus scutulatus*), a highly venomous pit viper species found in the deserts of the southwestern United States and central Mexico.

These heavy-bodied snakes grow up to four feet in length, and are easily recognized by the prominent rattles on their tails. The Mojave rattlesnake's belly is often gray, yellow, or brown, with light-bordered, dark diamonds or blotches in the scales along its body.

Mammals are their most common prey, but they also eat birds, lizards, frogs, and other snakes. It is often suggested that Mojave rattlesnakes are more dangerous than other species, primarily because the venom produced by most populations has repeatedly been shown to be more deadly to laboratory mice than the venom of other rattlesnakes. However, they aren't usually aggressive toward people unless they feel threatened.

Many of the animals on O'Reilly covers are endangered; all of them are important to the world.

The cover illustration is by Karen Montgomery, based on a black and white engraving from *Dover's Animals*. The cover fonts are Gilroy Semibold and Guardian Sans. The text font is Adobe Minion Pro; the heading font is Adobe Myriad Condensed; and the code font is Dalton Maag's Ubuntu Mono.

Milton Keynes UK
Ingram Content Group UK Ltd.
UKHW051818240724
445993UK00003B/4